S0-CVK-896

Prometheus Reimagined

Albert C. Lin examines technologies that will be among the most life-changing and controversial in coming decades: synthetic biology, nanotechnology, human enhancement, and geoengineering. While such technologies promise to address many of humanity's most serious problems, they also bring environmental and health-related risks and uncertainties. Moreover, such technologies can come to dominate global production systems and markets with very little public input or awareness. Existing governance institutions and processes do not adequately address the risks of new technologies, nor do they give much consideration to the concerns of persons affected by those technologies.

Far from demanding omniscient planning or a halt to technological development, Lin calls for a more thoughtful and democratic approach. Instead of treating technological development and environmental matters as two discrete fields, laws must acknowledge their fundamental relationship. Laws must become more forward-looking, anticipating both future technological developments and their potential adverse effects on human health and the environment. Laws must encourage international cooperation and the development of common global standards, and legal systems must allow for flexibility and reassessment.

Albert C. Lin is Professor of Environmental Law at the University of California, Davis, School of Law.

Prometheus Reimagined

Technology, Environment, and Law in the Twenty-First Century

Albert C. Lin

The University of Michigan Press • Ann Arbor

Published in the United States of America by
The University of Michigan Press
Manufactured in the United States of America
⊗ Printed on acid-free paper

2016 2015 2014 2013 4 3 2 1

A CIP catalog record for this book is available from the British Library.

Library of Congress Cataloging-in-Publication Data

Lin, Albert C.
Prometheus reimagined : technology, environment, and law in the twenty-first century / Albert C. Lin.
pages cm
Includes bibliographical references and index.
ISBN 978-0-472-11883-0 (cloth : alk. paper) — ISBN 978-0-472-02924-2 (e-book)
1. Technology—Social aspects. 2. Technology—Environmental aspects. 3. Technology and state. 4. Technology and law. 5. Technological innovations—Law and legislation. I. Title.
T14.5.L5535 2013
303.48'3—dc23

2013025446

TO LINH & ANDREW,

With gratitude for your patience and inspiration

Contents

Abbreviations

AEC	Atomic Energy Commission
APHIS	Animal and Plant Health Inspection Service
BIC	broader impacts criterion
BWC	Convention on the Prohibition of the Development, Production, and Stockpiling of Bacteriological (Biological) and Toxin Weapons and on Their Destruction
CBD	Convention on Biological Diversity
CDR	carbon dioxide removal
CEQ	Council on Environmental Quality
CNS-ASU	Center for Nanotechnology in Society at Arizona State University
CO_2	carbon dioxide
CPSA	Consumer Product Safety Act
CSR	corporate social responsibility
CTA	constructive technology assessment
DARPA	Defense Advanced Research Projects Agency
DIY biology	do-it-yourself biology
DNA	deoxyribonucleic acid
EA	environmental assessment
EIS	environmental impact statement
EPA	U.S. Environmental Protection Agency
EU	European Union
FCCC	United Nations Framework Convention on Climate Change
FDA	Food and Drug Administration
FFDCA	Federal Food, Drug, and Cosmetic Act
FIFRA	Federal Insecticide, Fungicide, and Rodenticide Act

GAO	U.S. Government Accountability Office
GHG	greenhouse gas
GM	genetically modified
GMO	genetically modified organism
GRAS	generally recognized as safe
iGEM	International Genetically Engineered Machine
LC/LP	London Convention/London Protocol
NBIC	nanotechnology, biotechnology, information technology, and cognitive science
NCTF	National Citizens' Technology Forum
NEPA	National Environmental Policy Act
NGO	nongovernmental organization
NIH	National Institutes of Health
NIOSH	National Institute for Occupational Safety and Health
NMSP	Nanoscale Materials Stewardship Program
NNI	National Nanotechnology Initiative
NRC	National Research Council
NSF	National Science Foundation
OECD	Organisation for Economic Co-operation and Development
OSHA	Occupational Safety and Health Administration
OSTP	Office of Science and Technology Policy
OTA	Office of Technology Assessment
PELs	permissible exposure limits
PIP	plant-incorporated protectant
PMN	premanufacture notice
ppm	parts per million
PTA	participatory technology assessment
R&D	research and development
SNUN	significant new use notice
SPS Agreement	Sanitary and Phytosanitary Measures Agreement
SRM	solar radiation management
TA	technology assessment
TSCA	Toxic Substances Control Act
USDA	U.S. Department of Agriculture
WTO	World Trade Organization

Preface and Acknowledgments

We live in an era of rapid technological change. Advances in biotechnology, nanotechnology, geoengineering, and other fields have the potential to address the most pressing challenges humanity faces and to transform countless aspects of our lives. Applications on the horizon range from waste-devouring bacteria and edible vaccines to engineered human tissues. These changes are occurring in an increasingly interconnected world characterized by the globalization of capital flows, trade, and labor. Although globalization may promote economic efficiency and improve living standards, particularly in the developing world, it also threatens to exacerbate the strains humanity places on the environment. Wide-ranging concerns such as rising commodity prices, growing food insecurity, and climate change remind us of the limits of our shared resources on planet Earth.

Technology can help to push back or even circumvent some of these limits. Improvements in energy efficiency can reduce dependence on fossil fuels. Newly engineered materials may emerge as superior substitutes for existing materials. And redesigned manufacturing processes may require less energy and generate less waste. Other limits, however, are more intractable: for example, there are no substitutes for clean water or clean air. Equally important, past experiences teach us that technologies often create new dangers. These dangers may involve novel hazards to human health and the environment as well as threats to valued cultural practices and social structures. We can identify some of these dangers from the outset, but others reveal themselves only with time.

Law is an essential tool for dealing with the dangers that technologies may pose. Through law, societies can attempt to impose order and shape the world around them. Narrowly defined, law refers to rules and statutes that govern conduct, backed by governmental authority.[1] A broader conception of law, however, encompasses not only these rules and statutes but also the institutions and processes involved in implementing them as well as less formalized norms that influence behavior.[2] Rooted in the legal process school of thought, this view acknowledges law as a complex and dynamic system for designing the society and environment in which we live. Law influences the ways in which we understand and interact with the world. Put more succinctly, law is "a purposive activity, a continuous striving to solve the basic problems of social living."[3]

A critical way in which law shapes society and the environment is through its treatment of technology. Regulations, subsidies, taxes, tort liability, and exemptions from liability are common legal tools to directly stimulate or inhibit the development of new technologies. Research policies and intellectual property regimes established through law likewise affect the course of technological innovation. In addition, legal requirements to collect and disclose information about new technologies can generate and inform public debate about how to handle those technologies. Law, particularly public law, "stands opposed to the unplanned market"[4] in offering an alternative regime for distributing goods, services, and *disservices*—the undesired effects of social and economic activity. At its best, law embodies and effectuates democratically determined public choices regarding social objectives, including choices about technology.

New technologies pose a particular challenge for law, however. These technologies often give rise to problems and circumstances that existing laws were not meant to address. Historically, law has been a laggard, reacting to the hazards of new technologies only after harms to life, health, or the environment materialize and become too significant to ignore. The rapid pace of modern technological change and the pressures of international competition threaten to exacerbate the problem. Technologies now emerging, such as nanotechnology and synthetic biology, demand special attention because of their swift evolution, global reach, and uncertain yet potentially catastrophic risks. Conventional regulatory regimes, which rely heavily on quantitative risk data, will struggle to respond in the face of the uncertain hazards and unknown consequences that accompany these technologies. Waiting for definitive evidence of harm, as is the predominant course under these regimes, may leave little opportunity to avoid disastrous effects or to undo technologies that have become entrenched.

In addition to challenging existing laws, emerging technologies can call into question and reshape the basic principles that underlie our laws. Emerging technologies have enabled the creation of stem cells and genetically modified organisms and might someday give rise to enhanced humans. These new biological entities, which do not fall neatly within existing legal boundaries, promise to alter long-held norms regarding those who have legal rights and the specific rights they hold.[5] As our technical powers expand, technological choices will increasingly become ethical choices about who we are, who we want to be, and what matters to us.

This book presents a vision for reorienting our legal institutions in a way that acknowledges the transformative power of technology, recognizes the physical and nonphysical consequences associated with its use, and engages the public in technology management. To meet this vision, existing legal approaches must change in at least three fundamental ways.

First, our laws must address more directly the relationship between technology and the environment. As currently written and implemented, laws tend to treat technology development and environmental matters as discrete fields, at least until the environmental consequences of a new technology become obvious. Yet the concept of technology—the use of machines and techniques to control and adapt to the environment—presupposes a fundamental relationship between the two. Our laws must acknowledge this relationship directly and become more forward-looking. We must build legal institutions and structures that enable the anticipation of future technological developments and facilitate the identification of adverse effects on health, the environment, and society.

Second, our laws must become more attuned to the dynamics of globalization. As the problem of climate change demonstrates, the environmental risks associated with technological development are increasingly global. More generally, the development, implementation, and effects of technology are not confined to national boundaries. In response to globalization, international law in areas such as trade, investment, and the environment has grown in scope and stature. International cooperation and development of common standards for technology management will become even more necessary in the future. Ostensibly domestic concerns may require international cooperation because the efficacy of one nation's legal response to such concerns may depend on other nations' policies.

Third, laws and institutions meant to oversee technology and its environmental consequences must reflect greater democratic control. Average citizens have little input regarding most of the technological changes affecting their lives. For example, genetically modified foods and food ingredients have become widely prevalent in our society, notwithstanding widespread discomfort with the process of genetic engineering. Regulators and the biotechnology industry have resisted even the disclosure of genetically modified ingredients, further fueling public distrust. This is but one prominent instance in which public concerns regarding technological change have been disregarded. Of course, satisfying all individual preferences regarding technology is impossible in a diverse and complex society. Effectively integrating public input into technology development and management will be challenging. Our social and legal institutions nonetheless must provide more information about technological risks and offer citizens more varied and meaningful opportunities to participate in managing technology.

This manuscript benefited from the comments and suggestions of numerous colleagues, students, and friends. I am especially grateful for feedback from Eric Biber, Bret Birdsong, David Dana, Terry Davies, Holly Doremus, Todd LaPorte, Peter Lee, John Nagle, Gene Rochlin, Jake Storms, Doug Sylvester, Margaret Taylor, David Victor, and David Winickoff; participants at workshops at the Arizona State University College of Law, the UNLV School of Law, and the 2011 Annual Meeting of the Law and Society Association; and to anonymous reviewers of the proposal and draft manuscript for the book. I have had the benefit of excellent research assistance from law students Aylin Bilir, Ronny Clausner, Theresa Cruse, Emily Gesmundo, Pearl Kan, Liz Kinsella, Lynn Kirshbaum, Atticus Lee, Autumn Luna, Christopher Ogata, Bao Vu, and Nick Warden and from librarians Aaron Dailey and Erin Murphy of the Mabie Law Library. Finally, I am grateful to dean Kevin Johnson, associate dean Vik Amar, and the University of California, Davis, School of Law for supporting this project and to Melody Herr of the University of Michigan Press for her assistance in the publication process.

Chapter 1 is based on Albert C. Lin, "Technology Assessment 2.0: Revamping Our Approach to Emerging Technologies," *Brooklyn Law Review* 76 (2011): 1309–70, and portions of chapter 3 were derived from Albert C. Lin, "Size Matters: Regulating Nanotechnology," *Harvard Environmental Law Review* 31 (2007): 349–408.

Introduction

What Is Technology?

Defining technology is fundamental to understanding its relationship to the environment and to seeing how law might better mediate that relationship. For many, the word *technology* brings to mind the Internet, iPhones, vacuuming robots, or other gadgets produced by the digital high-tech revolution. But to fully appreciate the relationship between technology and the environment, we must expand our view beyond this subset of technology. Technology is ubiquitous. It affects what we do and how we do it, how we live, and even who we are. Indeed, the term *technology* appears in virtually every context imaginable. We speak, for example, of industrial technology, information technology, military technology, biotechnology, nanotechnology, and even reproductive technology.

A historical perspective on the development of human civilization reinforces the notion that almost everything around us involves technology. Technological artifacts such as stone tools and clay pottery facilitated the rise of early civilizations, illustrating that technology encompasses even the most mundane objects. Technological objects have no significance, however, without the knowledge of how to make and use them. Thus, the concept of technology must also include the know-how regarding the construction and use of these objects. Drawing on these observations, we can develop a working definition of *technology* as the tools, techniques, and knowledge that humans use to mediate their environment.[1]

As this definition recognizes, an inherent relationship exists between technology and the environment. Through technology, we modify or manipulate aspects of the world around us to improve our lives. The technology-environment relationship is most obvious in the case of rudimentary technologies. For example, humans constructed shelter to shield themselves from the elements. To improve the reliability and abundance of food supplies, humans planted crops and domesticated animals. These were the first steps in an ongoing process of environmental modification that continues today with the genetic engineering of plants and livestock. Even modern communication and information technologies operate on the environment by shrinking the elements of space and time, expanding our memory and data-processing capacities, and facilitating new understandings of the world around us.[2]

Humans have done so much to modify the environment that it is now somewhat misleading to refer to a "natural environment." Braden Allenby's declaration that the "Earth has become an anthropogenic planet"[3] aptly describes the scale of human activity and its effects. Take, for example, the forests of North America. Possessing hundred-foot trees towering over lush undergrowth, some of these forests may appear untouched by humans. Nevertheless, their current state—in terms of appearance, location, age distribution, and species they contain—is a product of deliberate land management policies. Through decisions regarding land use, timber cultivation and harvest, fire suppression, and wildlife management, humans have shaped those forests. Similarly, humans have built dams on rivers to generate power, store water, and protect against floods. These dams constitute critical pieces of the technology that shapes our rivers, valleys, and floodplains to facilitate the uses humans have selected.

These examples involve the deliberate use of technology to engineer our environment. Many technologies, however, affect the environment in unintended ways.

Climate change provides an illustration of the drastic and unintended consequences that technology may bring. Since the Industrial Revolution, we have burned huge quantities of fossil fuels. Our goal in so doing was not to release large quantities of greenhouse gases (GHGs) or other pollutants. Rather, these fuels have run our factories, heated our homes, and transported us and our goods. Nonetheless, the most lasting legacy of fossil fuel combustion may well be global climate change. While inadvertent, anthropogenic climate change cannot be characterized as unexpected. As early as the late 1800s, scientists theorized that elevated carbon dioxide concentrations in the atmosphere would result in significant global temperature increases.[4]

Another example of technology's unintended environmental consequences involves the use of the insecticide DDT. DDT was introduced to control mosquito populations that spread malaria and subsequently was applied to control other insect populations as well. From the outset, biologists raised concerns that the chemical could harm humans and wildlife. Those concerns, however, were ignored.[5] Only after three decades of widespread use irrefutably demonstrated that the insecticide was highly detrimental to bird populations and probably humans as well did the United States ban its use.[6] Despite persistent concerns about human health risks, many developing countries continue to employ DDT as a relatively cheap and effective means of protecting humans from malaria.[7]

The DDT and fossil fuel examples illustrate two important points. First, technology often generates widespread costs and benefits. Technology has effects on users, society at large, and the surrounding environment, and those effects can extend well into the future. Second, despite the uncertainties that may surround a new technology's environmental hazards, we often have reasonable suspicions about what some of those hazards might be. We may choose to ignore those potential hazards, but we do so at our own peril.

These and similar experiences have sometimes led to simplistic portrayals of technology and the environment as diametrically opposed. Famously, the Luddites of 19th-century England opposed industrialization and mechanization as threats to a decentralized, more traditional way of life. Their present-day counterparts, the neo-Luddites, critique the predominant ethos equating technology with progress. The neo-Luddites do not object to all forms of technology, but they do lament the ecological destruction and community disruption that technology often brings.[8] The neo-Luddites call for a broader evaluation of technology that considers more than its immediate utility. Technology's long-term effects, economic ramifications, and political meanings also matter, as does the sociological context in which technology changes occur. Many of these sorts of concerns have been reflected in public reactions to the introduction of modern technologies, including biotechnology. As chapter 2 discusses, the revelation that genetically modified foods have become widespread in the food supply has led to ethical objections and fears of unpredictable and detrimental effects on human health and the environment.

It would be a mistake, however, to issue a blanket condemnation of technology and processes of technological change. First, technology is not monolithic.[9] It encompasses a wide range of tools and techniques that vary in the risks, uncertainties, and ethical concerns they raise. Even research and development within a specific field, such as nanotechnology, may involve sufficiently distinct activities to merit discriminating approaches to oversight. Second, technology is embedded in our lives and undoubtedly brings us many benefits. Modern society depends on multiple layers of technology to enable global commerce, instantaneous communication, and food production on a scale sufficient to feed billions. Technology has enabled higher living standards, new choices, and greater leisure. Thanks to the capabilities and ubiquity of technology, we are more powerful than ever before—as well as more vulnerable to technological failure. To address the economic, social, and environmental challenges we face today, further innovation is essential. Third, technology is in some ways inevitable. As inquiring and creative beings, we necessarily develop tools and knowledge to mediate our environment. At times, certain technologies may even seem to take on a life of their own. The particular tools we develop and the way we use them, however, are not inevitable. We make technological choices and ultimately must take responsibility for them.

The Prometheus Myth

The ancient Greek myth of Prometheus provides a useful metaphor for reflecting on the complex relationship between humans and technology. Prometheus is perhaps best known for stealing fire from the gods and giving it to humankind.[10] In leading interpretations of the myth, fire symbolizes "the spirit of technology, forbidden knowledge, the conscious intellect, political power, and artistic inspiration."[11] Humanity's use of fire is a double-edged sword, with both creative and destructive aspects. The Greek dramatist Aeschylus emphasized in *Prometheus Bound* that fire is a "precious gift which hath become the mistress of all arts and crafts."[12] Nonetheless, fire is a *stolen* gift. This aspect of the myth underlines the gift's dark side—specifically, "the limits of the human ability to meddle with the divine spark of creation."[13] The fire metaphor cautions against hubris in the aspiration for knowledge and technological achievement. More concretely, the metaphor serves as a warning to humanity to wield technologies carefully, consider potential effects, and recognize that technologies may bring on consequences that cannot be anticipated.

These warnings and the Prometheus myth in general remain relevant today. Indeed, the lessons of Prometheus are particularly worth heeding in light of the potency and scope of modern technologies. More than ever before, humans possess the ability to influence global climate processes, synthesize new forms of matter, and create new forms of life. The Prometheus myth brings to mind not only technology's tremendous potential for improving human welfare but also the attendant risks. The title of this book underscores the point that in an era in which we routinely hail technological progress, we need to reexamine our relationship with technology, rediscover technology's many dimensions, and reimagine how we develop and manage technology.

Further, less familiar aspects of the Prometheus myth are also instructive. The name *Prometheus* refers to "the one who thinks in advance": *pro* meaning "before," and *metis* meaning "clever intelligence."[14] Prometheus, the forethinker, stands in stark contrast to his dim-witted and lesser-known brother, Epimetheus, "the one who thinks afterward." In Hesiod's telling of the myth, Epimetheus, despite Prometheus's warning to the contrary, accepted the gift of Pandora from Zeus. As a result of this impetuous decision, various ills beset humankind.[15]

The contrast between Prometheus and Epimetheus corresponds to another major theme of this book: the need for more forethought regarding whether and how we develop and use our technologies. For too long, we have taken an Epimethean approach to technology. Societies often promote the widespread adoption of a promising new technology without seriously considering its broader consequences for society, individuals, or the environment. This approach fails to envision future developments, anticipate adverse effects, or reduce uncertainties. Such an approach is particularly troubling if the harms that may result from using a technology are serious and irreversible. When we acknowledge the problems arising from a technology's use only after it has become entrenched or difficult to modify, it may well be impossible to undo or address the damage that has already occurred.

Why has the Epimethean approach predominated? Economic factors are partly to blame. Capitalist market economies demand growth and provide intellectual property protection to innovators. Accordingly, those who develop and commercialize a technology reap financial rewards. Though researchers and developers are usually in the best position to identify a technology's potential hazards, the profit motive discourages them from developing information regarding possible negative consequences.[16] Cultural factors also play a role. Scientists are rewarded by society and their peers for research breakthroughs and technological innovations, not for their forbearance.[17] Moreover, those who perform research work in support of regulatory agencies receive less respect and recognition than do their counterparts in the private sector.[18] Finally, human nature, in the form of hyperbolic discounting, is to blame as well.[19] Drawn to the immediate and obvious benefits of a technology, we tend to discount future events relative to those in the present. In contrast, latent, indirect, and less certain effects, which often include environmental and social costs, receive little weight in our decision-making processes. Indeed, uncertain harms often are not treated as harms at all. Deferring consideration of these uncertainties can seem rational, at least in the short run, because harms may never arise or may not amount to a serious problem until a technology is deployed on a large scale.

Although it is impossible to eliminate all uncertainties with respect to a particular technology, we can nevertheless be more forward-looking. Law can powerfully facilitate prediction and detection of the consequences of technological change, and it can equip society to better handle problems as they arise. Simply requiring companies to disclose the use of certain technologies would assist users and other affected persons in identifying adverse effects. Going further, mandating that manufacturers perform life-cycle assessments of new products—from the extraction of raw materials to disposal of used products—would generate more complete information regarding overall environmental impacts.[20] Similarly, requiring government agencies to conduct ongoing assessments of government-funded research would help predict technological developments and their ramifications. These sorts of analyses will be imperfect but nonetheless can enable emergency preparedness, subsequent study of adverse effects, and more informed decision making.

Tools such as technology assessment and environmental impact assessment can bring greater foresight to the development and implementation of new technologies. Even when performed systematically and thoroughly, however, such tools cannot provide sufficient data to make fully informed decisions about adopting or managing a technology. The question then becomes how to proceed in the face of lingering uncertainty and ignorance. U.S. laws generally have treated the absence of evidence of harms as presumptive proof that a given technology is not harmful.[21] Instead of imposing limitations on a technology only when harm is manifest, however, we can act in a manner more consistent with the precautionary principle. A leading articulation of the principle provides, "When human activities may lead to morally unacceptable harm that is scientifically plausible but uncertain, actions shall be taken to avoid or diminish that harm."[22] The precautionary principle does not dictate the specific course to be followed in a particular situation. Rather, the principle leaves decision makers with broad discretion regarding how to respond to uncertainty and risk. Applied to new technologies, the precautionary principle can support a wide range of policies, including studies of adverse effects, measures to minimize exposure or mitigate potential harm, or restrictions on further development or use until society has had adequate opportunities to consider environmental and ethical implications. Unlike conventional approaches that focus on manifest harm, the precautionary principle explicitly acknowledges the problem of uncertainty.

Finally, the fact that Prometheus gave the gift of fire to *all* mankind inspires an additional set of insights regarding how we might manage technology. Technology affects not only direct users and beneficiaries but also the surrounding community and broader society. Technology's pervasive influence, positive and negative, has important implications for how technology management decisions should be made. That is, principles of democratic governance counsel that those whose lives may be substantially affected by a technology should have a voice in its management.

Current policies on technology, however, often develop with little meaningful public participation and input. Indeed, technology is sometimes characterized as beyond human control. The worry that technology can acquire a life of its own and defy human efforts to direct it is not unique to our era, as literary classics such as *Frankenstein* illustrate.[23] The irony in this characterization is that technology, far from being an autonomous force, consists of tools and techniques that humans choose.[24] Complaints about technology run amok nevertheless do raise legitimate questions regarding technological innovation, commercialization, dissemination, and control: Who decides these matters, and how are such decisions made? Underlying these questions is a sense that narrow interests dominate technology policy matters and that as a result, public concerns are ignored.

Private enterprise and governments play critical roles in technological change. Industry sponsors an ever increasing share of technological research and development.[25] In addition, private insurance can facilitate technology development by spreading risk. Private involvement in general can harness private resources and individual initiative to produce important technological benefits for society, but the interests of private actors do not necessarily align with society's interests. Moreover, private activity does not occur in a vacuum but exists against the backdrop of government policy and regulation. In particular, Congress funds research and development, sets tax and investment policies, defines intellectual property protections, and establishes regulatory standards.[26] Governments view technology policy primarily as a matter of economic competitiveness, and thus perceive their primary role as facilitators of new technologies.[27] Their role as regulators is often secondary. With respect to the $1.5 billion expended each year by the federal government on nanotechnology research, for example, less than 5 percent is directed to the study of health and environmental risks associated with such technology.[28]

Disregard of public concern about emerging technologies can be traced not only to private incentives and government policies but also to the difficulty of timing meaningful public input. When a technology is in development, public awareness of the technology and its potential implications may be low. To policymakers, lay perceptions and concerns may seem irrelevant or uninformed. Yet once a technology is commercialized and begins to affect a significant portion of the public, the views of laypersons may have little effect because of investments and commitments already made.

The Need to Adopt a Reimagined Promethean Approach

The value of taking a proactive and participatory approach to managing emerging technologies is reflected in various aspects of the Prometheus myth. Two long-term trends underscore the urgency of adopting such an approach, which this book refers to as a Promethean approach. First, the widespread use of technologies detrimental to the environment is testing the Earth's ability to sustain us as a species.[29] Together, population growth, rising global consumption, and our technologies are straining our natural support systems in unprecedented ways. Second, the rapid development of transformative technologies such as nanotechnology, accompanied by various unknown risks, demands greater agility in recognizing and responding to the technologies' strengths, weaknesses, and unintended consequences.

Testing the Earth's Limits

Signs increasingly indicate that we are approaching the physical limits of the Earth's ability to sustain us. The manifestations of climate change are becoming more obvious and frequent. Worldwide, fisheries are in decline. In many places, water supplies are shrinking. Competition for scarce resources ultimately could lead to armed conflict, reduced living standards, displaced populations, and environmental degradation.[30] Systemic factors contributing to these difficulties include population growth, high consumption rates in developed countries and rapidly rising consumption rates in developing countries, and increasingly powerful technologies with drastic environmental effects. Though our encroachment on the Earth's limits may not immediately threaten human survival, other species may decline and become extinct as a consequence of our current course.[31]

Climate change provides an incredible illustration of the magnitude of harm that aggregate human activity can produce. We are generating GHGs in volumes sufficient to alter radically the chemistry of the Earth's atmosphere and oceans. Although the Earth's climate is the product of complex interactions among the atmosphere, land, oceans, and other factors, its basic mechanics are well understood and undisputed.[32] The sun's energy drives the climate system and is either reflected back into space or absorbed by the Earth. From the beginning of the Holocene epoch some 11,000 years ago until recent times, the Earth's climate has remained relatively constant because energy is reflected into space at approximately the same rate that it is absorbed. Carbon dioxide and other GHGs naturally present in the atmosphere act as a blanket, preventing some heat from radiating back into space. This "greenhouse effect" maintains the Earth's surface temperature at temperate levels, sustaining life as we know it.

Human activities such as fossil fuel combustion and mass deforestation, however, release excess GHGs into the atmosphere and magnify the greenhouse effect. Since the beginning of the Industrial Revolution, human activity has raised the atmospheric concentration of carbon dioxide from 280 parts per million in 1750 to approximately 390 parts per million today.[33] The various phenomena that experts have attributed to increased GHG levels include elevated temperatures, changed precipitation patterns, retreat of polar icecaps and glaciers, and declining ecosystem health. Moreover, effects predicted for the coming decades include reduced water supplies, longer and more intense heat waves, more frequent droughts and wildfires, more widespread tropical diseases, decreased agricultural production, and flooding of island and coastal communities.[34] Such effects clearly will prove costly, if not deadly, for millions of people.

Climate change is but one example of how we are bumping up against the Earth's physical limits. More generally, various indexes suggest that humans are consuming natural resources in unsustainable ways.[35] Humanity's ecological footprint—a comparison of the planet's regenerative capacity with human demands placed on its ecosystems—suggests that resource consumption levels must be reduced by 25 percent to be sustainable.[36] Despite a tripling of world grain production between 1950 and 2000, shortages occurred in 2008 as a result of increased demand for animal feed, conversion of cropland to biofuel production, and climate-related stresses on production.[37] Many of the world's fisheries are in decline or on the verge of collapse, thanks to increasingly powerful harvesting technologies, ineffective management, pollution, and other factors.[38] And in many parts of the world, freshwater supplies are under pressure from industrial demand, irrigation, domestic consumption, and climate change. By 2025, the United Nations predicts that 1.8 billion people will be living in areas of absolute water scarcity.[39]

People have looked to technology as a cure-all for these and various other concerns. Technology offers both promise and pitfalls, however, in the struggle to live within the Earth's physical constraints. Some technologies have the potential to increase production while reducing environmental impacts and the consumption of raw materials.[40] Genetic engineering promises increased crop yields, nutritionally enhanced livestock, and plant-based production of pharmaceuticals and industrial chemicals. Nanotechnology offers the prospect of bottom-up manufacturing processes that create improved products, use less energy, and generate less waste. Synthetic biology may generate biofuels that can mitigate climate change by reducing our dependence on fossil fuels. We are an ingenious species, capable of finding and developing alternative materials and methods for carrying out old tasks and accomplishing new ones. Indeed, thanks largely to technological innovations, dire predictions about the Earth's inability to sustain human populations growing at exponential rates have repeatedly proven inaccurate.[41]

Nonetheless, our experiences with technology to date should also give us reasons for pause. Technologies frequently have unanticipated adverse effects. Furthermore, the introduction of new technologies may result in greater rather than lesser impacts on the environment.[42] Developed countries, which tend to have the most widespread and latest technologies, also consume the most resources and have the greatest environmental impacts on a per capita basis. By one estimate, the one-eighth of the world's population that lives in North America and Western Europe is responsible for 60 percent of private consumption worldwide.[43] Put another way, the average American consumes the same amount of resources as 32 of his or her counterparts in developing nations.[44] Understandably, much of the developing world aspires to a standard of living similar to that enjoyed by Americans. Even the partial achievement of this aspiration, however, could put unbearable strains on the environment. We need an approach to technology that encourages the alleviation of global poverty and inequality yet guards against the destruction of the natural systems that sustain us.

Coping with Rapid Technological Change

Compounding these concerns about the Earth's ability to sustain current and projected levels of human activity, rapid technological developments threaten to overwhelm conventional approaches to technology management. Anecdotally, the spread of Internet-based social networking tools and the surge of innovations in nanotechnology and consumer electronics support the notion that technological change is accelerating.[45] Improvements in the rapidity of global communications, the rate of scientific advancement and innovation, and the speed, power, and memory capacity of modern computers provide quantitative evidence of quickening technological change.[46]

Ongoing technological changes may have particularly broad and long-lasting impacts. The effects of emerging technologies will likely extend to society and the environment generally. Some emerging technologies, such as nanotechnology, are expected to revolutionize manufacturing techniques and find application in almost all areas of human activity. Other emerging technologies, such as geoengineering, will necessarily have global and enduring impacts. Still other technologies, such as genetic engineering and synthetic biology, may lead to consequences that are especially difficult to register and control because the organisms created by these technologies have the capacity to reproduce. In addition, developments in seemingly disparate technological fields may converge to radically reshape manufacturing processes, communication techniques, human capabilities, and even our understanding of what it means to be human.[47] Each of these emerging technologies involves great promise as well as tremendous uncertainty.

As a society, we should pay particular heed to transformative technologies—those with the potential to radically change production processes, communities, or ways of life. While incremental technological advances also can have considerable cumulative impacts, those impacts are more foreseeable and less likely to be disruptive. The challenge of managing emerging technologies is made more difficult by the fast pace and increasingly interconnected environment of the 21st century, where modern means of communication facilitate rapid technology dissemination, innovation, and information exchange.[48] Although we cannot always predict at the outset which technologies will be transformative, applying a Promethean approach can help us to identify such technologies and address their effects on human health, society, and the environment.

Looking Ahead

This book examines various technologies that are likely to be among the most important and controversial in the coming decades. To set the stage for this discussion, chapter 1 considers in some detail the practice of technology assessment. In its conventional form, technology assessment examines technologies that have already been developed and provides policymakers with information to facilitate the management of those technologies and their consequences.[49] An important component of technology assessment is risk assessment, an analytical process aimed at quantifying the hazards posed by an activity. Risk assessments are necessarily incomplete, however, because it is impossible to resolve all uncertainty regarding future effects of an activity. Overreliance by policymakers on risk assessments, moreover, can lead to the neglect of qualitative considerations that are pertinent to technology management decisions.

Many of the concerns relevant to risk assessment also apply more generally to technology assessment, at least in its rudimentary form. Consisting of the expert evaluation of fully formed technologies, conventional technology assessment contributed to a false sense of security that technology could be readily controlled. In addition, it did little to involve the public or account for social values. Recent innovations in technology assessment, however, have acknowledged its limitations and sought to open up the process to public input. These innovations include constructive technology assessment, which seeks to inject technology assessment feedback into the process of designing new technologies, and lay citizen participation, which can educate the public regarding developing technologies, provide a forum for public concerns, and facilitate consideration of public attitudes in technology decisions.[50] Further innovations in technology assessment and public engagement ultimately will be necessary to achieve more thorough and inclusive evaluation and management of emerging technologies.

Case studies of emerging technologies will advance our understanding of how better technology management can occur. Chapters 2–6 survey the technologies of genetic engineering, nanotechnology, geoengineering, synthetic biology, and human enhancement. These fields merit close consideration because of their wide-ranging effects and their potential to transform how we live. These fields also present valuable case studies in managing technologies in varying stages of development. Oversight at earlier stages presents practical difficulties but can facilitate incorporation of public concerns and more careful consideration of social and environmental consequences. The analysis of relevant legal regimes in each chapter will focus on domestic-level regulation, which serves as the primary means of technology management today. International cooperation or regulation may also be necessary and will also be considered.

Chapter 2 considers genetic engineering, focusing specifically on the development and use of genetically modified organisms (GMOs) in agriculture. Genetic engineering first came to prominence in the 1970s. Thanks in part to very limited oversight, GMOs have come to dominate the major commodity crops grown in the United States. Today, genetic engineers continue to develop new types of GMOs, including livestock and fish engineered for human consumption. Although GMOs remain controversial, the field is fairly well established as an economic matter and thus is unlikely to come under wholesale reconsideration. Genetic engineering provides an example of a technology that has already emerged and continues to evolve. Our experience in this field, including missteps in its development and management, provides valuable lessons not only for future genetic engineering efforts and oversight but also for other emerging technologies.

Chapter 3 examines the presently emerging field of nanotechnology. Nanotechnology refers to a suite of technologies that manipulate matter at a very tiny scale (measured in nanometers—one-billionth of a meter). Nanotechnology offers an even broader array of potential applications than genetic engineering, with current and projected uses in medicine, defense, electronics, and personal care products, to name just a few. At the same time, nanotechnology is surrounded by great uncertainty. Despite reasonable grounds for concern, we know little about the health and environmental effects that exposure to nanomaterials may cause. Developing an approach to safely manage nanotechnology without unduly impeding valuable uses presents a daunting and pressing challenge. Even though numerous nanotechnology applications are already in use, nanotechnology as a field remains largely in the developmental phase. The current state of affairs thus may allow a limited opportunity to involve the general public and other stakeholders in technology assessment and management before the technology becomes entrenched within economic and social systems.

Chapter 4 turns to geoengineering, which is attracting growing attention as a possible response to climate change. Geoengineering refers to a variety of risky, controversial, and untested techniques to "engineer" the Earth's climate at a planetary scale. These techniques seek to counter the elevated temperatures and other climatic consequences of higher GHG concentrations. Because these techniques necessarily would have global effects, geoengineering particularly raises questions of international technology governance. Although the global dimensions of governance will complicate geoengineering oversight, early recognition of the need for governance offers some hope for the adoption of a proactive approach to this group of technologies. At present, geoengineering is in its infancy. Relatively few private interests have a substantial stake in its promotion, and little field experimentation has taken place.

Chapter 5 discusses the latest iteration of the biotechnology revolution, the emerging field of synthetic biology. Building on techniques developed by genetic engineers, synthetic biologists seek to develop new and more efficient drugs, chemicals, and biofuels by crafting and manipulating novel genetic sequences. Synthetic biologists may one day even create entirely new organisms from genetic codes wholly designed by humans. Hazards of synthetic biology include health and environmental risks as well as biosecurity risks from deliberate misuse. The excitement and risks are magnified by the broad accessibility of synthetic biology technology: Amateurs can engage in rudimentary do-it-yourself biology experiments at a relatively low cost and with minimal technical training. The accessibility of synthetic biology suggests the need for creative governance mechanisms to complement more conventional regulatory regimes.

Finally, chapter 6 briefly considers "converging technologies for improving human performance." These anticipated technologies raise serious ethical concerns, thereby necessitating ethical sensitivity among scientists and broad societal debate. The subject of human enhancement technologies also serves as a springboard for reflecting more generally on the challenges posed by emerging technologies. The chapter discusses common shortfalls in our approaches to emerging technologies and suggests various reforms that would foster greater consideration of public values and better equip society to address the uncertainty and changes that accompany new technologies.

In light of the transformative nature of emerging technologies as well as the various pressures we continue to exert on the environment, a reactive technology management determined with little public input is no longer tenable. We should adopt an approach that peers into the future to identify potential risks, reevaluates a technology as it develops, acknowledges lingering uncertainty, and actively involves the public in critical questions of technology development and management. Indeed, a 2011 Obama administration memo concerning the oversight of emerging technologies acknowledges the importance of developing adequate information, encouraging public participation, and recognizing uncertainty but provides little detail on how to achieve these objectives.[51] As a practical matter, what might a Promethean approach to technology entail? And what role should law play in instituting and implementing such an approach? These questions have no simple answers. We nonetheless can identify the tools available through law to direct technology development and dissemination, diagnose flaws in how we handle technology today, and articulate principles to guide the course ahead.

CHAPTER I

Existing Law and New Technologies

Our society is divided, if not schizophrenic, in its attitudes toward regulation. The regulation of emerging technologies is no exception. While some Americans support stronger regulatory protection, many others expound the virtues of limited government and decry excessive state control. This latter view stresses the costs of regulation and attributes our nation's success in large part to the freedom enjoyed by entrepreneurs, innovators, and inventors.[1] Indeed, Americans generally have great faith in technology and see it as a powerful solution to many of the challenges we face. This faith is reflected both in the popular embrace of new gadgets and in government policies that promote the development of new technologies through grants, tax breaks, and intellectual property protections.

Notwithstanding these attitudes, many Americans also assume that government is protecting us from the excesses of the market and the risks that new technologies may pose. We expect emerging technologies and the products that incorporate those technologies to be thoroughly tested and reasonably safe. We also expect clean air, clean water, and a healthy environment, and we expect that emerging technologies will not injure these resources. Such expectations, however, often are not met. Toys contaminated with lead, baby bottles tainted by endocrine-disrupting chemicals, and the Deepwater Horizon oil spill are just a few recent instances in which human health and environmental regulatory efforts have fallen short.

Technological change does not occur on its own, of course. Technologies are a product of human discovery, choice, and policy.[2] Accordingly, managing technology not only is possible but also is a critical social endeavor. An essential step in sound technology management is technology assessment, which attempts to predict the course and consequences of technological development. Using the information generated through such assessments, society can regulate or reshape a technology or seek to direct its development. This is not to suggest, however, that society can exert complete control over technologies and their effects. Even technologies subject to close scrutiny during development and implementation may give rise to unanticipated consequences. The spread of a technology, moreover, may make its control costly and difficult. Ultimately, technologies may transform societal values and society itself.[3]

The predominant approach to managing technology has been reactive. In the United States, the production and use of new chemicals, for example, is generally allowed until there is hard evidence indicating significant and adverse health or environmental effects.[4] Similarly, government oversight of nuclear reactor safety was minimal until the Three Mile Island and Chernobyl accidents,[5] and regulators did not seriously address the risks associated with deep-ocean oil drilling until the recent Deepwater Horizon disaster.[6] The development of safety requirements in response to the risks laid bare by such incidents is of course rational and appropriate. Addressing harms after the fact, however, can be problematic. In many instances, the harms that materialize are difficult, if not impossible, to remedy. Responses crafted after the fact often are limited by existing infrastructure, prior commitments, and the policy preferences of powerful vested interests. In addition, such responses tend to target the risks associated with the specific events contributing to a harm, rather than provide for a broad and comprehensive evaluation of the underlying technology.

The widespread and potentially irreversible effects of emerging technologies warrant prompt attention and proactive management. As these technologies generally promise great benefit, knee-jerk moratoriums should be avoided. But these technologies also have the potential to adversely affect critical aspects of people's lives, and the general public should be kept informed of technological developments and have meaningful opportunities to be involved in their management.

Proactive technology management is not a new concept. Policymakers developed the tools of technology assessment and environmental impact assessment decades ago. However, the implementation of these tools has been generally ineffective and has failed to involve the public in meaningful ways. This chapter explores past technology assessment practices and thus sets the stage for the case studies in subsequent chapters and for considering how society might better manage technology. One option would be to more effectively implement existing tools of technology assessment. Though efforts along these lines have been initiated, their impact is likely to be limited. More fundamental changes are needed to bring about participatory and effective management of emerging technologies.

The discussion in this and subsequent chapters focuses on technology assessment and regulation in the United States. The United States is a leader in technological innovation, and it effectively invented the concepts of technology assessment and environmental assessment. Scientific and technological advances often begin here and then spread abroad. Because much of the current research and development activity in nanotechnology, synthetic biology, and other emerging technologies is occurring in the United States, this country offers a suitable context for considering the challenges of emerging technologies. Moreover, governance standards and techniques adopted by the United States frequently serve as models for other nations. The United States does not have a monopoly on technological innovation or technology management, of course. Efforts by other nations to assess and manage technologies also can be instructive. Furthermore, international dimensions of technology management are increasingly important thanks to globalization and global problems such as climate change. International governance thus merits serious attention as well.

Goals of Technology Assessment: Assessment and Public Engagement

Assessment

The practice of technology assessment arose in the 1970s with the ambition of predicting and analyzing the full range of consequences—social, environmental, and otherwise—of a given technology.[7] As then conceived, technology assessment was to involve objective analysis that drew on the natural and social sciences; subjective value judgments were to be left to democratically elected officials.[8] A basic premise behind technology assessment was the belief that society could use and manage technology effectively through the application of comprehensive rationality—that is, that such assessments could be "precise, value-neutral, and exhaustive of relevant concerns" and could thereby enable fully informed policy decisions.[9]

Conventional technology assessment presumed, moreover, that technology develops in a linear fashion: Basic research would be followed by applied research and then production. As such, technology assessment focused on the latter part of that process and sought to generate an objective analysis of virtually finished technologies. At this stage, the ability to refashion a technology or to shape it to account for concerns identified in an assessment was often limited. Additionally, in providing technical analysis, technology assessors did little to involve the public. Societal values could presumably be incorporated later, when politically accountable actors made decisions on whether and how to regulate a technology.[10]

Technology assessment's faith in information gathering, rational analysis, and human control was overly optimistic, as was the assumption that societal values could adequately be folded in at the end stages of technology production. Technology development is not linear, and there are limits to our capacity to predict and direct technological futures.[11] But the underlying motivation of developing and managing technologies in a more deliberate and publicly accountable manner remains critical. Recent refinements to technology assessment, discussed later in this chapter, have sought to incorporate assessment earlier in technology development, encourage greater reflectiveness among scientists about the ramifications of their work, and integrate public input into the assessment process.

In addition to these shortcomings of conventional technology assessment, there are further concerns regarding risk assessment, which is a central component of technology assessment.[12] Risk assessment is a type of technical analysis that seeks to produce quantitative estimates of the probability and magnitude of potential harms from an occurrence.[13] That occurrence might involve a commonplace activity, such as driving a car, or a complex policy decision, such as the adoption of a new technology. Risk assessment for a new chemical substance, for example, consists of a process to identify potential hazards, estimate the probability and magnitude of injury at different exposure levels, and analyze the likelihood of exposure. The risk assessment process ultimately generates a risk characterization, which ideally includes a range of estimates to quantify identified hazards. A risk characterization may also include discussion of uncertainties, underlying assumptions, and the degree of confidence with which estimates are made.

Risk assessment and other technical analyses cannot provide all the data necessary to make perfectly informed decisions, however. Risk assessment can identify some hazards, but the quantification of risk often involves rough probability estimates whose confidence intervals may be so wide as to render such risk quantification useless.[14] Certain hazards, moreover, simply cannot be quantified because insufficient data exist or the ability to perform useful experiments is limited.[15] Beyond that, other hazards—"unknown unknowns"—cannot be identified because of limitations in modern scientific understanding, random processes, and the inherent unpredictability of interactions among technology, society, and the environment.[16] For emerging technologies, unknown unknowns loom large because we are often dealing with novel mechanisms beyond our expertise. Decisions regarding technology nonetheless must take uncertainties and unknowns into account, even if these indefinites cannot be quantified or readily described.

These points lead to a more fundamental criticism of risk assessment and by extension technology assessment. The quantitative analyses typically generated by risk assessments tend to hinder the consideration of qualitative factors and other pertinent concerns less amenable to scientific characterization.[17] In the climate change arena, for example, a focus on the expected costs and benefits of reducing carbon emissions has sometimes led to the disregard of low-probability catastrophic hazards and of extreme events that cannot yet be identified, let alone quantified.[18] Rational policy making in this area should consider such extreme events, however, as well as concerns of intragenerational and intergenerational equity. Important factors that are often overlooked in the assessment of new technologies—in addition to uncertain and unknown hazards—include loss of economic security, compromise of traditional practices, and ethical matters. In other words, the "technical rationality" of risk assessment, which focuses on scientific measurement, may differ quite dramatically from the "cultural rationality" often reflected in the attitudes of nonscientists.[19] Popular views on risk matters, informed by cultural rationality, consider quantifiable effects as well as contextual factors such as personal experience and social values. Thorough technical analysis is an important part of sound technology management, but incorporating social values and other concerns is equally critical. A more comprehensive assessment of technology would build on the exploration of these concerns to consider broadly the purposes of technology and the conditions we might place on its development and use.

Public Participation

In addition to analyzing technological consequences, technology assessment should engage the public in technology decisions. Participation by the broader public and incorporation of social values in technology management are essential because of the pervasive influence of technology. At a general level, technology shapes society and its institutions. And at an individual level, technology shapes people's lives and affects their vital interests, including health, relationships, employment opportunities, and cultural practices. However, as science and technology policy scholar Daniel Sarewitz observes, "[T]he pursuit of technological transformation is largely exempted from formal democratic processes of eliciting value preferences and adjudicating value disputes about desired future states, even though technological innovation strongly expresses those very things."[20]

Public participation in decisions regarding new technologies rests on several basic rationales: instrumental, normative, and substantive.[21] The instrumental rationale for participation is a relatively narrow one of facilitating support for technological innovation and acceptance of new products. Instrumentally oriented participation treats the public as object rather than as partner in technology development and largely reinforces existing power structures.[22] The participation envisioned in this book, however, would have broader aims of advancing normative and substantive goals. Normatively, public participation in technology management is inherently valuable in that it reflects principles of democratic governance, social justice, and equality.[23] Democratic ideals of autonomy and freedom from excessive government control call for citizens to have an active and meaningful role in the management of powerful technological forces.[24] Moreover, under contemporary democratic notions of public reason and discourse, policymakers should engage the broadest possible array of societal interests.[25] Such engagement seeks to empower citizens and thereby ensure that policy decisions having fundamental effects on society—including decisions regarding new technologies—are not reduced to the agenda of dominant institutions or a small group of elites. A further rationale for public participation is that it can contribute to substantively better outcomes. At a basic level, the public sometimes provides useful insights or suggestions that might otherwise be overlooked. More important, public input into technology management decisions can inform the policy-making process with public values and preferences, such as those regarding tolerance for risk and uncertainty, desire for change, and willingness to make trade-offs.

Who is "the public" that should be involved in decisions on technology? Under one conception of the public, set forth by philosopher John Dewey, the public encompasses all persons substantially affected by the consequences of an activity.[26] Several observations follow from this description. First, there may be multiple publics, depending on the activities that are at issue. Although different technologies may have different associated publics, emerging technologies increasingly have such broad and sweeping scope that their publics largely overlap. Indeed, since such technologies affect nearly all persons, the public impacted by a specific technology may not differ in any meaningful way from what we understand as "the general public." Second, publics are or can be created. Emerging technologies often generate effects or potential effects on persons who are not yet aware of these effects. Genetically modified organisms, for example, are widely present in processed foods, yet few consumers are aware of this fact. In such instances, the affected persons may be shaped into an engaged public by raising awareness of technological developments and encouraging affected persons to apply their foundational values and beliefs to new circumstances. Third, given the impracticality in our complex society of creating a citizenry that is fully participatory on all issues of public significance, the idea of public participation necessarily implies representatives who speak on behalf of others having shared interests.[27] In addition to persons formally elected, these representatives may include those who lack formal power but nonetheless speak for others and not simply themselves.

Public input is warranted not only when explicit policy decisions are made regarding a technology affecting the public but also in the technical, expert-dominated process of technology assessment. Technology assessment is riddled with value-based judgments that should be attuned to the values of the public and not merely the values of the experts conducting the assessment. Take, for example, the practice of risk assessment as applied to chemical substances. Although risk assessment is often characterized as a purely scientific, values-free undertaking, the values of those performing risk assessments necessarily influence the assumptions made, inferences drawn, and calculations performed.[28] The subjective judgments involved in assessing chemical risks include judgments about what exposure levels are likely, which health consequences should be analyzed, how uncertainties should be interpreted, and how risk information should be framed.

Each of these judgments wrestles with problems of uncertainty. This is especially the case with respect to emerging technologies, where data on hazards are often sparse and where future applications may be unknown. Dealing with the uncertainty inherent in risk assessments and technology assessments is not a scientific issue; rather, it is fundamentally a policy issue that turns on subjective beliefs, values, and experiences.[29] For this reason, incorporating public input and discourse into even the technical aspects of risk assessment and technology assessment is essential. Although lay citizens may have limited knowledge on technical matters and are sometimes misled by popular understandings of science and technology,[30] experience has shown that laypersons can communicate effectively with experts and contribute greater breadth and depth of knowledge and experience to technology assessments.[31] Rational discourse involving citizens, experts, and other interested parties ultimately can lead to an expansion of relevant considerations and to the development of consensus regarding the common good.[32]

This emphasis on involving the public more directly and at earlier stages in technology management responds to a number of flawed assumptions of conventional technology assessment. First, the view that technology assessment was a purely objective process failed to take into account the subjective factors that inevitably enter into such assessments. Second, the assumption that technology assessments could focus on finished technologies overlooked the difficulty of controlling or modifying technologies that are often backed by substantial investments and powerful interests. Third, the conventional approach incorrectly assumed that technology policy decisions by elected officials would adequately reflect public values. In theory, political representatives are competent to make decisions consistent with the public interest and public values. Although this theory is foundational to our representative government,[33] there is a sound basis for concluding that elected officials inadequately represent the public with respect to technology matters.

Several factors systematically bias lawmakers against focusing on the potential risks of new technologies. These factors include the specialized body of knowledge necessary to understand science and technology issues as well as electoral pressures on politicians to respond to current headlines.[34] Few political rewards result from addressing issues having distant time horizons, such as emerging technologies. Not surprisingly, our government's general policy toward uncertainties, such as those surrounding the health and environmental effects of emerging technologies, is one of willful ignorance rather than precautionary action.[35] In addition, as public choice theory predicts, lawmakers tend to cater to the interests of an organized and vocal minority at the expense of a majority whose interests are more diffuse.[36] This tendency is especially pronounced with respect to emerging technologies, which often have vigorous corporate or institutional advocates but little opposition from a public largely unaware of a technology or its potential risks.[37] All of these obstacles to the expression of public values through ordinary channels of representation warrant more direct and active public participation in emerging technology management.

In instances where the government does turn its attention to emerging technologies, the empirical techniques of risk assessment and risk-benefit analysis often dominate its policies to the detriment of public preferences.[38] Technically, *risk management* (the values-driven, policy-making process of deciding how to respond to risk data) can be distinguished from *risk assessment* (the expert-driven process of identifying, analyzing, and quantifying risks).[39] Notwithstanding this distinction, risk managers often defer to the quantitative data of risk assessment (or the lack of such data) in deciding how to respond to the hazards posed by new technologies.[40] Because empirical techniques fail to account for all of the factors relevant to social decision making, however, an approach centered on quantifiable risks hides difficult, values-based choices under a veneer of objectivity. Indeed, the act of engaging in the discourse of risk through the risk management process reinforces the questionable assumption that risk can be understood, comprehensively measured, and effectively managed.[41]

In sum, technology assessment, once characterized as the exclusive province of experts, can and should be opened up to public participation. This broader and more open assessment should occur *during* and not just *after* technology development, thus bringing public concerns into research decisions and technology design. Care must be taken in the design of participatory processes to guard against the potential for simply reinforcing existing power relations or creating meaningless exercises in outreach. Nonetheless, if incorporated into properly designed processes, ongoing public participation can improve the quality of analysis and make the decision process more democratic.[42]

Our Experience with Technology Assessment

Past and present efforts in the United States to analyze and manage technology's consequences fall into three basic categories: (1) formal technology assessment, which was practiced by the Office of Technology Assessment (OTA) until 1995 and has since been undertaken occasionally by other government agencies; (2) environmental impact assessment, which evaluates the potential health and environmental effects of policy choices and serves as a form of technology assessment when applied to technology policy decisions; and (3) nongovernmental assessment, which is carried out by technology developers themselves or by other nongovernmental entities. An examination of these efforts reveals that they fall short in advancing the analytical and participatory functions that technology assessment should achieve.

Formal Technology Assessment

The Office of Technology Assessment

The concept of technology assessment arose as "critical voices within science began calling for preassessment before committing society to innovations such as supersonic transport and nuclear weapons."[43] In the United States, Congress created the OTA in 1972 to provide nominally objective analyses that would inform policy decisions on technology matters. While the OTA earned high marks for the quality of its work on issues ranging from the feasibility of the Star Wars missile defense system to the cleanup of nuclear weapons laboratories, it ultimately played only a modest role in influencing U.S. handling of new technologies.[44]

Congress created the OTA at a time of increasing government spending on scientific research and development and growing social unease regarding the negative effects of science and technology. In the words of the statute establishing the OTA, modern technologies were "increasingly extensive, pervasive, and critical in their impact, beneficial and adverse, on the natural and social environment."[45] Against this backdrop, Congress recognized the need for an independent source of technical information and objective analyses on which to base its decisions. The OTA's mission was "to provide early indications of the probable beneficial and adverse impacts of the applications of technology and to develop other coordinate information which may assist the Congress."[46]

Although the OTA originally was expected to perform long-term analyses of entire technological fields, in practice, technology assessment at the OTA was fairly constricted. Congress's requests for information, which tended to emphasize specific subjects and require rapid responses, were typically met with short-term, narrow policy analyses.[47] In an effort to maintain credibility and avoid alienating legislators who might disagree with its conclusions, the agency generally sidestepped recommendations and confined itself to reports on the technical components of the issues.[48] More significantly, technology assessment came to be understood not as a process but as a product: the reports generated by the OTA.[49] These developments were not surprising, given the OTA's subordinate role and limited mandate. The OTA had no regulatory power and limited practical advisory power. The OTA ultimately never fulfilled its potential to study technologies in a broader and more systematic way or to transform how Congress and society related to technology.

The OTA eventually became unable to maintain the difficult balancing act of providing objective information to Congress while securing political support for its survival. Shortly after the 1994 Republican takeover, Congress eliminated the agency as part of the new majority's promise to enact its "Contract with America."[50] Among the primary reasons given for eliminating the agency was its slow pace relative to congressional timetables. The OTA's assessments, though limited in scope, often took one to two years to complete and thus often failed to meet the more pressing political needs of the members of Congress. Also contributing to the OTA's elimination were accusations of bias—whether justified or not—commonly made by parties whose interests were undermined by the OTA's reports.[51]

America's experiment with formal technology assessment through an agency dedicated to the task came to an end with the demise of the OTA. The needs for technology assessment and for deliberate management of technology, however, have not diminished. To the contrary, those needs are even greater today, as rapidly developing technologies will have widespread and long-lasting impacts and raise serious social and ethical concerns.

Institutional Alternatives to the OTA

In the OTA's absence, the function of technology assessment has devolved to a number of other government institutions. Such assessments, however, have occurred on a narrow and ad hoc basis, if at all, and their scope and impact have been limited by traditional and legal constraints. The Government Accountability Office (GAO), for example, has conducted a few technology assessment pilot projects at Congress's direction.[52] With the exception of a recent report on geoengineering, these analyses have focused mostly on narrow topics related to counterterrorism, such as the use of biometric technologies for border security and the use of cybersecurity measures to protect infrastructure. The credibility of the GAO's work, moreover, is open to question, given the agency's traditional expertise in performing audits, its limited experience with predictive assessments, and the past use of its reports for partisan objectives.[53] Another institution, the Office of Science and Technology Policy (OSTP), advises the president on science and technology issues. The OSTP has the authority to "initiate studies, including technology assessments, to resolve critical and emerging problems."[54] With its fairly limited resources, however, the OSTP has focused primarily on serving as a channel of communication between the president and the scientific community and coordinating science and technology policy across the federal government.[55] Finally, the National Research Council (NRC) issues reports on science and technology topics in response to congressional and agency requests. Committees of leading experts prepare these authoritative reports through an extensive, peer-reviewed study process. While well respected, these reports require significant time and resources to produce, and they are designed to generate expert recommendations in response to specific questions.[56] The reports do not raise issues independently, address broader policy questions, or seek to foster public debate—activities that technology assessment should promote.[57]

NEPA as Technology Assessment

Technology assessment also can take place in the context of environmental impact assessment. In particular, the National Environmental Policy Act (NEPA) requires federal agencies to assess the environmental impacts of their actions.[58] NEPA contemplates that this assessment be done in a public process, and its regulations provide for extensive public involvement in that process.[59] To the extent that federal agency actions involve the development or implementation of technology, NEPA analysis could serve as a form of technology assessment with respect to health and environmental effects. The federal government's extensive role in sponsoring research and development, establishing technology policies, and possibly regulating technology's adverse impacts underscores NEPA's potential reach.[60] Nevertheless, narrow interpretations of NEPA by agencies and courts have minimized the statute's value in technology assessment.

The Basics of NEPA

Sometimes described as a "Magna Carta" for the environment,[61] NEPA was enacted with the purpose of integrating environmental values into national policies and planning processes. The central requirement of NEPA is the environmental impact statement (EIS), a document that must be prepared for "major Federal actions significantly affecting the quality of the human environment."[62] In the EIS, federal agencies must describe the environmental impacts of the proposed action, alternatives to the proposed action, and any irreversible and irretrievable commitments of resources that the proposed action would involve. As the Supreme Court has recognized, the EIS requirement is meant to ensure

> that the agency, in reaching its decision, will have available, and will carefully consider, detailed information concerning significant environmental impacts; it also guarantees that the relevant information will be made available to the larger audience that may also play a role in both the decisionmaking process and the implementation of that decision.[63]

For federal actions deemed to have environmental impacts that are less than significant, agencies still must prepare an environmental assessment (EA). An EA is a more limited inquiry aimed at identifying the environmental consequences of a proposed action and documenting an agency's determination that those consequences are not significant.[64]

NEPA was not meant merely to impose a requirement on agencies to document environmental impacts. Rather, NEPA was intended to achieve a wholesale reorientation of governmental actions and values, sensitizing agencies to environmental effects.[65] Through preparation of a programmatic EIS, for example, an agency could evaluate environmental impacts of entire programs instead of conducting piecemeal assessments of individual projects. Overarching and comprehensive analyses presumably would motivate agencies to reorient programs in light of overall impacts.

Implementing and Interpreting NEPA

As implemented by federal agencies and interpreted by courts, however, NEPA has not fulfilled the goals embodied in its broad statutory language and legislative history. NEPA's primary mandate, the requirement that federal agencies prepare EISs, has turned out to be less powerful than some of its crafters might have envisioned. First, that mandate applies only to "major federal actions"; it does not apply to private actions or to the majority of legislation enacted by Congress.[66] Moreover, the Supreme Court has construed the EIS requirement as imposing only a procedural duty rather than as mandating substantive results.[67] This does not necessarily make the EIS requirement toothless: Federal agencies, having analyzed the environmental ramifications of a proposed action and considered public concerns, may make more environmentally sound decisions than would otherwise be the case. Nonetheless, agencies sometimes view the EIS as little more than a paper-pushing hurdle to be overcome before the agency proceeds with a predetermined course of action.[68]

Supreme Court decisions have played a critical role in constraining the statute's reach and effectiveness. The Court first addressed NEPA in 1975 in *Aberdeen & Rockfish R.R. v. Students Challenging Regulatory Agency Procedures (SCRAP)*.[69] There, the Court rejected the argument that an EIS must be integrated into the earliest stages of an agency's decision-making process. Rather, an EIS must be prepared only when an agency "makes a recommendation or report on a proposal for federal action." The Court subsequently held in *Kleppe v. Sierra Club* that neither the contemplation of a project nor the preparation of studies of possible development triggers the obligation to prepare an EIS absent a proposal for federal action.[70] Together, *SCRAP* and *Kleppe* allowed agencies to defer analyzing environmental impacts until after the overall policy-making process is well underway. These decisions did not specifically involve new technologies, but they offered little hope to those expecting that NEPA would compel agencies to conduct effective technology assessments.

The Supreme Court made its most direct pronouncements regarding the analysis of new technologies under NEPA in *Vermont Yankee Nuclear Power Corp. v. Natural Resources Defense Council, Inc.*[71] In this challenge to licenses granted to specific nuclear facilities, the Court rejected the notion that NEPA provides a forum for the wholesale consideration of the desirability of new technologies:

> Nuclear energy may some day be a cheap, safe source of power or it may not. But Congress has made a choice to at least try nuclear energy, establishing a reasonable review process in which courts are to play only a limited role. The fundamental policy questions appropriately resolved in Congress and in the state legislatures are *not* subject to reexamination in the federal courts. . . . NEPA does set forth significant substantive goals for the Nation, but its mandate to the agencies is essentially procedural.[72]

In this passage, the Court correctly recognized that democratic institutions—Congress and state legislatures—should resolve fundamental policy questions. Nonetheless, in holding that NEPA imposes no substantive obligations, the Court failed to appreciate NEPA's potential to inform legislative policy decisions and to guide agencies in the exercise of delegated policy-making authority. Rather, the Court continued to steer agencies toward the narrow, technocratic approach to NEPA analysis that has prevailed ever since. The Court's opinions have not ended the preparation of programmatic EISs—the sort of overarching analyses that could serve as a useful instrument of technology assessment. Many agencies, however, try to defer such programmatic assessments as long as possible, if not completely.[73] NEPA ultimately has not lived up to its potential as a tool for analyzing environmental risks and uncertainties of major policy developments, such as the development of new technologies. In the words of law professor Oliver Houck, "NEPA is missing the point. It is producing lots of little statements on highway segments, timber sales, and other foregone conclusions; it isn't even present, much less effective, when the major decisions . . . are made."[74]

Nongovernmental Mechanisms

Technology assessments can be carried out not only by government agencies but also by those directly engaged in technology development. Researchers, for example, may look to personal morals and professional ethics to guide, shape, and constrain their research pursuits. Under the predominant model of technology development, however, scientists and engineers regard their research endeavors as morally neutral. Technology, in this view, is not inherently good or bad; rather, it is the particular uses that society chooses for a technology that are subject to normative judgments.[75] This model does not encourage researchers to consider their efforts in broader contexts or to pursue the ethical implications of their work. Furthermore, researchers—including those in academic institutions—face strong economic incentives to tailor their efforts in favor of technologies with commercial potential.[76]

Potential legal liabilities may give firms a stronger incentive than individual researchers to carry out health and safety assessments on the technologies they develop and market. In addition, companies may lose their social license to operate should they fail to live up to social expectations regarding environmental behavior.[77] As a result, corporations sometimes go beyond what the law explicitly requires.[78] With respect to new technologies, for example, corporations might assess potential consequences with an unusual degree of care, solicit public concerns, or choose not to develop hazardous applications despite the potential for profit. Such actions can translate into reduced risk exposure, avoided regulation, and higher product premiums.[79] Whatever firms' motivation may be, however, efforts along these lines are unlikely to lead to adequate technology assessment. Firms are profit-seeking entities whose primary obligations are to their shareholders. The analyses that firms do undertake often are internal assessments focused narrowly on marketing potential. Such analyses generally do little to consider negative externalities, inform policymakers, or involve the public in fundamental decisions about technology.[80]

Revitalizing the Basic Tools of Technology Assessment

Neither the government nor technology developers are doing enough to assess the health and environmental effects of emerging technologies or to integrate the public into decisions about technology. The rest of this chapter considers options for addressing these inadequacies. Reconstituting the OTA or reinvigorating NEPA would be relatively straightforward yet modest ways to increase our understanding of new technologies. Because such steps alone are insufficient to achieve substantive change and meaningful participation, however, more far-reaching reforms must be considered.

Reconstituting the OTA

Reestablishing the Office of Technology Assessment would be a first step toward building societal capacity to make informed and accountable decisions about new technologies. Indeed, since the OTA's elimination, there have been several proposals to either create a new technology assessment office or authorize funding to reestablish the OTA.[81] The primary benefit of a reestablished OTA would be to address Congress's ongoing need for objective advice on scientific and technical matters. Few current members of Congress have had any experience with the OTA, however, and none of the proposals has gained much traction.[82] More important, a reconstituted OTA would be subject to similar time and political constraints as its predecessor. Indeed, a truly independent technology assessment office might raise concerns that are not politically expedient to address. Furthermore, it is unlikely that such an entity would generate new information on the hazards of emerging technologies, nor would it bring about broad and meaningful public engagement. To achieve these ends and bring about robust technology assessment that extends beyond narrow analyses of risks and benefits, we will need more than a reconstituted OTA.

Recapturing NEPA's Lost Potential

More vigorously implementing NEPA would be another modest step to improve societal decision making on technology. Overall, current agency practices under NEPA are a pale shadow of the possibilities embodied in the statute.[83] As already discussed, those practices entail little wholesale consideration of technological developments and their ramifications for society and the environment. Rectifying this shortcoming in NEPA implementation would help address concerns regarding emerging technologies. NEPA ultimately can provide no more than a partial response, however. NEPA's scope is confined to those developments that involve federal agency action, and its lack of substantive teeth would limit its effect on resulting policies.

NEPA's Concern with Technology

Reexamination of NEPA and its legislative history demonstrates the statute's potential to serve as a starting point for better technology assessment. NEPA requires that agencies adopt a long-term orientation toward considering environmental consequences for present and future generations.

Specifically, each EIS must discuss "the relationship between local short-term uses of man's environment and the maintenance and enhancement of long-term productivity," as well as "any irreversible and irretrievable commitments of resources."[84] Moreover, the statute explicitly recognizes "the profound influences of . . . new and expanding technological advances" on the environment and strives to "promote efforts which will prevent or eliminate damage to the environment and biosphere."[85]

NEPA's legislative history describes even more emphatically the dangers posed by technology. In floor debates and committee hearings, various members of Congress expressed their individual concerns about damage to the environment caused by technology.[86] Quoting an editorial that appeared in the *New York Times*, the House committee responsible for considering NEPA legislation identified technology as the greatest threat to the environment:

> By land, sea, and air, the enemies of man's survival relentlessly press their attack. The most dangerous of all these enemies is man's own undirected technology. The radioactive poisons from nuclear tests, the runoff into rivers of nitrogen fertilizers, the smog from automobiles, the pesticides in the food chains, and the destruction of topsoil by strip mining are examples of the failure to foresee and control the untoward consequences of modern technology.[87]

The counterpart committee in the Senate likewise recognized the dangers posed by a "growing technological power which is far outstripping man's capacity to understand and ability to control its impact on the environment."[88] Indeed, an influential report prepared for that committee singled out technology as the root cause of the perceived environmental crisis:

> Technology . . . has greatly increased environmental stress in general. The net result has been enormously increased demands upon the environment in addition to the increase in population. . . . Unfortunately, our productive technology has been accompanied by side effects which we did not forsee [*sic*]. . . . It is now becoming apparent that we cannot continue to enjoy the benefits of our productive economy unless we bring its harmful side effects under control.[89]

That report called for a "pay-as-you-go" approach to the development and use of technology, under which the cost of environmental harms would be internalized, with "provision . . . made for the protection, restoration, replacement, or rehabilitation of elements in the environment before, or at the time, these resources are used." Unfortunately, current implementation of NEPA is not achieving the statute's objective of transforming the relationships among human activity, productivity, and long-term effects on the environment.

The EIS as Technology Assessment

NEPA's EIS requirement was intended as the central tool for identifying and better managing the effects of emerging technologies on the environment.[90] A federal appellate court's analysis in *Scientists' Institute for Public Information, Inc. v. Atomic Energy Commission* (*SIPI*), issued three years after NEPA's enactment, is particularly instructive on how EISs might enable a more proactive and comprehensive approach to new technologies.[91] In *SIPI*, the plaintiffs alleged that the Atomic Energy Commission (AEC) was required to prepare an EIS for its breeder reactor program, which was in the research and development stage. Conceding that an EIS would be required before the construction of individual breeder reactors and facilities, the AEC argued that NEPA analysis was not required for the research and development program as a whole.

Citing "NEPA's objective of controlling the impact of technology on the environment," the court rejected the AEC's arguments.[92] The court explained that environmental analysis of the research and development program was necessary because the program would facilitate subsequent use of breeder reactor technology by private parties that in turn would affect the environment. As the court declared, "[T]he decisions our society makes today as to the direction of research and development will determine what technologies are available 10, 20, or 30 years hence."[93] The court further noted that consideration of a technology's environmental impacts would be far more meaningful at the research and development stage than at the point when specific facilities are being constructed. Once planning for a specific project or facility has begun, substantial resources are already committed in developing the technology, and parties with vested interests in that technology—perhaps including the regulatory agency itself—will undermine objectivity in the decision-making process. In the court's view, only a programmatic EIS at the research and development stage would fulfill NEPA's purpose of adequately informing Congress, the executive branch, and the public of the environmental effects of new technologies and thereby ensure informed decision making.

The *SIPI* court's interpretation of the EIS obligation, which has since been largely abandoned,[94] is consistent with NEPA's text, purposes, and implementing regulations. As noted earlier, an EIS must be prepared for "major Federal actions significantly affecting the quality of the human environment." NEPA does not define "major federal actions," but U.S. government decisions and policies regarding new technologies often significantly affect environmental quality. Moreover, NEPA's regulations provide that "major federal actions" "include new and continuing activities, including projects and programs entirely or partly financed, assisted, conducted, regulated, or approved by federal agencies."[95] The regulations also indicate that agencies may conduct their evaluations by "stage of technological development including federal or federally assisted research, [and] development or demonstration programs for new technologies."[96]

In light of these regulations and the *SIPI* analysis, federal research funding decisions should be treated as major federal actions subject to NEPA. Of course, the nature of the environmental analysis and level of detail will depend on the stage of technological research and development. Where the funding is meant to support basic research in a new field, discussion of potential environmental effects may necessarily be general and couched in uncertainty. Such discussion can nonetheless identify potential applications of prospective research findings. In contrast, where the supported research involves technology nearing commercialization or deployment, environmental effects should be analyzed more thoroughly and specifically. Funding agencies should not be expected to have perfect foresight, but they also should not be allowed to disregard the potential consequences of the research they sponsor. In sum, the EIS is well suited for prompting agencies to study and consider an emerging technology's potential consequences throughout a technology's development, well before a particular technology becomes entrenched.

Other NEPA Tools for Technology Assessment

NEPA contains additional mechanisms that could help to manage new technologies better. Section 101 of the statute expresses Congress's concern about the environmental impacts of new technologies and suggests a more thoughtful approach to the relationship between technology and the environment:

> The Congress, recognizing the profound impact of man's activity on the interrelations of all components of the natural environment, particularly the profound influences of . . . *new and expanding technological*

> *advances* . . . , declares that it is the continuing policy of the Federal Government, . . . to use all practicable means and measures, . . . to create and maintain conditions under which man and nature can exist in productive harmony, and fulfill the social, economic, and other requirements of present and future generations of Americans.[97]

Though the Supreme Court has held Section 101 to be a judicially unenforceable policy statement, the provision represents a broad, government-wide commitment to address potential dangers posed by new technologies.[98] Section 101, in other words, directs and authorizes federal agencies to make discretionary choices in ways that reduce environmental damage and maximize public input. Such choices may concern, among other things, funding of research or investigation into the effects of emerging technologies.

NEPA also established an institutional mechanism to study and address the environmental ramifications of new technologies: the Council on Environmental Quality (CEQ). The CEQ's functions were to include the gathering of information concerning conditions and trends in environmental quality—including the environmental impacts of new technologies.[99] Congressional debate on the provisions to establish the CEQ repeatedly emphasized the council's authority to conduct research and provide policy advice regarding the often-overlooked environmental consequences of new technologies.[100] And shortly after NEPA's enactment, President Richard Nixon issued an executive order directing the CEQ to foster "investigations, studies, surveys, research, and analyses relating to (i) ecological systems and environmental quality, (ii) the impact of new and changing technologies thereon, and (iii) means of preventing or reducing adverse effects from such technologies."[101] Just as the Council of Economic Advisers provides economic advice to the president, the CEQ was to oversee the execution of NEPA's declared policy of protecting the environment and provide advice on how to further that policy.[102]

The CEQ's role, however, has turned out to be far more circumscribed than originally envisioned. The CEQ has issued regulations and guidance documents that have played a significant role in implementing the EIS requirement.[103] But the agency's broader mission of anticipating environmental problems and providing policy advice has been largely neglected. Perpetually underfunded, the CEQ has since 1980 operated through a single member rather than a full council of three.[104] The CEQ nonetheless retains the statutory authority to analyze environmental trends, study environmental impacts of new technologies, and help formulate policy. With robust support from Congress and the president—support that it has not received to date—the CEQ could play a key role in technology assessment. In particular, the CEQ could identify emerging technologies and their potential hazards, thereby laying the groundwork for more detailed assessment, policy initiatives, and legislative action.

Technology Assessment 2.0

A reconstituted OTA and revitalized NEPA could improve our understanding of emerging technologies and better inform relevant policies. Genuine transformation of our relationship with such technologies, however, requires redesigned technology assessment practices and new mechanisms to incorporate public participation.

In the conventional form of technology assessment practiced by the OTA (referred to here as "conventional TA"), the OTA submitted reports to Congress on the potential impacts of a relatively finished technology. This form of technology assessment was quite limited: Its purposes were relatively circumscribed, its intended audience was narrow, and its impacts were uncertain. Rather than attempting to involve the public or to project the extended consequences of technological developments, the OTA took on the more manageable task of summarizing existing knowledge on narrowly focused issues.

An important factor in the relative ineffectiveness of conventional TA was the long-standing separation of the development of technology from its control and regulation. Conventional TA, in other words, had little effect on whether and how a technology developed. Whatever modest influence it exercised came about by acting as an "after-the-fact gatekeeper."[105] Furthermore, conventional TA tended to treat technology decisions as static rather than ongoing processes. Assessing a technology, however, should be a continuous and integral aspect of technology management. Depending on who participates in TA and how it is carried out, these processes themselves can have a critical influence on the shapes and uses of a technology. The separation of the OTA's technical analysis from the congressional decision-making process meant that there was little assurance that the agency's analysis would make a difference in policy decisions. And because the conventional TA process was limited largely to experts, it was relatively distant from democratic control and unresponsive to public values.

In an effort to address the deficiencies of conventional TA, various countries have experimented with modified forms of TA, including participatory technology assessment (PTA) and constructive technology assessment (CTA). These more recent forms of TA represent important advances but would benefit from further reforms to improve the analysis of emerging technologies and better incorporate public values.

Participatory Technology Assessment

Unlike conventional TA, PTA actively seeks to incorporate public values into the assessment process by involving a wide range of actors.[106] Participants in PTA include lay members of the public as well as technology stakeholders and experts. By widening the scope of participation, PTA may educate the public, stimulate public debate, set the political agenda, and inform decision makers of public opinions and values. PTA, in other words, enhances the expert assessment function of conventional TA by bolstering the credibility of expert assessments and providing more information on the social acceptability of a technology. Its strength ultimately lies in its ability to project public values into decisions concerning technology.

PTA taps into both participatory and representative conceptions of democracy. Citizens who take part in PTA engage directly and extensively with critical issues, and their views may be presented to decision makers. Participants are not accountable to constituents, nor do they formally represent particular interests, but they frequently reflect varying viewpoints.[107] PTA differs from the more familiar participatory format of town hall meetings in this representational aspect as well as in the depth of citizen involvement. Unlike polls or focus groups, which are comparatively superficial means of gauging public opinion, PTA strives to elicit diverse views in a policy-making context after giving citizens the opportunity to learn about and reflect on an issue.[108] And because panel membership is drawn from a pool of randomly selected individuals, PTA is less subject to political capture or grandstanding than are advisory commissions.[109]

PTA techniques include consensus conferences, citizens' juries, and planning cells. Each of these techniques organizes laypersons into panels that consult over several days and consider the input of experts and others.[110] In a consensus conference, for example, a multiday public meeting is convened to foster a dialogue between a panel of fifteen to twenty-five citizens and a group of experts.[111] Citizens invited to serve on the panel are drawn from a random sample of the population and then selected through an application process that seeks to identify persons sufficiently dedicated to participate. Assisted by a facilitator trained in communication skills and cooperative techniques, the citizen panel conducts two preparatory meetings prior to the consensus conference. In these meetings, panel members learn basic information about the technology at issue and formulate questions to be addressed at the conference. At the conference itself, experts (selected in part by the citizen panel) present their answers to the panel's questions and respond to cross-examination by citizens. After a discussion period, the citizen panel prepares a consensus-based report presenting its conclusions and recommendations. The report has no binding effect but is available to the public, experts, and politicians for their consideration.

Other PTA techniques may differ in the details of implementation, but the underlying purposes are the same: to involve citizens in technology assessment and incorporate public input more effectively into technology decisions.[112] Citizens' juries, for example, generate lay findings and recommendations on focused policy questions in a process akin to a jury trial, often in a local or regional context.[113] Planning cells also involve randomly selected citizens in a trial-like process but allow participants more freedom to design policy options and consider a range of concerns.[114]

Constructive Technology Assessment

Innovation often occurs within private laboratories outside of public scrutiny, and external input and regulation are contemplated only after a technology appears in commercial applications.[115] In contrast to this approach of after-the-fact assessment and oversight, constructive technology assessment (CTA) seeks to influence technology design itself by promoting interaction among stakeholders throughout a technology's development process.[116] Here, the concept of stakeholders is defined broadly to include technology developers, regulators, workers, end users, and the potentially affected public. Incorporating various interests through such interaction, it is hoped, will generate more widely accepted outcomes and fewer adverse effects. Although CTA may involve public participation, such participation has a substantive aim of shaping technology and identifying risks and is not a normative goal in and of itself.[117]

Advocates identify three key features of CTA: anticipation, reflexivity, and social learning.[118] Anticipation of potential technological interactions and adverse effects can be promoted by involving a broad range of stakeholders and by conducting appropriate experiments early in the design process. Because not all concerns can be identified at the outset, however, the technology development process should be flexible and iterative, allowing for periodic reexamination of interactions and effects. Reflexivity refers to an appreciation of the potential social effects of different technology design options. Rather than focusing narrowly on technical goals, technology developers should take into account social, environmental, and other consequences as they design and evaluate new technologies. Finally, social learning refers to a mutual process in which stakeholders learn from each other in the course of technology development and use. Through social learning, companies learn about consumer preferences and regulatory requirements and can design products accordingly. Variants of CTA incorporate public opinion polling, content analysis, and other tools to complement the basic CTA process.[119]

Denmark and the Netherlands have been at the forefront of the development of PTA and CTA. In Denmark, the consensus conference technique has shaped policy on food irradiation, genetically engineered animals, and the use of knowledge about the human genome.[120] In the Netherlands, CTA techniques were applied to the creation and design of novel, environmentally friendly alternatives to meat. Meetings among various stakeholders generated agreement on minimum standards for such products and led technology developers to devote more attention to issues of taste and texture.[121]

Concerns

Although PTA and CTA offer valuable tools for managing new technologies more effectively and democratically, commentators have raised a number of concerns in regard to these more recent forms of technology assessment. These concerns underscore the difficulty of changing established practices and bringing about meaningful public involvement.

One obvious issue involves the competence and willingness of laypersons to weigh in on emerging technology issues. The average citizen is likely to have limited understanding of the technical matters that undergird modern technologies. Laypersons may defer to experts or rely on simplifying heuristics when assessing new technologies.[122] Citizen participation may even cripple technology development if unwarranted and irrational fears come to dominate the assessment process.[123] When a possible outcome evokes strong emotional responses, for example, people tend to focus on adverse outcomes and to neglect probabilistic calculations.[124] Furthermore, low levels of civic engagement, as reflected in modest U.S. voter turnout, suggest that laypersons might not take much interest in technology assessment.

If given the opportunity to participate in a meaningful way, however, citizens have proven willing and competent to engage in matters of public debate. Layperson interest is suggested not only by anecdotal evidence, such as vigorous participation in policy blogs, but also by studies finding an eagerness to participate and resentment at exclusion.[125] In evaluating citizen competence, it is critical to keep in mind that lay participation is primarily meant to provide a read on public values, particularly with respect to concerns and risks, and not to replace technical expertise.[126] Of course, laypersons occasionally will identify technical issues, problems, and contextual factors overlooked by experts.[127] But the main question is whether laypersons can become informed on the technical issues to the extent required to meaningfully apply their personal experiences, belief systems, and values to new technological situations of risk and uncertainty.

As an initial matter, public opinions regarding emerging technologies do not appear to be driven by irrational fears. A survey of public views regarding synthetic biology, for example, reported a 2:1 majority in favor of continued research rather than a ban, even after respondents were informed of potential risks and benefits.[128] Similarly, scientists who participated in public engagement exercises on nanotechnology in the United Kingdom found the public to be more open-minded and accepting of the technology than they had expected.[129] PTA, moreover, enables informed participation by providing citizens with the time and access to expertise required to learn about the issues in some depth.[130] As "values consultants," citizens participating in past PTA exercises have vigorously and critically questioned experts, regulators, and stakeholders.[131] Examples of citizen activism in technical controversies likewise suggest that laypersons can come to informed views, particularly if assisted by experts.[132] This is not to say that laypersons necessarily will agree with experts or policymakers after discussing technology matters. Public opposition to the deployment of a new technology ultimately may reflect not an inability to comprehend scientific matters but rather a distrust of the scientific and government institutions that are invested in them.[133]

Another criticism of participatory approaches to TA concerns the legitimacy of citizen participation.[134] Participation in TA is typically limited to a few selected citizens, yet these lay participants do not constitute democratic representatives of the public will. Not needing to answer to an electorate, participants may endorse unrealistic or politically infeasible policy options.[135] Such criticism, however, overlooks the fact that PTA does not replace representative decision-making procedures. Rather, PTA serves a consultative role to support decision making by politically accountable elected representatives.[136] PTA thus incorporates elements of direct participatory democracy yet remains rooted in a representative democratic system. The method of selecting lay participants nevertheless does matter to the credibility and utility of the process. Random selection of participants does not ensure that citizen deliberations will be representative, but it can facilitate the expression of a wide range of views, promote participant independence, and ameliorate concerns that interest groups will rig the process.[137]

To the extent that PTA and CTA techniques tend to emphasize the formation of consensus, they may mute the expression of alternative viewpoints or dampen values conflicts that are in fact healthy for social risk management.[138] Conversely, achieving consensus may not be possible where value differences are great, as is often the case with new technologies. Indeed, one review of PTA efforts suggests that PTA "increases the complexity of decision making by taking into account different values to assess impacts of technology, [and] by supplying all information and knowledge available and conveying uncertainties or deficiencies of knowledge."[139] Moreover, the time constraints inherent in some participatory techniques may not allow adequate opportunity for the deep normative deliberation that democratic idealists might desire.[140] Such concerns underscore the limitations of PTA and CTA. These techniques nonetheless can serve as a good starting point for identifying public concerns and provide a useful road map for future research and analysis.

Mirroring concerns about the willingness of ordinary citizens to participate in PTA are concerns that private industry will be reluctant to participate in processes, such as CTA, that risk divulging trade secrets. Indeed, perhaps the most serious objections to PTA and CTA involve the relatively limited effect that such assessments may have on actual decisions by technology companies and policymakers. Simply providing the results of assessments to decision makers is unlikely to overcome vested interests' domination of political decisions.[141] Policymakers may be inclined to dismiss such results as nonrepresentative or lacking reliable expertise. And public participation may be of limited value if experts or vested interests control the framing of a problem and policy options. Rather than changing the relationship between society and new technologies, TA might serve as little more than an academic exercise, defusing opposition by creating the appearance of an open, participatory process.[142] At the same time, powerful actors might escape accountability for their role in promoting risky technologies.

Indeed, the effect of citizen deliberations on the formation of technology policy, while difficult to measure, generally appears modest.[143] Defenders of PTA attribute its limited policy impact to its consultative nature;

PTA is intended to inform rather than dictate policy decisions.[144] Critics, however, contend that PTA fails to bring about the multilateral communication among lay citizens, stakeholders, and decision makers that is needed.[145] PTA's effectiveness can be enhanced by ensuring openness in the process, providing for widespread dissemination of results, and timing citizen deliberations to facilitate incorporation of their output into an ongoing decision-making process.[146] Likewise, integrating the results of CTA into actual technological development has proven to be a daunting challenge.[147] Technology developers often hesitate to open up their processes to outsider scrutiny, let alone outsider participation. Absent a regulatory mandate, the effect of CTA activities may be at best indirect, serving primarily as an external critique or source of pressure on technology developers.

Beyond Current PTA/CTA Efforts

Attempts to use modified TA techniques in the United States during the 1980s and 1990s were limited in scope and generally did not consider emerging technologies.[148] However, the growing field of nanotechnology is now serving as a testing ground for these methods in the United States and other countries, as chapter 3 discusses. Questions nonetheless persist regarding how to engage the public in emerging technology issues and how to interject public values more effectively into technology decision-making processes. This section considers several options for doing so and suggests possible applications.

A Role for Elected Officials

One mechanism to connect PTA more directly to the policy-making process involves the participation of elected officials and other policymakers directly in technology assessment efforts. In the 1980s, several citizens' juries were convened in the United States on an experimental basis.[149] Seeking to overcome the political and cultural resistance to these forms of public participation, organizers convinced two congressional representatives to participate in portions of citizens' jury hearings. Widespread legislator participation in PTA is unlikely, however, in light of the numerous demands on legislators' time and the limited payoff for legislators who do participate.

A more realistic approach might involve public leaders raising the profile of emerging technology issues to stimulate public debate. A prominent example of this approach involves federal funding of research on stem cells derived from human embryos. In 2001, President George W. Bush issued restrictions on such funding after personally deliberating on the issue in a very public way.[150] Of interest here is not the substance of that decision, which was criticized on a number of grounds and later reversed, but rather the process leading up to it.[151] President Bush publicly declared that he was reviewing the policy and announced his decision in a prime-time address after consulting with scientists, ethicists, elected officials, and others.[152] This high-profile consideration of the issue appropriately expanded to the general public a debate that had been previously limited to a "small, professionally invested elite."[153] More recently, President Barack Obama used the news that scientists had created a synthetic bacterial genome as an opportunity to order a bioethics commission to report on the implications of synthetic biology research.[154] Although the commission's report stimulated debate within the policy community, greater presidential attention to the report could have elevated the subject to public prominence.

Technology Referendums

Nonbinding national technology referendums are a potentially more systematic mechanism to increase public involvement and deliberation. Potential subjects of such referendums could be identified by Congress, the president, or the CEQ. Ideally, such referendums would take place on a periodic basis, and they would follow consideration of an emerging technology by citizens' juries or similar panels. The deliberations and recommendations of citizens' juries could provide valuable background, promote rational discourse, and help to counter the effects of superficial media campaigns that might arise.[155] In preparation for the referendum, an appointed panel of experts, social scientists, and citizens would develop a limited number of questions to be considered on the ballot, each having several possible responses. As nonbinding instruments, such referendums would supplement rather than replace representative decision making.[156] A technology referendum nonetheless would make emerging technology issues salient and bring the broader public into discussions about technology. Moreover, the resultant opportunity for societal debate would enable the airing of concerns that might otherwise be overlooked.[157] Through technology referendums, citizens would gain a direct voice in critical developments that shape their lives.

Referendums are an imperfect tool for gauging public sentiment or promoting public deliberation, of course. Like voters in general, referendum voters tend to be older, wealthier, and more educated than the average person.[158] The results of a nonbinding referendum could be viewed as analogous to the results of a national public opinion poll. Because these results are meaningful only to the extent they reflect informed views, well-designed efforts to educate the public are essential. In addition, referendum questions must be framed with care to ensure their comprehensibility and neutrality. Referendums may be especially suited for gauging public views on technologies whose potential applications are fairly concrete and less appropriate for measuring public views on abstract questions. Finally, referendums should be used judiciously, as the electorate has limited time and attention to devote to the issues.

Notwithstanding these caveats, nonbinding referendums could bring needed attention to emerging technology issues and prove more effective than polls in engaging and educating the public. A nonbinding referendum would essentially open up the public hearing phase of legislation—a process typically dominated by lobbyists and insiders—to the entire electorate while leaving the lawmaking details and final policy judgments to Congress.[159] The question nonetheless remains whether the results of such referendums would make a practical difference. As an empirical matter, representative bodies frequently do follow the results of nonbinding referendums, a fact that suggests their persuasive effect on elected officials.[160] More generally, public policy follows public opinion more often than not, particularly when an issue is salient and public opinion is clear.[161] Carrying out technology referendums in conjunction with presidential elections could promote relatively high voter turnout, stimulate public interest, and limit the potential for strategic timing of referendums.[162]

Broadening Research Grant Criteria

In terms of bringing about effective and open technology assessment, technology developers must face stronger incentives to consider health and environmental risks as well as other public concerns. First, technology assessment and public involvement should be included as criteria for awarding research grants. The National Science Foundation (NSF), for example, considers both the "intellectual merit" and "broader impacts" of proposed research in reviewing research grant proposals.[163] Under the "broader impacts criterion" (BIC), the NSF considers whether a proposal would (1) promote teaching, training, and learning; (2) broaden the participation of underrepresented groups; (3) enhance the infrastructure for research and education; (4) enhance scientific understanding through broad dissemination of results; and (5) provide benefits to society.[164] In theory, the BIC could be a means of evaluating a research proposal's incorporation of public participation as well as the social and environmental implications of such research. However, application of the BIC has run into deep-rooted resistance among researchers and reviewers in the scientific community. In practice, the BIC is often treated as a relatively unimportant criterion that can be fulfilled merely by hiring professional educators to disseminate information to the general public.[165] Such a limited conception of the BIC fails not only to encourage scientists to involve the public in technology research and development but also to transform how scientists think about their research and its broader societal implications. Including social scientists on proposal review panels would facilitate an expanded understanding of the BIC and strengthen its role in government research funding decisions.[166] Another way to make the BIC more effective would be for the NSF to amend its grant proposal guidelines to explicitly require PTA or other forms of public consultation and to require consideration of a broader range of impacts. Just as institutional review boards have become an accepted mechanism for incorporating ethical concerns into human subjects research, modified versions of TA might someday be broadly institutionalized within scientific research processes.[167]

Research Oversight

Greater governmental oversight of research activities, including privately funded research, may be needed in some instances. Most basic research should proceed without pause. Research activities that involve significant risks or uncertain hazards to society, however, and research that raises serious ethical issues should not proceed without mechanisms to ensure consideration of these concerns. Many emerging technologies, including synthetic biology and human enhancement technologies, involve at least some research activity in this category. Moreover, publicly sponsored research in support of specific projects with far-reaching and potentially controversial effects likewise requires careful review and deliberation. Such goal-directed research, exemplified by the geoengineering research discussed in chapter 4, arguably implies a conditional societal commitment to implement the project in the future.

Other Incentives for Assessment and Responsible Behavior

Incentives to hold technology developers responsible for their products' adverse health and environmental effects also are essential. An important step toward creating such incentives would be to require greater public disclosure of the use of emerging technologies. Labeling genetically modified foods or products that incorporate nanotechnology, for example, can enable more informed consumer choices, increase public awareness of new technologies, and reduce barriers to holding manufacturers liable for their products' adverse consequences. Additional measures will likely be necessary, however, to create stronger incentives further upstream in the technology development process. One proposal to incentivize more thorough private assessment of potential externalities would impose tort liability for "foreseeably unforeseeable" consequences of new technologies.[168] The downsides of this proposal include legal uncertainty and potentially crippling liability for technology developers. Alternatively, as chapter 3 explains in connection with nanotechnology oversight, environmental assurance bonding can be a suitable policy tool to address situations where substantial uncertainty surrounds health and environmental risks. Requiring companies to post bonds in such circumstances helps to assure the existence of funds to pay for subsequently discovered damages without blocking new technologies from entering the market.[169]

Global Governance

Finally, because technological innovation, use, and consequences are not confined within national borders, mechanisms for addressing technology at a global level also will be necessary. All of the emerging technologies considered in this book raise substantial issues of global governance: Genetically modified organisms—and the choice of some countries to reject them—have given rise to international trade disputes; nanotechnology and synthetic biology have largely escaped regulation, thanks in part to concerns that domestic oversight might drive research and development activities to other jurisdictions; and geoengineering purports to address the global problem of climate change through controversial techniques that could pose distinct hazards for different parts of the world. The fundamental insight of ecology—that we are all interconnected—applies as well to today's emerging technologies and their regulation.

The global nature of technology can be beneficial: Cooperation in research and development can hasten breakthroughs, and joint efforts to identify potential hazards can yield synergistic benefits. But the obstacles to technology assessment and public participation, as imposing as they may be domestically, are multiplied at the global level. Here, efforts should focus on harmonizing standards, when appropriate, to avoid a regulatory race to the bottom that can undermine domestic regimes. Complete harmonization will be impossible, as national technology policies will reflect differing social preferences and value judgments. Nonetheless, common baseline standards of health and environmental protection will reduce incentives for technology developers to seek out less restrictive regimes. Chapter 3 considers this issue further in the context of nanotechnology oversight.

Finding a way to operationalize public participation on global issues poses a substantial challenge as well. The emphasis here should be on developing decision-making structures that broaden representation, particularly of disadvantaged persons who could suffer adverse health, environmental, or social effects as a consequence of technological change. The possibility of geoengineering the Earth's climate, which has sometimes been likened to a thermostat for the planet, raises these issues in an especially pointed way, as chapter 4 discusses.

Conclusion

Transforming our approach to emerging technologies to integrate more thorough assessment and greater public participation into the ongoing process of technology management will not be easy. By their nature, emerging technologies often defy prediction regarding their developmental paths, applications, and adverse consequences. Furthermore, a more proactive and participatory approach to technology management runs counter to cultural norms that celebrate innovation, scientific paradigms that emphasize freedom of inquiry and expert peer review, and political discourse that demands scientific certainty as a prerequisite for regulatory oversight. Yet to continue with our current approach would be narrow, shortsighted, and unrepresentative. Transforming our relationships with emerging technologies will require new processes and mechanisms to incorporate citizen participation and compel thoughtful and responsible technology development. In addition, citizens must take an active role in the management of emerging technologies by informing themselves on relevant issues, participating in available assessment and management processes, and demanding that public officials exercise effective oversight.

The following chapters explore what more participatory and effective technology management might look like in the context of specific technologies. First, our experience with GMOs provides a nice example of how technology assessment should *not* be done. As chapter 2 explains, GMO development and oversight were left largely to scientists and industry, no programmatic NEPA analysis of the technology's impacts took place, and the public was—and continues to be—deprived of basic information about the prevalence of GMOs. Exclusion of the public from the process has contributed to controversy over GMOs, and that controversy persists notwithstanding their widespread use. Second, ongoing efforts with respect to nanotechnology illustrate some of the limitations of current technology assessment techniques. Chapter 3 relates efforts to study nanotechnology's health and environmental risks, which are lagging behind the expanding presence of nanotechnology. The chapter also describes efforts to engage the public, which are having minimal effects on nanotechnology policy. Reforms that might enable more proactive and participatory management are available and should be incorporated promptly. Third, geoengineering offers an opportunity to identify hazards and integrate public values early on, before any geoengineering technologies are developed. Compared to other emerging technologies, geoengineering presents a relatively discrete issue: use of a specific set of tools in response to climate change. One pivotal challenge here, considered in chapter 4, is the global nature of geoengineering, which ultimately will require technology management decisions to be made by the international community. Finally, synthetic biology and human enhancement, additional emerging technology fields in their infancy, provide further opportunities to do technology management properly. But as chapters 5 and 6 explain, ethical issues are central to determining how to proceed, and these issues make the task especially challenging.

CHAPTER 2

Biotechnology

Emerging Technology Past and Present

This chapter reviews the development and regulatory history of genetically modified organisms (GMOs)—organisms produced by combining DNA from different living things. Given their widespread presence in American agriculture today, GMOs represent more an established than an emerging technology. Even after two decades of commercialization, however, agricultural biotechnology still has not gained full acceptance, and its consequences remain uncertain. The long-term health effects of consuming GMOs are largely unstudied, for example, and adverse effects on agriculture, such as the development of pesticide resistance, are just becoming apparent. In addition to the uncertainty of such effects, consumer unease with GMOs reflects a fundamental discomfort with genetic manipulation and modern, industrialized methods of food production. Forthcoming and expanded biotechnology applications, such as genetically modified (GM) salmon and plants engineered to produce pharmaceuticals, raise further concerns and will warrant close attention as producers seek to introduce these products into the market. Most important for the purposes of this book, the rise of GMOs and the accompanying controversy and history of GMO regulation offer general lessons in how society should—and should not—handle emerging technologies.

Introduction

Biotechnology can refer to conventional genetic engineering, where scientists transfer existing genetic material from one organism to another to produce desired traits, and to synthetic biology, where researchers synthesize novel genetic material coding for desired traits. Conventional genetic engineering has given rise to medical and agricultural applications. In the medical field, genetic engineering has been used to manufacture synthetic insulin, human growth hormone, and other desired proteins for treatments. In agriculture, genetic engineering has been used to modify crops and other plant and animal species. This chapter focuses on these agricultural applications, which are particularly controversial. In contrast to conventional genetic engineering, synthetic biology, considered in detail in chapter 5, is at an early stage of research and has not yet achieved practical applications.[1]

Over the past two decades, genetic engineering has gained widespread use in crop modification in the United States and other countries. Genetic engineering involves the introduction of transgenes—genetic material isolated from one organism and transferred to another—and thereby offers two significant advantages over traditional plant breeding techniques.[2] First, genetic engineers are able to incorporate a wider range of desired traits into a crop, including traits originally found only in unrelated species. Second, genetic engineering allows desired traits to be incorporated more quickly into a crop than through traditional cross-breeding. To date, genetic engineers have focused primarily on incorporating pesticidal traits or herbicide resistance into commodity crops.[3] These modifications can lower production costs and increase crop productivity. They can also reduce chemical pesticide use and improve soil and water quality by decreasing the need to till fields.[4] Genetically modified varieties of commodity crops have become widely prevalent, accounting for 88 percent of corn, 93 percent of soybeans, 94 percent of cotton, and 93 percent of canola planted in the United States.[5] Furthermore, an estimated 80 percent of processed foods in American grocery stores contain GMOs of some sort.[6] In other words, genetically modified foods are virtually unavoidable in the marketplace. Genetic engineers are now shifting their attention to noncommodity crops and to a wider range of crop traits, including drought and frost tolerance, nutritional content, and flavor.[7] If successful, such efforts could be vital in combating malnutrition and adapting to climate change. Genetic engineers are also seeking to develop plants and animals that can produce industrial chemicals, pharmaceutical compounds, and even organ and tissue replacements.[8]

GMOs have become commonplace in U.S. agriculture and on American supermarket shelves despite surprisingly low public awareness and support. Only about one-fourth of the American public favors the use of GMOs in human food.[9] The widespread presence of GMOs is perhaps made possible by many people's unfounded belief that they have never ingested foods containing GMOs. When asked to explain their views, opponents of GMOs identify health and environmental concerns, effects on small-scale farming, and general concern about the "unnaturalness" of GMOs.[10]

The consumption of GM foods on a broad scale has to date resulted in little evidence of adverse human health effects. Nonetheless, concerns about hazards to human health persist. These concerns generally involve toxicity and potential allergic reactions. Toxic effects from the consumption of GMOs may result from pesticides or other compounds produced by engineered crops or from naturally occurring toxins whose levels have inadvertently increased as a result of genetic manipulation.[11] With respect to allergenic effects, genes are sequences of DNA that code for proteins; proteins, in turn, are potential food allergens. Thus, the transfer of a gene from one plant to another may transfer to or induce in the host plant allergenic properties not otherwise present.[12] Disagreements regarding the potential for these and other unintended effects arise from contrasting models of how genetic engineering works. Proponents portray the insertion of new genes into living cells as a precise and mechanistic endeavor. This view assumes that each gene codes for exactly one protein and thus that the transfer of a gene from one organism to another leads to predictable effects.[13] Critics contend, however, that genetic manipulation can affect the expression of various genes in unpredictable ways. This latter, more complex understanding is supported by recent discoveries that many genes overlap each other and that genes and the DNA sequences regulating their activity are sometimes located on distant parts of a chromosome.[14]

The environmental risks of GMOs include the potential for gene transfer, ecosystem disruption, and the development of pesticide resistance. Specifically, gene transfer refers to the inadvertent movement of transgenes through cross-pollination from an engineered crop to relatives of that crop.[15] Those relatives may include other varieties of the same crop, which can acquire new traits without our knowledge. For example, an heirloom variety of a crop might acquire pesticidal properties through cross-pollination with a genetically engineered strain of that crop.[16] Gene transfer may also occur between a GMO and its wild relatives. As a result, a rare wild species might become extinct. Or conversely, wild relatives that acquire transgenes might demonstrate increased weediness and consequently crowd out competing species and disrupt ecosystems in unexpected ways. Ecosystem disruption also may occur as a result of the establishment of wild populations of GM plants or of the loss of beneficial insects (such as pollinators) that are exposed to crops that have been engineered to contain pesticides.[17] Finally, the use of crops genetically engineered to incorporate pesticidal traits or herbicide resistance has led to the development of pesticide-resistant insects and herbicide-resistant weeds.[18] Of particular note, the widespread introduction of crops resistant to glyphosate, accompanied by extensive use of this herbicide, has led to the rapid evolution of glyphosate resistance among weed species. Farmers have responded by increasing herbicide use, resorting to more toxic chemicals, and tilling the soil more frequently, thereby undermining much of the potential environmental benefits from GMO use.

Social and ethical concerns also contribute in important ways to public discomfort with GMOs. First, widespread use of GMOs can lead to monopolies that harm small farmers, as economic power is concentrated in the biotech companies that own patent rights to GMOs.[19] Powerful corporations such as Monsanto have aggressively enforced these rights against growers.[20] The monopoly control of seeds may not only undermine small farmers in both industrialized and developing countries but also have detrimental effects on food security and consumer choice.[21] Second, some persons view the consumption of foods containing GMOs as repugnant, and others express religious or ethical objections to the transfer of genes between species. These objections are neither surprising nor irrational: Food has always played a central role in human history, culture, and identity.[22]

GMOs presently under development raise further concerns. Scientists are in the process of genetically modifying traditional food crops to produce drugs and industrial chemicals. The introduction of potentially toxic substances into food crops could pose health and environmental hazards more significant and unpredictable than those posed by the current generation of GM crops.[23] In addition, researchers are developing and seeking to market GM animals for human consumption. This work warrants careful scrutiny, as transgenic animals may act as agents for transmitting disease, and escaped transgenic animals may outcompete wild animal species and damage ecosystems.[24] Consumers have expressed particular unease with genetically engineering animals for food and with the idea of mixing animal and plant genes.[25] In general, further development and commercialization of GMOs could intensify social, economic, and ethical concerns.

(Non)Regulation of GMOs

Despite public unease about the presence of GMOs in the food supply, GMOs have become widespread in the United States thanks to policies designed to foster the growth of the biotechnology industry. GMOs are loosely governed in the United States under the Coordinated Framework for Regulation of Biotechnology (Coordinated Framework or Framework), a policy that organizes the regulatory responsibilities of the federal agencies in this area.[26] The federal government established the Coordinated Framework ostensibly to regulate health and environmental risks that might result from the development, commercialization, and consumption of GMOs.[27] However, another motivation behind the policy—apparently the primary one—was to "minimize the uncertainties and inefficiencies that [could] stifle innovation and impair the competitiveness" of the nascent biotechnology industry.[28] Indeed, although the Coordinated Framework purports to be a "comprehensive federal regulatory policy,"[29] it is better understood as a patchwork of existing laws pieced together to deflect calls for legislation that would have directly addressed the potential hazards of GMOs. In a nutshell, the Framework rests on legal authorities that were not designed to address the concerns unique to GMOs, and it relies heavily on industry to comply voluntarily with those authorities. A product of the Reagan administration's overall deregulatory philosophy, the Coordinated Framework remains the foundation of U.S. GMO policy today.

Self-Regulation Prior to the Coordinated Framework

The Coordinated Framework was a logical outgrowth of long-standing efforts by researchers and the nascent biotech industry to avoid stringent government oversight. Indeed, much of the history of GMO governance reflects a systematic attempt to promote genetic engineering and to exclude the public from its management.

GMO governance efforts were initiated during the early 1970s by the scientific community, which was excited about the prospects of research breakthroughs yet concerned about potential risks. The rules the scientists developed to govern themselves deflected external criticisms and enabled research to proceed with little interference. These self-regulatory efforts merit closer examination not only because they continued to influence policy even as GMOs were commercialized but also because they demonstrate the potential significance of initial governance regimes and hence the importance of opportunities for early public input.

A vital first step in the scientists' self-regulatory efforts was the establishment of a voluntary and temporary moratorium on recombinant DNA experiments in 1974.[30] The moratorium set the stage for the 1975 Asilomar Conference, an international gathering of scientists to address concerns such as the potential for inadvertently creating new pathogenic organisms.

A set of recommendations from the Asilomar meeting anchored U.S. policy on genetic engineering for years to come. The recommendations classified experiments into four categories, with containment measures for each category corresponding to rough estimations of the risk posed by each type of experiment.[31]

Touted as an example of successful self-regulation by the scientific community, the Asilomar recommendations were widely followed.[32] Scientists were commended for their self-restraint. The effort, however, by no means involved open and democratic deliberation on the future of the new technology. For example, participation in the Asilomar Conference was limited largely to scientists working in the field, many of whom were eager to move forward with their research projects. These scientists understood that crafting credible research guidelines on their own would be their best defense against potential government regulation.[33] Few scientists in attendance worked in the health or environmental sciences, critics of recombinant DNA research boycotted the conference, and members of the general public were not invited.[34] The limited scope of participation is not surprising, given the prime motivation behind the conference. Conference organizers were seeking to end the moratorium on recombinant DNA research and to do so in a way that avoided outside intervention in scientists' research efforts.[35] This motivation permeated various aspects of Asilomar, including the selection of participants, the crafting of the agenda, and conference discussions. In light of the excitement surrounding advances that already had been made, a permanent halt to recombinant DNA research simply was not an option.[36]

Discussions at Asilomar focused narrowly on the development of guidelines to address the biological hazards associated with experimentation.[37] At the outset of the conference, issues regarding how research results might be applied—in gene therapy, genetic engineering of animals, or biological warfare, for example—were taken off the table. Occasional attempts to raise such questions were ignored or rejected. Broader concerns, such as the effects that GM crops might have on rural social organization, were left unmentioned. And discussion of legal and ethical issues, left largely to a panel tacked on at the end of the conference, failed to heighten participants' ethical sensibilities; rather, such discussions primarily raised scientists' worries about personal liability. By marginalizing social and ethical matters from serious consideration, organizers were able to frame conference discussions in terms of technical issues that they asserted were suitable for determination by scientists.

The recommendations that emerged from Asilomar nonetheless served as the basis for more detailed guidelines issued by the National Institutes of

Health (NIH) in 1976.[38] NIH oversight was no accident. The scientists responsible for the 1974 moratorium, though wary of outside regulation, recognized that a government imprimatur would confer legitimacy on the research guidelines developed by the scientific community. Accordingly, they suggested that an agency naturally disposed to favor scientific research, the NIH, take the lead in formal government oversight of genetic engineering.[39] Under the guidelines, an NIH committee reviewed proposals for conducting NIH-funded recombinant DNA research to ensure compliance with the guidelines.[40] Research funded by other sources, however, was not subject to the guidelines, although scientists involved in such research could follow the guidelines voluntarily.

The scientists' desire to minimize outside influence over the guidelines was reflected in the NIH's halfhearted efforts at NEPA compliance. Though the NIH recognized that NEPA required it to prepare an EIS for the guidelines, the NIH failed to issue even a draft EIS until after the guidelines were already in place.[41] As a result, the NEPA process, which is intended to influence agency decision making through public input and thoughtful environmental analysis, had no effect on the initial guidelines.[42] The NIH did update its belated EIS when it revised the guidelines.[43] In addition, the agency—under court order—later prepared NEPA documentation before approving individual field tests of GMOs.[44] The courts nevertheless rejected attempts to require the NIH to conduct a more general programmatic evaluation of the environmental effects of genetic engineering field experiments.[45]

The NIH guidelines enabled the scientific community to sidetrack congressional efforts to regulate recombinant DNA research.[46] Scientists carried out experiments under the guidelines without apparent harm. Meanwhile, concerns grew that the guidelines put NIH grantees at a competitive disadvantage relative to private researchers as well as those in other countries. Over the next few years, the NIH relaxed and then essentially undid the guidelines by shifting from a somewhat precautionary approach to one that placed the burden on proponents of oversight to demonstrate the existence of a hazard.[47] This occurred even though no systematic risk assessment had demonstrated that recombinant DNA research was safe.[48]

The Coordinated Framework

Legal developments and research advances in the early 1980s set the stage for the commercialization of GMOs and for more active government involvement. The U.S. Supreme Court held that living organisms could be patented under federal law in a 1980 decision, *Diamond v. Chakrabarty*.[49] The ruling established the legal foundation for GMO researchers to profit from their labors and served as a green light for economic investment.[50] Meanwhile, scientists continued to progress toward field testing. The controversy and litigation surrounding field test proposals, however, helped expose the NIH guidelines' inadequacies in the face of changing circumstances. For example, in 1985, a federal court of appeals enjoined a proposed experiment to release a genetically modified bacterium that would increase the frost resistance of plants.[51] The court ruled that the NIH had neglected its obligations under NEPA by "completely fail[ing] to consider the possible environmental impact from dispersion of genetically altered bacteria."[52] The decision raised questions about the NIH's oversight and underscored the discrepancy between NIH-funded research, which was subject to mandatory NIH review, and other research, which was not.[53] In addition, a House subcommittee examining the environmental implications of genetic engineering recommended that the Environmental Protection Agency (EPA) regulate the environmental risks of GMOs through its authority over chemical substances under the Toxic Substances Control Act.[54] The EPA, however, struggled in the face of political opposition to determine how it could apply that authority to GMOs.[55]

Amid this growing uncertainty, the Reagan administration, through the Office of Science and Technology Policy (OSTP), established the Coordinated Framework in 1986. The Framework sets out the federal government's overall policy for evaluating GMOs. It identifies three primary agencies with authority under preexisting laws to regulate the potential risks of biotechnology products: the Animal and Plant Health Inspection Service (APHIS) oversees organisms that could pose plant pest risks; the Food and Drug Administration (FDA) regulates safety and labeling of foods; and the EPA regulates pesticidal substances. The Framework directs these agencies to focus on the characteristics of the biotechnology product at issue and to regulate only those risks they deem unreasonable.[56] The Framework, however, neither requires comprehensive analysis of the health and environmental effects of individual GMOs nor compels systematic consideration of the desirability of GMOs in general. Indeed, the Framework involves little actual oversight.

The Framework rests on two fundamental yet questionable assumptions. First, the policy assumes that genetic engineering techniques are merely an "extension" of traditional plant breeding and that therefore only the products of biotechnology, and not the processes that generate those products, warrant regulatory attention.[57] The policy thus disregards the fact that many consumers care about how their food is produced and whether it contains GMOs. Second, the policy assumes that laws enacted prior to the emergence of genetic engineering are adequate to address the potential health and environmental hazards of biotechnology products.[58] The possibility that genetic engineering technology might undermine the premises under which these laws operate is ignored. These two assumptions have proven critical in enabling the widespread adoption and use of GM plants in U.S. agriculture. Just as the NIH guidelines had shielded recombinant DNA research from congressional action and public deliberation, the Coordinated Framework came to serve as a shield for the commercialization of GMOs by the biotech industry.[59] A closer examination of the Framework agencies' authority and practice reveals the limited nature of oversight under the Framework.

APHIS

The Coordinated Framework gives APHIS primary authority over the assessment of environmental effects of GM plants, with the exception of plants modified to produce pesticides. APHIS is an agency within the U.S. Department of Agriculture (USDA), whose general mission is to promote American agricultural interests. The allocation of primary regulatory authority over GM plants to APHIS rather than the EPA—the federal agency generally charged with protecting the environment—underscores the pro-industry bent of the Coordinated Framework. APHIS's authority over GM plants derives from the Plant Protection Act, the purpose of which is to prevent the introduction, spread, or establishment of plant pests.[60] Thus, only GM organisms that have the potential to be plant pests are deemed "regulated articles" subject to APHIS oversight; other GM plants are not regulated by APHIS.[61]

Before the release of regulated articles into the environment, as in a field test, one must either provide a "notification" to APHIS or obtain a permit. Notification is a streamlined process for species that meet specified criteria and are not noxious weeds.[62] The criteria require, inter alia, that the introduced genetic material be stably integrated into the crop's genome and that expression of the genetic material not result in plant disease. Activities conducted under a notification must incorporate precautions to reduce the likelihood that regulated articles will persist in the environment or be mixed with nonregulated materials. Within 30 days of receipt of a notification, APHIS provides acknowledgment "that the environmental release is appropriate."[63]

Plants engineered to produce pharmaceutical drugs, industrial compounds, or toxic substances do not qualify for notification.[64] For these GMOs, a permit is required prior to field testing. APHIS reviews permit applications individually for potential plant risks. Permits may contain specific conditions to prevent the dissemination and establishment of plant pests. To fulfill its NEPA obligations, APHIS conducts an environmental assessment for a small percentage of permit applications; APHIS exempts the rest from environmental analysis under a categorical exclusion.[65] For releases beyond field testing—that is, commercial plantings—a developer typically files a petition for deregulation.[66] APHIS grants the petition if it concludes that the plant does not pose a plant pest risk. Once deregulated, a GM plant may be freely moved and planted without a permit or any regulatory supervision.

APHIS's oversight of GMOs is subject to various criticisms. First, because the agency relies on its authority over plant pests as its basis for regulating GMOs, APHIS's jurisdiction is incomplete. While most GM plants developed to date use DNA sequences from plant pests and thus fall within the definition of regulated articles, a growing number of GM plants do not use such sequences and thus may evade APHIS scrutiny.[67] Similarly, GM vertebrate animals are not regulated by APHIS because they are excluded from the regulatory definition of a plant pest.[68] Second, where APHIS does have regulatory authority, its oversight focuses narrowly on plant pest risks to agriculture. As a result, APHIS tends to give less attention to broader concerns such as human health and environmental risks.[69] Given the USDA's mission of promoting U.S. agriculture, APHIS's narrow focus is unsurprising but nonetheless troubling in this context.

Third, in practice, APHIS's oversight has been shallow. The vast majority of GMO field tests are performed pursuant to notifications rather than permits.[70] Notifications are the subject of cursory review involving no environmental assessment, external scientific review, or public input.[71] Supervision of field testing under a notification is similarly minimal. APHIS inspects only about one-third of field trials conducted under notifications.[72] Moreover, such inspections do not entail testing surrounding areas to determine whether GMOs have escaped into the environment or whether transgenes have been transferred to other species.[73] Indeed, APHIS has sometimes lacked information on the precise location of field tests, and a 2005 internal audit criticized the agency's failure to impose measures restricting public access or requiring prompt destruction of GM plants after testing.[74] Even when field testing occurs pursuant to a permit, APHIS's analysis of environmental effects is sometimes negligible. For example, in a series of tests involving corn and sugarcane genetically modified to produce pharmaceutical products—that is, novel proteins that should have triggered immediate concern about toxicity and allergic reactions—APHIS simply issued permits for testing without preparing an EIS or even an EA.[75]

Fourth, APHIS's routine practice of deregulating GM crops prior to commercialization leaves the public and the environment vulnerable to unanticipated effects from their widespread use. Limited field trials of GM plants can detect only large-magnitude effects and are poor predictors of important traits such as whether a modified species will be invasive.[76] Accordingly, postcommercialization monitoring and testing are essential to identify effects not predicted by prior testing.[77] Such monitoring also can be valuable in evaluating the effectiveness of premarket risk assessments. Once APHIS grants a petition for deregulation, however, it no longer exercises oversight authority on the GMO in question or its descendants.

Indeed, recent lawsuits challenging APHIS's deregulation of various GM crops highlight the inadequacies of the agency's analyses of postderegulation environmental effects. In *Geertson Seed Farms v. Johanns*, the plaintiffs contended that the introduction of GM alfalfa would cross-pollinate non-GM alfalfa grown nearby, a significant environmental impact that would necessitate preparation of an EIS.[78] Such "biological contamination," argued the plaintiffs, would impair organic alfalfa farmers' ability to market their seed as non-GM and the ability of organic livestock farmers to obtain non-GM alfalfa for their livestock. Agreeing with the plaintiffs, the trial court found a significant environmental impact because deregulation would potentially "eliminat[e] a farmer's choice to grow non-genetically engineered alfalfa and a consumer's choice to consume such food."[79] A coalition of consumer groups, environmentalists, and farmers raised an analogous challenge to the deregulation of GM sugar beets in *Center for Food Safety v. Vilsack*. The trial court there reached a similar conclusion to the *Geertson* court and likewise ordered APHIS to prepare an EIS.[80]

Finally, opportunities for meaningful public involvement in APHIS's review and decision-making processes are limited. Much of the data and information developers submit to APHIS is claimed to be confidential business information and thus is not available for public review.[81] Furthermore, though APHIS publishes proposed decisions in the Federal Register for public comment, the agency's responses to public comments have often been perfunctory.[82]

FDA

The FDA's role in regulating GM plants similarly reflects the piecemeal and relatively narrow oversight that is characteristic of the Coordinated Framework. The FDA views its authority over GM plants as limited to those found in foods or food products.[83] That authority derives from the agency's jurisdiction over adulterated foods—in particular, its power to regulate foods containing unsafe additives.[84] Thus, objections to GMOs in food based on ethical, cultural, and other concerns lie outside the FDA's regulatory purview. Moreover, GM plants used to produce industrial chemicals receive no FDA review, and the FDA does not concern itself with environmental risks associated with the growing of GM crops.[85]

The FDA's actual oversight of foods containing GMOs is negligible, as a careful parsing of the agency's policy reveals. The FDA generally must approve food additives prior to use.[86] A genetic modification to a food crop may result in the presence of a food additive.[87] Nonetheless, FDA allows food additives to be exempt from regulation if they are "generally recognized as safe" (GRAS).[88] FDA policy presumptively applies the GRAS exemption to GM foods and as a result, almost all GM foods escape any requirement of premarket approval.[89] This policy, announced in 1992 without any NEPA analysis or public comment and over objections from several agency scientists, is based on the doctrine of "substantial equivalence."[90] Pursuant to this doctrine, the FDA presumes that substances added to foods via genetic modification are "the same as or substantially similar to substances commonly found in food" and thus GRAS.[91] The FDA therefore treats GM foods no differently than foods developed through traditional plant breeding. The FDA does not require that GM foods be labeled, nor does the agency mandate or conduct postmarket monitoring for adverse health effects.[92]

The FDA's minimal oversight is of particular concern because producers of GM plants themselves make the somewhat subjective determination of whether an added substance is GRAS.[93] Moreover, producers are not required to report their GRAS determinations to the FDA, and the FDA does not conduct any systematic monitoring to police the GRAS determination process.[94] Compounding the FDA's lack of formal oversight is the agency's refusal to mandate GM food labeling. This refusal undermines public awareness of the widespread presence of GMOs in the food supply, therefore hindering consumers' ability to avoid or object to GMOs. Weaknesses in the FDA's regulatory scheme ultimately render it largely voluntary, ineffective, and opaque.

EPA

Finally, the Framework's role for the EPA, the federal agency charged with protecting the nation's environment, is extremely constrained. The EPA's oversight of GMOs is limited to plants that are genetically engineered to express pesticidal substances. With respect to these plants, the EPA relies primarily on two statutes as the basis for regulation: the Federal Insecticide, Fungicide, and Rodenticide Act (FIFRA) and Section 408 of the Federal Food, Drug, and Cosmetic Act. All other GM plants, including herbicide-resistant GM plants, disease-resistant GM plants, and plants engineered to produce pharmaceuticals or industrial chemicals, receive no EPA review.

FIFRA authorizes the EPA to regulate pesticides and mandates the registration of pesticides sold or distributed in the United States. Prior to registering a pesticide, the EPA requires that an applicant submit data demonstrating that the pesticide "when used in accordance with widespread and commonly recognized practice, . . . will not generally cause unreasonable adverse effects on the environment."[95] The EPA interprets the term *pesticide* to include substances produced by GM plants for protection against pests as well as the genetic material necessary to produce these substances, if intended for use in "preventing, repelling or mitigating any pest."[96] Accordingly, the EPA applies FIFRA's requirements to crops genetically engineered to produce pesticides (which the EPA refers to as "plant-incorporated protectants" [PIPs]). Although the EPA acknowledges that PIPs differ from traditional chemical pesticides in "their ability to spread and increase in quantity in the environment," the EPA applies essentially identical regulatory standards to PIPs and to ordinary chemical pesticides.[97] Specifically, the EPA mandates that companies obtain an experimental use permit to conduct field trials of plants containing PIPs on 10 or more acres and that companies register a PIP prior to commercialization.[98]

The basic framework of FIFRA is ill suited to address many of the potential health and environmental risks posed by the GMOs within the EPA's limited jurisdiction. For example, unlike chemical pesticides, GMOs can reproduce, evolve, and exchange genetic material with other species.[99] The resulting progeny may take on characteristics quite different from those of the original GM plants that were subject to FIFRA's registration requirements. In addition, FIFRA regulation relies heavily on product labeling requirements to reduce health and environmental risks.[100] Such requirements, designed with chemical pesticides in mind, may be ineffective or insufficient for pesticidal GMOs.[101] Directions accompanying an initial shipment of GMO seed, for example, are likely to be of little use if the crop planted from that seed exchanges pesticidal genes with conventional crops or wild relatives.[102] And such directions may be disregarded or forgotten if a farmer saves seeds from a GM crop and plants a subsequent crop from the saved seed.[103]

Moreover, a fundamental objection to relying on FIFRA to regulate GMOs involves the suitability of applying the statute's standard for granting a chemical pesticide registration—whether the use of a pesticide will cause "unreasonable adverse effects." This relatively lenient standard requires that the EPA balance the economic, social, and environmental costs and benefits associated with the use of a pesticide.[104] As a general matter, legal standards that involve a balancing of costs and benefits tend to overlook uncertain consequences, which are often inherent in emerging technologies. For GM plants, determining and quantifying environmental costs is exceptionally difficult because of the lack of information regarding the potential for the exchange of genetic material and other risks.[105]

The EPA also regulates GMOs—at least nominally—under Section 408 of the Federal Food, Drug, and Cosmetic Act, which authorizes the agency to regulate pesticide residues in food. Under that statute, the EPA first determines tolerance limits for pesticide residues, and the FDA then monitors foods for compliance with those tolerance limits.[106] The setting of tolerance limits and granting of exemptions from such limits are subject to limited public comment.[107] To date, all currently registered PIPs have received tolerance exemptions from the EPA, based presumably on the agency's finding of "a reasonable certainty that no harm will result from aggregate exposure" to these PIPs in food.[108] Thus, PIPs generally have not been subject to tolerance limits. The EPA's routine practice of granting tolerance exemptions is based on limited data regarding toxicity, however, since the EPA does not require testing of PIPs for the chronic toxicity that might result from the long-term consumption of GM foods.[109]

Unauthorized releases of GMOs into the food supply and the environment reinforce concerns about the effectiveness of the precautions relied on by the EPA and other regulators.[110] In the most widely publicized incident, StarLink, a genetically modified strain of corn approved for nonhuman consumption only, was detected in human food in 2000. As a condition of FIFRA registration, the EPA had required StarLink's developer to adopt various measures in cultivation, harvest, storage, and transport to isolate the strain from other corn varieties.[111] For example, the developer was supposed to ensure that StarLink plantings were separated from other cornfields by a 660-foot buffer zone and that farmers purchasing StarLink seeds signed a contract agreeing to various restrictions. These measures, as it turned out, were inadequate, poorly implemented, and weakly monitored.[112] StarLink turned up in taco shells and other food products, leading to broad product recalls and reduced corn exports. Fortunately, the U.S. Centers for Disease Control found no evidence of allergic reactions resulting from human consumption of StarLink corn.[113] To prevent a recurrence of the incident, the EPA no longer grants FIFRA registrations of GM food crops solely for nonhuman consumption.[114]

Future unauthorized releases of GMOs are nonetheless likely. Such releases are virtually unavoidable in modern food production, since the same cultivation, storage, and transportation equipment is used for GM and non-GM crops.[115] In addition, inadvertent releases can result from simple human error, as has occurred on several occasions subsequent to the StarLink incident.[116] Future unauthorized releases could have serious consequences. For example, plants engineered to produce chemicals and pharmaceuticals pose far greater dangers of toxic contamination. Before such crops are commercialized, effective and enforceable containment measures—such as a prohibition on open-air cultivation or close monitoring of growers by independent auditors—must be utilized to prevent gene transfer and commingling of such plants with their counterparts destined for human consumption.[117] Unless the Coordinated Framework is changed, however, these particularly hazardous GMOs may be subject to *less* regulatory scrutiny than most GM crops already on the market because these new GMOs do not constitute pesticides or food additives that would be regulated by the EPA or FDA.[118]

In sum, premarket oversight of GMOs under the Coordinated Framework is ineffective, and postmarket oversight is nonexistent. Weak implementation has contributed to these inadequacies, but the primary culprit is the flawed design and legal limitations of the Framework itself. The Framework puts APHIS—an arm of the agency tasked with promoting U.S. agricultural interests—in charge of identifying and addressing the environmental effects of GMOs. APHIS's oversight is focused narrowly on plant pest risks to agriculture, and even those risks are managed mostly in a cursory manner. Once a GM plant is commercialized, APHIS exercises no further oversight. The FDA and EPA address human health risks from consuming GM foods, at least in theory. The FDA's approach, however, focuses on nutritional equivalence rather than safety and entrusts producers with regulatory determinations. The EPA's role is similarly circumscribed by its narrow authority and liberal use of tolerance exemptions. The absence of systematic monitoring by any regulatory agency once a GMO is commercialized is particularly troubling because of the patchwork nature of the regulatory oversight prior to commercialization.[119] While comprehensive, long-term monitoring would be challenging to implement, the absence of efforts to study adverse effects on the consuming public means that we have little way of knowing if the Framework is effectively identifying and addressing risks.[120]

International Aspects

No discussion of GMO regulation would be complete without noting the international dimensions of the subject. GMOs have provoked opposition in numerous countries. As demonstrated by a drop in U.S. corn exports resulting from the StarLink corn incident, failure to provide adequate oversight of GMOs could have broad impacts on U.S. trade and economic activity.[121] Moreover, the potential environmental hazards of GMOs, such as ecosystem disruption, need not respect national boundaries. Food contamination by GMOs not suitable for human consumption likewise can escalate into global problems due to high levels of international trade.

National policies are not formed in a vacuum, and it would be unwise to disregard the international implications of domestic policy choices. The U.S. desire to maintain a competitive edge in biotechnology development was a major driver behind the establishment of the Framework.[122] Major agricultural exporters such as Canada and Brazil have taken a similarly permissive approach to GM crops and products.[123] By contrast, the European Union (EU), Japan, and Korea have adopted more precautionary approaches that the United States criticizes as protectionist.[124] Many developing countries also have supported restrictions on the trade of GM products despite their potential benefits in combating hunger and malnutrition. These countries often express concerns that powerful Western corporations will secure monopoly control of essential seed technology.[125]

EU GMO policy in particular has diverged from U.S. policy. A brief examination of EU and U.S. approaches, including the resulting trade dispute, not only sheds light on the cultural differences that can inform policies on emerging technologies but also points to the possibility of reconciling rational technology regulation with public values.

EU policy, recognizing that genetic engineering technology may give rise to unique hazards, has focused on the process of genetic modification as well as its products.[126] Responding to public controversy and member-state concern regarding uncertain health and environmental hazards, the EU's more open and precautionary legal regime mandates premarket authorization, traceability, and labeling of GMOs.[127] In deciding whether to authorize a particular GMO, regulators take into account not only effects on health and the environment but also "consumer interests" as well as other considerations that may include ethical concerns.[128]

A number of factors explain the divergence between EU and American policies: greater opportunities for public deliberation on GMO technology in Europe, which led to the performance of farm-scale field trials; European cultural attitudes more wary of technological innovation, particularly in food production; and the occurrence in Europe of large-scale agricultural disasters, such as the outbreak of mad cow disease, that undermined public trust in regulators and the agricultural industry to address risks adequately.[129] These factors underscore the potentially significant role of public values in technology policy, with the first factor, public deliberation, particularly noteworthy. In contrast to the United States, where adoption of the Framework essentially short-circuited public deliberation, the EU has fostered a continued societal debate through advisory groups, open meetings, online consultations, and the like.[130] That debate has revealed public concerns and provided important informational feedback.

Conflicts between EU and American GMO policies came to a head in the *EC-Biotech* trade dispute, which was initiated by the United States, Canada, and Argentina before the World Trade Organization (WTO) in 2003.[131] Though a detailed discussion of the case is beyond the scope of this chapter, *EC-Biotech* highlights the conflict between a regulatory approach based narrowly on quantitative risk assessment and a more comprehensive approach that recognizes the critical role of public values in informing risk assessment and risk management.[132] In *EC-Biotech*, the United States contended that EU delays in authorizing the importation of products containing GMOs constituted a de facto moratorium that violated free trade principles and the Agreement on Sanitary and Phytosanitary Measures (SPS Agreement).[133] The SPS Agreement recognizes the authority of member states to adopt measures that protect human life or the environment from, among other things, invasive species and food safety hazards, as long as such measures are based on risk assessments.[134] In response to the allegation that EU actions were not properly grounded in risk assessments, the EU argued that its policies were permitted under Article 5.7 of the SPS Agreement, which authorizes the adoption of provisional protective measures in "cases where relevant scientific evidence is insufficient."[135] The EU contended that its measures were also authorized by the Cartagena Protocol on Biosafety, an international legal agreement that explicitly affirms the authority of sovereign states to adopt GMO risk management decisions that reflect a precautionary approach.[136]

The WTO's decision in *EC-Biotech* held that the EU had engaged in undue delay in its GMO approval process. Because the decision sidestepped numerous substantive issues, however, it has had little effect on the EU's overall regulatory regime.[137] *EC-Biotech* left undecided issues such as whether the EU's actions were based on a risk assessment and whether those actions were authorized by Article 5.7.[138] Had the WTO confronted the merits of these issues, it might have addressed fundamental questions about the relationship between science and democracy as well as the tension between global standards and state autonomy. In light of the WTO's free-trade mission, the proper role of the WTO on these issues should be limited to policing against protectionist measures. The WTO has no legitimate authority to impose particular conceptions of risk assessment and risk management that lack general acceptance.[139]

In theory, international harmonization of GMO regulation could prompt further innovation, facilitate trade, and help to address concerns about transboundary hazards. The coexistence of continued GMO innovation with diverse national approaches to GMO regulation, however, suggests that international harmonization may not be necessary at this time. States should retain the ability to adopt GMO policies reflecting domestic values as long as those policies are not motivated primarily by protectionism. Opposition to GMOs in Europe is ultimately rooted in social and cultural values surrounding food, food production, and risk tolerance, and these values should be respected. Moreover, regulatory harmonization is unlikely to substantially reduce transboundary hazards from GMOs because the United States and other major agricultural exporters would make certain that any harmonized standards are weak.

Concerns about international harmonization and competitiveness are not unique to GMOs and appear with other emerging technologies as well. For example, proponents of nanotechnology contend that excessive regulation will drive research and development efforts and any accompanying commercial benefits out of the United States and into other countries.[140] EU policy with respect to GMOs, however, suggests that it is possible and economically feasible for societies to make technological choices that reflect public values in the face of countervailing economic pressures. The GMO experience also illustrates that asserted economic benefits of technological change may not fully materialize in the absence of adequate public support.

Future Approaches to GMO Regulation

Fixing the Framework

Without effective oversight of GMOs, the American public must rely primarily on industry and genetic engineers to protect it from biotechnology's hazards. Defenders of this approach point to the absence of any incidents in which people have suffered significant health effects from consuming GM foods as well as to the absence of studies establishing human health or environmental harms.[141] Such arguments, while technically accurate, are misleading. The long-term effects of GM food consumption in humans have not been studied.[142] Furthermore, the fact that regulators and the biotechnology industry conduct minimal postmarket monitoring for adverse consequences makes it less likely that such consequences will be discovered.

In general, the potential for adverse effects from cultivating and consuming GM crops warrants closer attention. As chapter 1 explains, new technologies often have caused health, environmental, and social harms that became apparent only with careful study or the passage of time. Our changing understanding of what genes are and how they operate undermines the Framework's assumptions about the precision of genetic engineering techniques as well as the potential for unintended modifications.[143] Rather than assuming the adequacy of premarket review, health and environmental regulators should constantly monitor for such modifications and their effects. Moreover, even intended modifications may have unanticipated and detrimental effects. For instance, in vivo animal studies examining the effects of consuming GM foods, particularly those engineered to produce pesticides, suggest the potential for tumor formation and tissue and organ damage.[144] Though the PIPs that have become widespread in the food supply are believed to be relatively benign, they have not previously been consumed in the quantities now present in the modern American diet.[145] Moreover, animal feeding studies cited by industry to demonstrate GMO safety generally involve short-term trials that do not evaluate chronic toxicity.[146] GM foods should be studied further to evaluate potential long-term effects from their consumption on human health.

Carrying out such studies would not be easy.[147] First, GM foods are largely unlabeled, making it difficult to establish a control group of the population that does not consume GMOs at all. Second, distinguishing the health effects of GM foods from non-GM foods may be next to impossible because diet and food consumption patterns naturally vary over time. Nonetheless, more careful monitoring for allergenicity, increased use of food-tracking systems, and other improvements in postmarket surveillance efforts can enhance the reporting and detection of adverse effects.[148]

Labeling of GM foods has been a contentious issue. GMO developers and the food industry object to proposals to require labeling on the grounds that the additional effort needed to segregate GM crops from non-GM crops would raise costs.[149] They also worry that labeling would lead consumers to believe that GM foods are less safe than non-GM foods. Siding with industry, the FDA has refused to require labeling.[150] In fact, the agency has even discouraged voluntary efforts to include information on labels indicating that a food contains no genetically modified ingredients. The FDA disapproves, for example, of using the terms *not genetically modified* and *GMO* on food labels, and it has warned that inclusion of the term *GMO-free* "may be misleading on most foods."[151]

Nonetheless, labeling of GM foods can promote market efficiency, consumer autonomy, and manufacturer responsibility. Labeling fosters market efficiency by providing information that enables consumers to express their preferences more accurately.[152] Labeling encourages personal autonomy by recognizing consumers' right to have pertinent information about the food they consume, regardless of safety concerns.[153] And labeling helps the public and the government to hold the biotech industry responsible for damages that may result from consuming GM foods.[154]

To suggest that labeling of GM foods would mislead consumers is presumptuous, to say the least. American consumers overwhelmingly support labeling of GM foods,[155] and thus it is likely that many consumers would deem the presence of GMOs relevant to their purchasing decisions. Whether a food contains GM ingredients may be relevant to consumers who desire to avoid uncertain hazards, reduce environmental externalities, express particular religious or ethical beliefs, or support traditional farming methods.[156] Regardless of the specific grounds that shape the effect of GMOs on consumer choice, such decisions should be left to consumers. Granted, not all consumers will notice or care about the information that might be contained on a GMO label. But the rising interest in sources and quality of food—as reflected in the local food movement, the growth of the organic food market, and mounting state initiatives to require labeling—suggests that many consumers would take such information into account.[157]

Present labeling policies deprive members of the public the opportunity to independently evaluate the benefits and concerns of GMOs and ultimately preclude the public from adopting a precautionary approach to uncertain risks and from expressing ethical choices through their purchasing decisions. As law professor Douglas Kysar has argued, because the Coordinated Framework ensures that no single governmental decision maker performs a comprehensive evaluation of GMOs, consumers are "the only decision-makers in a position to evaluate new technologies such as GM agriculture in their totality."[158] In the absence of labeling, consumers simply cannot perform such an evaluation.

With respect to costs, the implementation of a national organic food labeling scheme indicates that GMO labeling would be practicable.[159] Various countries have successfully mandated labeling of GM foods.[160] Although any cost determination is undoubtedly complex, several studies conclude that costs are likely to be modest.[161] In the end, concerns about lost market share rather than about implementation costs drive industry opposition to mandatory labeling. Some consumers, of course, may reject GM foods. Even more worrisome to the GMO industry is the prospect that food processors, fearing consumer rejection, will reformulate their products to avoid GMOs and that retailers will similarly alter their product lines.[162] These fears underscore the lack of transparency that characterizes the management of genetic engineering technology in the United States. From the outset, the general public had little say in setting the guidelines for research. The public's concerns were brushed aside under the veneer of regulation known as the Coordinated Framework, and even today the public is being kept in the dark about the prevalence of GMOs in the food supply. Various stakeholders acknowledge that this approach has fostered an atmosphere of distrust that is likely to persist until there is greater transparency within the industry, the market, and the regulatory structure.[163]

Several possible measures would improve the oversight of GMOs. Such measures include vigorous premarket review, postmarket monitoring, and labeling. Importantly, these measures would not entail wholesale reconsideration of genetic engineering technology, nor would they necessarily involve the broader public in restructuring GMO policy. The modest nature of these measures reflects a realistic assessment that, in the absence of significant and demonstrable harms, wholesale reconsideration of the technology is unlikely. GMOs have yielded greater productivity and some environmental benefits, and they are supported by powerful biotechnology companies that have billions of dollars at stake and farmers who benefit from reduced costs and easier weed control.[164] GMOs now dominate the leading commodity crops grown in the United States, and the associated technology has become entrenched in American agriculture.

While GM crops are yet to gain broad and unquestioning public acceptance, they are nearing what science and technology scholars describe as a state of closure—the point at which the social controversy surrounding a technology ends and the scientific facts about a technology become stabilized in society's view.[165] The widespread use of GM crops undermines opponents' efforts to portray them as novel, foreign, and unnatural. Similarly, the absence of data demonstrating serious adverse effects undercuts safety objections to GMOs. Before closure occurs, however, closer scrutiny of GMOs and greater public awareness regarding their use can help to ensure that any closure is deliberate, is informed, and accounts for values-based concerns about social, economic, and environmental effects.

Next Up: GM Animals

The next generation of GMOs, particularly GM animals, presents circumstances warranting greater caution. In contrast to GM crops, GM animals are not commonplace and have not yet been approved for human consumption in the United States. Their introduction into the market is not a foregone conclusion and if it were to occur would likely be highly controversial. Now is the time for thorough, careful, and public deliberations on GM animals.

A number of GM animals are currently awaiting government approval for commercialization. AquaBounty Technologies has genetically engineered salmon to grow more rapidly to market size, and this animal is expected to be the first to receive FDA approval for human consumption.[166] Other GM animals presently under FDA consideration include cows resistant to mad cow disease and pigs that produce more environmentally friendly manure.[167] The FDA, which is taking the lead in regulating GM animals, asserts its authority to do so under its rules covering veterinary drugs.[168] The FDA explains that it considers the recombinant DNA inserted into the genome of a GM animal to be a "new animal drug" because it is intended to affect the structure or function of the body of the animal. As with GM plants in food, the FDA apparently does not intend to require labeling of food from GM animals as long as it is substantially equivalent to food from non-GM animals.

Unfortunately, by relying on preexisting regulatory authority designed to govern other products, the FDA is utilizing the same flawed paradigm under which GM plants are governed, and its regulatory scrutiny will likely be inadequate to address relevant concerns. For example, the FDA's evaluation of new animal drug applications focuses on whether a drug is "safe and effective for its intended use."[169] Presumably, those criteria are suitable for determining whether run-of-the-mill veterinary drugs are safe for the animals that receive the drugs and for the humans that consume those animals. Unlike ordinary veterinary drugs, however, GM animals can reproduce and thus pose a far greater potential for adverse environmental effects.[170] Ecologists worry, for example, that GM salmon will escape into the wild, potentially disrupting fragile ecosystems or interbreeding with and devastating wild fish populations.[171] Especially troubling, the FDA's authority to consider environmental concerns in approving a GM animal is unclear because those concerns are not obviously relevant to the issue of whether a particular GM animal is "safe and effective for its intended use."

Though the FDA has promised to analyze environmental consequences and comply with NEPA in the course of approving GM animals, doubts remain about the quality and thoroughness of its environmental analyses.[172] In reviewing new drug applications, for example, the FDA relies heavily on applicants to produce underlying safety data and to prepare environmental assessments.[173] This practice is contrary to the spirit of NEPA, which was intended to compel federal agencies to analyze environmental impacts in the hopes of sensitizing agencies to environmental concerns. It also presents a potential conflict of interest and is a departure from the general practice at other agencies, which either prepare such documents themselves or hire third-party contractors to draft them.[174] Moreover, the FDA instructs applicants that the environmental assessments they prepare are to focus on "environmental issues relating to the use and disposal" of pharmaceuticals.[175] These issues comprise only a small fraction of the environmental concerns that GM animals raise, however. Concerns regarding broader ecological effects, which arguably lie beyond the FDA's expertise, are likely to receive short shrift.[176]

The FDA's reliance on its veterinary drug authority to govern GM animals also results in a lack of transparency. The FDA treats the licensing of new drugs as a confidential regulatory matter, with minimal public disclosure, let alone public input, prior to approval.[177] Because the FDA considers the recombinant DNA in GM animals to be a new animal drug, its review of GM animals is subject to the agency's confidentiality rules. Indeed, the FDA treats even the existence of an application for approval of a GM animal as confidential.[178] Complaints regarding lack of transparency have led the FDA to promise to hold public advisory committee meetings before issuing any approval, at least for the first GM animals under consideration.[179] Public input is critical, but the precise role of that input in the overall approval process as yet remains unclear.

The introduction of GM animals to the human food supply raises new and serious public concerns. The American public is more uneasy about GM animals than about GM plants.[180] The public's concerns appear to involve not just worries about the safety of any particular GMO but also ethical concerns about the genetic engineering of animals in general, their use, and their consumption.[181] The pending applications before the FDA provide an opportunity for the industry and regulators to forge a more Promethean and participatory path than that followed for GM plants. Documents supporting AquaBounty's salmon application suggest that the FDA and GM animal developers are giving some consideration to potential adverse health and environmental consequences.[182] AquaBounty has promised to market only sterile, female fish that will be grown exclusively in contained facilities.[183] Moreover, the FDA conducted a public meeting to consider AquaBounty's application and solicited public comment on whether the GM salmon should be labeled as such.[184] Nevertheless, a panel of outside experts was generally persuaded that GM salmon is safe for human consumption, and the FDA appears inclined to give its approval.[185]

It is clear that much more should be done to engage the public on the issue. The participatory mechanisms discussed in chapter 1 offer various options for doing so. Specifically, a number of factors make this an especially appropriate time to hold a nonbinding national referendum on the use of GM animals for food purposes. First, the technology is sufficiently advanced to provide concrete examples that the public can readily understand and discuss. The GM salmon application in particular has attracted the public's attention and thus would serve as an effective starting point for a broader public debate. Second, the technology has not become entrenched in the marketplace, and public debate thus could substantially inform critical policy decisions. Public input on the salmon application should be considered not only in the determination of this specific case but also in the development of other GM animals. While scientific expertise will be useful in estimating the risks posed by GM animals, the question of whether such animals should be commercialized is in large part a question of values. Holding a referendum, public forums, or consensus conferences on GM animals would help ascertain the public values that should guide policies in this area.

Lessons

Consideration of America's experience with GMO regulation not only reveals ways to improve oversight of GMOs but also suggests lessons for managing emerging technologies in general.

The Lasting Effects of Early Technology Decisions

First, decisions and policies adopted early in the technology development process can have tremendous and lasting influence, even when they are not formally incorporated into law. With respect to GMOs, for example, the Asilomar recommendations set the precedent that biotechnology scientists would largely govern themselves. Although research scientists have traditionally resisted outside regulation, risk management should not have been left solely to the scientific community. The Asilomar recommendations became the basis for the NIH guidelines, which reinforced the premise that management of this revolutionary technology was best left to those developing it. That premise persisted even with the advent of government regulation: the Coordinated Framework established a veneer of government oversight, but the nascent biotech industry was free to continue to develop and then to commercialize the technology essentially unhindered. As American farmers rapidly adopted GM crops, the increasingly formidable biotech industry gained powerful allies in its efforts to block effective regulation of GMOs.[186]

In the years since the government's establishment of the Framework, criticism has mounted.[187] Critics note that the Framework continues to dominate the regulatory approach to GMOs even though each of the three agencies within the Framework has raised doubts about its central assumption that the process of genetic modification is irrelevant to the need for regulation.[188] The Framework's persistence illustrates the phenomenon of path dependence—the ongoing influence and perpetuation of past choices resulting from institutional inertia, the costs of changing settled arrangements, and the operation of positive feedback mechanisms.[189] Path dependence can occur with respect to both technology development and legal regulation, and both types of path dependence come into play when new technologies are being regulated.[190] In technological path dependence, a decision to adopt a particular technology affects future decisions and investments and makes subsequent reconsideration of the technology unlikely.[191] In legal path dependence, a legal regime, once established, strengthens the influence of the parties that sought its adoption and leads to the rise of new interests vested in that regime. Early laws and policies may influence and lock in critical innovations, technological designs, institutional arrangements, and investment decisions before the adverse consequences of a technology are fully understood.[192] In the GMO context, the Coordinated Framework bolstered the biotech industry, and farmers who adopted GM crops subsequently became supporters of the Framework. Reversing U.S. policy at this point is virtually unimaginable, reallocating oversight authority under the Coordinated Framework is unlikely, and even a tightening of standards faces an uphill battle.

The Disproportionate Influence of Researchers and Industry

The second important lesson from our experience with GMOs is that researchers and industry will often have disproportionate influence on policies governing an emerging technology. That disproportionate control, which stems from the potential significance of early decisions and may persist even after a technology proceeds beyond initial research, will likely favor technology development and undervalue the interests of the general public. For biotechnology, a "scientific-industrial complex" comprised of joint ventures of research universities and private companies, private research efforts, and multinational corporations has served as a driving force behind the technology.[193] The seeds of that complex were sown during the Reagan years, as Sheldon Krimsky, a scholar of the social history of GMO regulation, has observed. Krimsky notes that the era's dominant "neoconservative political ideology . . . supported the breakdown of traditional sector boundaries between university and industry, which led to the adaptation of science toward private rather than public agendas."[194]

Historically, academic researchers have often served as consultants for government and private industry.[195] Furthermore, deferring to the expert judgment of scientists in assessing and managing the risks posed by new technologies is a natural response to complexity and uncertainty. Modern legal arrangements, however, have strengthened the ties between university researchers and industry and intensified academic scientists' vested interest in seeing technology research proceed.[196] Specifically, the 1980 Bayh-Dole Act gave universities control over intellectual property generated by government-funded research, thereby creating a strong incentive to commercialize new discoveries.[197] Tax incentives encouraged the formation of university-corporate partnerships.[198] In addition, the *Chakrabarty* decision, recognizing the ability to patent living organisms, spurred biotechnology scientists' efforts and stimulated investment by venture capitalists and the petrochemical and pharmaceutical industries.[199] Led by Monsanto, the biotechnology industry successfully pushed for the establishment of the Coordinated Framework to allay public fears and subsequently continued to wield outsized influence over regulatory policies that facilitated rapid commercialization.[200]

Despite the importance of early decisions regarding how or whether to proceed with research in a new field, the public typically plays no role in such decisions. The public is frequently unaware of early technological developments and has little time or opportunity to learn more about them. Researchers and sponsoring agencies, concerned that public fears about unfamiliar risks will lead to reactionary prohibitions on scientific research, may keep the details of their efforts and advances under wraps.[201] And even where early public involvement is contemplated, defining a useful role for the public may be difficult when the uses and implications of a technology are poorly understood.[202] With respect to GMO policy, the Asilomar Conference and subsequent developments effectively contained social debate in the United States and minimized the public's role in policy making.[203] Although lawsuits have occasionally challenged various aspects of that policy and popular resistance to GMOs has periodically surfaced, the Coordinated Framework has never been open to reconsideration. To this day, there has been relatively little broad public discussion of the role that GMOs should play in American agriculture, and most Americans remain unaware of GMOs' ubiquity in the food supply.

Technology developers contend that the public plays a role in technology management through the decisions its members make as consumers.[204] Consumers ultimately determine whether a technology is successful, so the argument goes, and are free to base their purchasing decisions on whatever factors they deem relevant, including religious beliefs, political preferences for specific production methods, or the satisfaction of personal needs.[205] Low levels of public awareness regarding the presence and use of a technology, as in the case of GMOs, however, undermine these claims. Under such circumstances, consumers lack the information required to make the informed choices on which such an argument relies. The absence of GMO labeling thus not only has undermined public trust but also has precluded the public from obtaining that information without substantial personal effort.

This is not to say that researchers and industry will inevitably dominate policy in favor of technology development. In the early stages of a technology, few interests may be willing to fund research having distant prospects of any economic payoff, and industry roles may not yet be established.

In addition, different commercial interests may not share a common perspective. Industries threatened by technological change may seek to stifle technology development, and manufacturers of consumer products may pressure suppliers not to incorporate a new technology that they expect will encounter consumer resistance.[206] Food-processing companies and restaurant chains, for example, refused to use GM potatoes because they feared a consumer backlash. Those fears ultimately drove the genetically modified potato strain from the market.[207] Notably, the public is absent from each of these decisions. The public should be made aware of potential technological developments early on and should be at the table to have a meaningful voice before a technology becomes entrenched.

Regulatory Gaps and the Tendency toward Patchwork Policymaking

A third lesson from GMOs is that laws and policies regarding new technologies often do not result from comprehensive and open policy-making processes. The enactment of laws tailored to a specific technology often lags far behind technological developments. As a result, the policy governing a new technology may effectively be a product of legislative and regulatory inertia.[208] Problematically, new technologies often fall into regulatory gaps. GMOs, for example, do not obviously fit within the definition of a "plant pest," "food additive," or "new animal drug." In fact, until the biotech industry pushed for the establishment of the Coordinated Framework, the Reagan administration had been inclined to leave GMOs unregulated.[209] In the absence of affirmative steps by legislatures or regulatory agencies, new technologies may escape government oversight.

As a general matter, enacting legislation is not quick or easy. It is especially difficult to enact laws addressing public health and environmental hazards because the costs of such legislation are narrowly concentrated and the benefits are widely dispersed.[210] There are additional barriers to legislation governing new technologies, such as a lack of publicity and a deficit of scientific expertise in Congress. In the case of GMOs, the scientific community and the biotech industry warded off legislation that would have specifically governed genetic engineering research and commercialization.

Applying existing laws to new technologies often is politically more feasible than passing new legislation. As Adam Sheingate has observed, "[W]hen moments of regulatory uncertainty arise, the executive branch enjoys distinct advantages over Congress in matters of agenda setting and policy innovation."[211] Regulators can use or adapt established structures and procedures, and stakeholders may prefer to work with familiar regulatory systems. Indeed, existing laws sometimes are sufficiently broad and flexible to apply to a new technology. But as the history of the Coordinated Framework illustrates, relying on existing laws and regulations to govern the risks of new technologies can result in inadequate and haphazard governance. From the start, the laws invoked by the Coordinated Framework were inadequate for dealing with the unique issues surrounding GMOs. The hazards plausibly posed by GMOs extend beyond the plant pest risks, pesticidal hazards, and food safety concerns addressed by the statutes that underlie the Framework.[212] The central tenet underlying the Framework, substantial equivalence, inappropriately discounts the possibility that genetic engineering merits closer scrutiny because it so greatly differs from conventional plant breeding. Moreover, the Framework's division of authority among multiple agencies undermines the effectiveness of the limited regulatory oversight that does exist.[213] Adoption of the Framework ultimately facilitated the avoidance of value-laden questions regarding the risks and uncertainties society is willing to tolerate, the ethical aspects of genetic manipulation, and humanity's relationship to the natural world.

These observations warrant a healthy skepticism regarding assertions that existing law is sufficient to manage emerging technologies. However, new legislation is not always the answer. Additional statutes can compound regulatory complexity and foster bureaucratic balkanization. Moreover, the compromises common to the lawmaking process frequently result in laws that are suboptimal from a policy perspective.[214] But a sound analysis should consider all regulatory options and counter the systemic bias toward the use of existing legal regimes.

The Socially Constructed Nature of Technologies

A fourth critical lesson from the history of GMO regulation is that the definition, boundaries, and categories of a technological field are socially constructed. As philosopher Andrew Feenberg explains, technology is "a social object" and "a scene of social struggle."[215] GMOs have generated global controversy in a way that the genetic engineering of pharmaceuticals has not.[216] Likewise, the proposed use of GM animals for human consumption has aroused domestic public concern to a greater extent than the food use of GM plants. The categories and distinctions drawn by society clearly matter. Legally speaking, they frame understandings of policy problems and help determine who and what should be subject to regulation.[217]

GMO advocates characterize biotechnology as an "evolution[ary]" step along a "continuum" of genetic modification rather than a "revolution."[218]

Consistent with this characterization, the Framework declares that the genetic engineering of plants is merely an "extension" of traditional plant breeding techniques. The Framework accordingly treats genetically engineered changes in food composition like conventional food additives and genetically engineered pesticides like conventional pesticides. In contrast, opponents of GMOs portray genetic engineering as new, unnatural, and potentially hazardous.[219] The European Union has essentially adopted this view by imposing on GMOs close regulatory scrutiny that is absent from its oversight of traditional agricultural practices and products. Disputes over framing are prominent with respect to other emerging technologies, too. Subsequent chapters explore the issues underlying these disputes, including what constitutes nanotechnology and whether it is a "new" field; what techniques fall within the rubric of geoengineering and whether the term *geoengineering* should be used at all; and whether synthetic biology differs in meaningful ways from conventional genetic engineering.

No simple formula exists for sorting through such disputes. However, laws requiring information disclosure and risk analysis can help develop the information that might assist in the task. In addition, the recognition that technology is inherently social underscores the importance of public participation in technological matters. That is, there should be ample opportunities for meaningful involvement from a broad spectrum of society in defining technology, directing technology development, and determining regulatory policy. Expanded participation with respect to GMOs could lead to greater public acceptance, increased attention to potential health and environmental effects, and the steering of research funding toward GM crops offering greater benefits for consumers and society at large.

The Importance of Public Trust

A final lesson from the governance of GMOs is that public trust in the risk management and regulatory structure for an emerging technology is crucial to its acceptance. Despite their prevalence in the food supply, GMOs remain controversial in the United States nearly two decades after their initial commercialization, in large part because industry and government did little to earn the public's trust. The biotechnology industry made some attempts to inform the public about the new technology as farmers adopted GMOs, but those efforts failed to engage the public in a meaningful dialogue.[220] Rather, industry assumed that a deficiency in public understanding existed and thus focused only on educating the public about the benefits of GMOs. However, no dialogue took place sufficient to address the public concerns that persist to this day. These concerns exist not because of the public's inability to understand the science behind GMOs, but because of public distrust fostered by the biotechnology industry's successful campaign to block mandatory labeling requirements and the lack of an effective and transparent regulatory system.[221] The failure of industry and government to engage the ethical dimensions of genetic engineering has compounded the matter.[222] GM crops demonstrate that a new technology can establish a dominant market share without the public's knowledge. Having the technology ultimately accepted by society is another matter. The failure of the biotech industry to establish trust has cost it access to foreign markets as well as significant portions of the domestic market and has contributed to the technology's inability to achieve much of its original promise.[223]

These lessons from biotechnology provide useful insights to keep in mind as society addresses other emerging technologies. Although each new technology presents a unique set of opportunities and challenges, thoughtful and public oversight in each case is essential.

CHAPTER 3

Nanotechnology

Emerging Technology Present

Now found in nonstick cookware, spill-resistant fabrics, transparent sunscreens, and various electronic devices, nanotechnology is a leading example of a presently emerging technology. Though nanotechnology applications are becoming widespread, much of the field is in the research and development (R&D) phase. Nanotechnology is an important subject of consideration because of its growing presence and its expected effect on almost all sectors of the economy. Moreover, nanotechnology is of particular interest for this book because it has been the subject of various participatory technology assessment experiments. These experiments suggest both the potential and the limits of current TA approaches.

What Is Nanotechnology?

Broadly defined, nanotechnology refers to the science of manipulating matter at the scale of one to one hundred nanometers, with a nanometer being one-billionth of a meter.[1] There is some disagreement regarding the exact boundaries of nanotechnology, however, since the term has come to include a wide range of nanometer-scale advances in biology, chemistry, physics, and materials science. Regardless, many people perceive nanotechnology as having nearly boundless applications and predict that it will revolutionize various manufacturing processes.[2] In the United States, the federal government funds more than $2 billion in nanotechnology research per year, and global R&D funding surpassed $18 billion in 2008.[3] Nanotechnology was incorporated into approximately $225 billion worth of products in 2009, and the value of nanotechnology products worldwide could reach $2.5 trillion by 2015.[4]

Nanotechnology encompasses a wide array of developments and activities. Though observers have suggested various schemes to categorize nanotechnology, a fundamental distinction often is made between passive nanotechnology and active nanotechnology.[5] Passive nanotechnology, also referred to as nanoscale science and engineering, researches the unique properties of nanomaterials. Nanomaterials, defined as materials having a size range of one hundred nanometers or less, may be fixed as integral features of larger objects such as electronic components or deployed as free nanoparticles such as in cosmetics or pharmaceuticals. Active nanotechnology, by contrast, refers to the performance of more elaborate functions at the nanometer scale, such as the use of molecular nanosystems or self-assembly processes to construct materials and devices in a bottom-up, molecule-by-molecule fashion. This more elaborate version of nanotechnology could serve as the basis for cleaner and more efficient manufacturing processes but is estimated to be years if not decades away from commercial applications.[6] Active nanotechnology ultimately may even lead to the creation of devices capable of self-replication in a process akin to cell division, a prospect that has generated both excitement and fear.[7]

This chapter focuses on passive nanotechnology, which is responsible for the commercial applications of nanotechnology to date.[8] The engineered nanomaterials produced by passive nanotechnology are of interest to scientists because they often behave differently from the conventional materials from which they are derived.[9] For example, nanosilver, which is produced by a variety of physical as well as chemical techniques, has greater antimicrobial properties than ordinary silver.[10] Thanks to these properties, nanosilver is now widely incorporated as an antibacterial and antifungal agent in socks, pillows, washing machines, and other consumer products.[11] More generally, the small size and high surface area–to–mass ratio of nanomaterials enhance the mechanical, electrical, optical, catalytic, and/or biological activity of a substance.[12] These characteristics make nanomaterials desirable for an astonishing range of potential uses. Classes of engineered nanomaterials include carbon nanotubes, metal oxide nanoparticles, and quantum dots.[13] Carbon nanotubes are extremely strong and flexible molecules, and they can exhibit varying electrical properties depending on their structure. These properties have led to the use of carbon nanotubes in plastics, battery and fuel cell electrodes, water purification systems, adhesives, and electronic, aircraft, and automotive components.[14] Nanoparticles of titanium dioxide and other metal oxides, which have photolytic properties such as the ability to absorb ultraviolet light, are commonly incorporated into sunscreens, cosmetics, solar cells, paints, and protective coatings.[15] And quantum dots, semiconducting crystals possessing special optical properties, have proven useful in medical imaging, targeted therapeutics, solar cells, and photovoltaics.[16]

Future applications appear similarly boundless. A recent Government Accountability Office report envisions nanotechnology applications ranging from smaller and more powerful batteries and better targeted medical treatments to more surreptitious surveillance techniques and stronger yet more lightweight military uniforms.[17] Nanotechnology could also be applied in ways that directly benefit the environment. Because of their high chemical reactivity, certain nanomaterials may be deployed to rapidly clean up environmental contamination, for example. Other possible environmentally beneficial applications include fuel additives that increase engine efficiency, more portable and affordable water desalination and filtration systems, more efficient solar energy generation and storage, and better-controlled release of pesticides and fertilizers.[18]

Health and Environmental Concerns

Supporters of nanotechnology promise that its applications will lead to better health, improved environmental quality, and general abundance. It remains to be seen whether these promises will be fulfilled. Furthermore, there are grounds to expect that some of the benefits of nanotechnology will be accompanied by serious drawbacks. Surveillance uses of nanotechnology, for example, may lead to privacy concerns. More significantly, a wide range of nanotechnology applications may raise health and environmental issues. Past experiences with substances like asbestos—to which one class of nanomaterials, carbon nanotubes, bears a strong resemblance—are revealing. Asbestos was once touted as a "magic mineral" because of its strength, flexibility, insulative properties, and fire resistance.[19] Exposure to asbestos was later found to cause cancer, asbestosis, and other health problems, and the substance is estimated to be responsible in total for half a million deaths in the United States.[20] The asbestos problem provides a stark and cautionary reminder that the hazards of a substance may be serious yet not apparent for many years.

Indeed, precisely the properties that make nanomaterials useful, such as their small size, chemical composition, surface structure, solubility, and shape, may make them harmful to humans as well as other organisms.[21] Free nanoparticles, as opposed to nanomaterials integrated into larger objects, are of particular concern because they are most likely to enter the body, react with cells, and cause tissue damage.[22] The small size of nanoparticles, for example, corresponds to a greater surface area for a given mass of material and hence a greater number of reactive groups displayed at the surface.[23] Surface reactive groups, scientists believe, play an important role in toxic reactions by generating reactive oxygen species that may damage DNA, proteins, and cell membranes.[24] Consistent with this theory, experimental results suggest that tissue injury from exposure to nanoparticles is correlated with surface area rather than mass.[25] Small size also enables some nanoparticles to move into and within the body in ways that bulkier versions of the same substance cannot.[26] Nanoparticles that come in contact with flexed or damaged skin may penetrate the epidermis and pass into the body.[27] Nanoparticles may be inhaled and move deeply into the respiratory tract, evading defense mechanisms that trap larger particles.[28] Once in the body, nanoparticles may even cross the blood-brain barrier, unlike most contaminants.[29] This means that nanoparticles may enter the central nervous system through neuronal pathways leading from the respiratory tract to the brain.

Humans have long been exposed to ambient nanoparticles from forest fires and industrial pollution, and data regarding the hazards of such exposure provide clues regarding the potential risks of exposure to engineered nanomaterials.[30] Although the human body has developed various mechanisms for filtering out or removing some ambient nanoparticles,[31] other ambient nanoparticles pose health and safety concerns. For example, mineral dust particles, which are comparable in size to engineered nanoparticles, can cause pulmonary inflammation, heart attacks, cardiac rhythmic disturbances, and oxidative injury.[32] Studies comparing ambient and engineered nanomaterials suggest that some engineered nanomaterials may pose hazards at least as great as those associated with mineral dust.[33] There are limitations on the ability to infer the potential hazards posed by engineered nanomaterials from information about ambient nanoparticles, however.[34] Ambient nanoparticles often have a fairly short life span as nanoparticles because they tend to agglomerate or dissolve in water, for example.[35] In comparison, engineered nanomaterials may be designed to persist for longer periods of time and thus may pose greater hazards. It is also possible that the novel properties associated with engineered nanomaterials may lead to additional mechanisms of injury that do not apply to ambient nanomaterials.[36] These novel properties may enable engineered nanoparticles to evade or overstimulate the body's defenses, causing inflammation or allergic responses.[37]

Actual data regarding the toxicity of engineered nanomaterials are limited. Nanotoxicology research efforts have grown substantially in recent years but still trail far behind the pace of nanotechnology development and commercialization.[38] With more than a thousand nanotechnology-based consumer products already on the market, nanotechnology development essentially "got a 15 year head start" on research into its health impacts.[39] Furthermore, the modest nanotoxicology efforts to date have made little progress for several reasons. First, many features of a nanomaterial—including the method of manufacture, presence of impurities, nature of surface coatings, and degree of aggregation—appear to affect the risks posed.[40] Thus, findings from studies of one nanomaterial may not be applicable to a similar nanomaterial. Research results may even be influenced by the particular method used to prepare a given nanomaterial for an experiment.[41] Second, scientists currently do not possess all the measurement technologies relevant to describing the characteristics of nanomaterials that affect their toxicity.[42] As a result, the data generated by toxicology research may be incomplete or inaccurate. Finally, models do not yet exist for predicting the toxicity of untested nanomaterials based on data involving substances that have already been tested.[43] Such models, which play a critical role in conventional toxicology, could be even more important to risk-based regulation of nanomaterials, given the wide variation in types, sizes, and surface coatings of nanomaterials.[44]

The nanotoxicity information generated thus far is discomforting. Researchers have found that once inhaled, ingested, or otherwise taken into an organism, certain nanoparticles can enter individual cells, release toxins, and damage various cell components, including DNA.[45] One class of materials giving particular cause for concern is carbon nanotubes, which are being produced and used in substantial quantities today.[46] Carbon nanotubes, which are extremely biopersistent, physically resemble asbestos fibers and thus could trigger similar effects on the respiratory system.[47] Studies of carbon nanotubes have found that they can cause oxidative stress, inflammation, cell damage, and other pathological effects.[48] Titanium dioxide nanoparticles, another type of nanomaterial used widely in sunblocks, cosmetics, and paints, have produced genotoxic and carcinogenic effects in rodents.[49]

At present, however, a quantitative risk assessment of nanomaterials is not possible, and evidence suggestive of danger falls short of establishing that exposure at levels likely to be encountered by humans or other organisms is harmful.[50] The data gap will not be closed any time soon, if ever. Government agencies are only beginning to put nanotoxicity research policies in place, and the various efforts under way will provide only a fraction of the data required to conduct meaningful risk assessments. Existing governmental efforts include: (1) testing of "representative" nanomaterials by Organisation for Economic Co-Operation and Development (OECD) members, (2) federal research efforts under the National Nanotechnology Initiative (NNI), (3) the EPA's Nanomaterial Research Strategy, and (4) the EPA's Nanoscale Materials Stewardship Program (NMSP).[51] Private companies are also conducting internal toxicology assessments. A brief discussion of these research efforts provides a sense of the limited progress to date.

In 2006, the OECD established a working group to study the potential hazards of nanomaterials. Member countries have agreed to jointly develop data for a group of fourteen nanomaterials purportedly representative of materials currently circulating in commerce or nearing commercial use.[52] Initial research efforts focused on the threshold question of whether such materials can be successfully tested, and the OECD has concluded that methods used to test traditional chemicals are generally appropriate but may need to be adapted to specific nanomaterials. OECD testing on a limited subset of nanomaterials to determine their properties, toxicity, and environmental fate and behavior is ongoing.[53]

The NNI is a multiagency program established in 2001 to coordinate nanotechnology R&D across the federal government. Although the NNI research program includes the study of health and environmental risks, such research has been relatively neglected. The NNI's 2008 *Strategy for Nanotechnology-Related Environmental, Health, and Safety Research*,"[54] for example, was sharply criticized by the National Research Council (NRC), which found that much of the ongoing research cited in the NNI strategy as evaluating health and environmental risks actually involved projects "focused [only] on understanding fundamentals of nanoscience."[55] The NRC also found the NNI strategy itself to be flawed in that it lacked a plan of action for achieving research goals, mechanisms to evaluate research progress, and other critical elements. In response to the NRC's recommendations to establish a broader strategic plan for risk research sufficient to support risk assessment and risk management, the NNI issued a revised strategy in late 2011.[56] That strategy, which emphasizes risk assessment and product life-cycle analyses, identified research needs in six core categories: (1) nanomaterial measurement tools; (2) human exposure assessment; (3) human health responses; (4) environmental effects; (5) risk assessment and risk management methods; and (6) informatics and modeling. The strategy underscores the vast informational needs for effective risk management, and its execution will depend on sufficient funding from Congress and adequate coordination by various federal agencies.

The EPA's Nanomaterial Research Strategy, issued in 2009, outlines the research on environmental and health effects of nanotechnology that the EPA intends to support in the coming years.[57] Like the 2011 NNI strategy, the EPA's strategy identifies numerous research needs, reinforcing the immensity of the task facing regulators, and acknowledges the various obstacles to the informed risk assessment of nanomaterials. In addition to the Nanomaterial Research Strategy, the EPA is also responsible for the NMSP, which the agency introduced in an attempt to solicit materials data from industry on a voluntary basis. The data, it was presumed, would include information on health and environmental impacts, but the EPA has characterized the results of the program as disappointing.[58] Only 31 companies ultimately participated, providing information on 132 materials—only about 10 percent of the nanomaterials that the EPA estimated to be commercially available at the time.[59] Much of the data submitted to the EPA, moreover, was of limited use because it contained no information on exposure or toxicity.[60]

While extremely limited, the information collected through the NMSP indicates that industry is conducting some research on health and safety risks. Apparently concerned about the potential liability and adverse publicity that might follow from toxic exposure, major chemical companies such as BASF and DuPont have publicly stated that they are voluntarily undertaking toxicology research on nanomaterials.[61] The extent and results of such efforts, however, are unknown because nanotechnology companies have made few details available to the public.

Existing Legal Authority over Nanomaterials

Despite the concerns discussed thus far and the increasingly widespread use of nanotechnology, legal efforts to manage health and environmental risks have been negligible. No federal law explicitly regulates the health and environmental effects of nanotechnology. The 21st Century Nanotechnology Research and Development Act (Nanotechnology Act), the only federal statute specific to nanotechnology, concentrates on the development and promotion of nanotechnology.[62] State and local regulatory efforts have focused on limited requirements of information disclosure.[63] In the absence of legislative action, regulators are following the all-too-familiar path of cobbling together an oversight scheme based on existing authorities. Such a course runs the risk of repeating the errors and difficulties experienced in applying the Coordinated Framework to GMOs.

Existing statutes that regulators would likely try to apply to nanomaterials were not designed with nanotechnology in mind. In general, these laws require proof of harm or a quantitative assessment of risk as a precondition for regulation. This means that nanomaterials will likely be unregulatable under existing statutes for some time, notwithstanding significant safety concerns. Even as a body of risk information develops that may be sufficient to regulate some nanomaterials under existing statutes, regulators will likely find themselves unable to keep pace with the rapid development and commercialization of nanotechnology.

The Toxic Substances Control Act

The most obvious legal authority that might be utilized to address risks associated with nanomaterials is the Toxic Substances Control Act (TSCA).[64] The TSCA provides the EPA with regulatory authority in several key areas relevant to the management of nanomaterials' potential risk: testing of chemicals (Section 4), notification to the EPA prior to the manufacture of new chemicals (Section 5), regulation of chemicals that present health or environmental risks (Section 6), and notification to the EPA when a manufacturer learns of a substantial risk (Section 8(e)). In contrast to other environmental laws that govern only the release of pollutants into the environment, the TSCA gives the EPA broad authority to regulate a chemical substance at any point in its life cycle.[65] Moreover, the TSCA's primary focus—that exposure to chemical substances may pose unreasonable risks to humans and the environment—appears to encompass the concerns raised by nanotechnology. TSCA, however, is a weak statute whose flaws are widely recognized.[66] The statute's weaknesses are magnified in light of the wide variety of nanomaterials in production and development, uncertainty regarding the materials' safety, and the swift pace at which nanotechnology is developing.

Section 4 Testing

Section 4 of the TSCA authorizes the EPA to require health and safety testing of specific chemicals.[67] The statute itself does not, however, impose a self-executing duty on manufacturers to conduct such testing. Nor can the EPA simply order a chemical manufacturer to carry out the desired tests. Rather, the EPA must promulgate a rule to require testing, and it must make a statutory finding that a chemical either (1) "may present an unreasonable risk of injury to health or the environment" or (2) "will be produced in substantial quantities," resulting in substantial human exposure or entry of substantial quantities into the environment.[68] Pursuant to Section 4, the EPA planned to issue a test rule for 15 to 20 different nanomaterials in December 2010 but had taken no such action by the beginning of 2013.[69] Promulgating a Section 4 test rule, moreover, is a lengthy process subject to industry challenge that the EPA has not adequately demonstrated the requisite potential risk.[70]

Section 5 Premanufacture Notification

For new chemical substances, Section 5 of the TSCA requires manufacturers to provide a premanufacture notice (PMN) and to submit any available health and safety data to the EPA.[71] A "new chemical substance" is a chemical substance that is not identical to any substance already found on the TSCA Inventory of chemicals in commerce.[72] If the EPA takes no action on a PMN within 90 days, manufacture of the chemical can proceed.[73] The EPA, however, may restrict or prohibit the manufacture, distribution, or use of a new chemical upon finding a reasonable basis that the chemical presents an unreasonable risk.[74] Section 5 also gives the EPA the authority to evaluate significant new uses of chemical substances that are already in commerce. In this context, the TSCA places the burden on the EPA to promulgate a rule determining that a particular use constitutes a "significant new use." A manufacturer subject to such a rule must provide a significant new use notice (SNUN), which is similar to a PMN.[75]

Thus far, the EPA's efforts to apply the TSCA to nanomaterials have primarily involved Section 5. Nanomaterials that qualify as new chemical substances, of course, are subject to the PMN requirement.[76] It is disputed, however, whether a nanomaterial that has the same molecular identity as a macroscale substance listed in the TSCA Inventory—such as the nanoscale titanium dioxide used in sunscreens—should be deemed a new chemical substance requiring a PMN. Nanomaterials are often derived from common substances that are not new, but they are of special interest precisely because they possess properties different from their parent material. Indeed, the issuance of patents for nanomaterials undermines the contention that engineered nanoscale versions of macroscale substances should not be treated as new materials or new uses under the TSCA.[77]

In January 2008, the EPA formally adopted the position that no PMN should be required for nanoscale versions of existing chemicals.[78] The EPA reasoned that the TSCA Inventory does not distinguish between two forms of a chemical substance that differ only in particle size or that have differing properties resulting from a difference in particle size. The EPA's policy—comparable to the FDA's presumption that GM foods are GRAS and therefore exempt from premarket approval—has opened the door to the commercial production of various nanomaterials that are not subject to Section 5 regulation and has left regulators without a reliable means of tracking the identity or characteristics of these nanomaterials.

Under the Obama administration, however, the EPA has reconsidered its approach. The EPA announced that it intends to issue a significant new use rule that would apply Section 5 requirements to nanoscale versions of existing chemicals. The anticipated rule, which had not been issued in draft form as of early 2013, would require an SNUN for certain nanomaterials that have the same molecular identity as macroscale substances listed in the TSCA Inventory.[79] The rule would only apply prospectively, however: No SNUN would be required for nanomaterials that come into use before the rule is finalized.[80] More important, while the notices submitted pursuant to the rule will help keep the EPA informed, they generally will not provide sufficient health and environmental data to assess risks. In a PMN or SNUN, manufacturers need only provide *available* toxicity data; they need not generate any data on toxicity unless the EPA specifically demands that they do so.[81] In fact, only about 15 percent of PMNs filed with the EPA include health or safety test data, and most lack test data of any type.[82] Without such data, the EPA must rely solely on modeling to estimate chemical hazards. Yet such models are of limited value in evaluating the hazards of nanomaterials.

Finally, the EPA's Section 5 regulations contain exemptions that could apply to some nanomaterials. For example, the regulations include an exemption for chemicals produced in volumes of 10,000 kilograms or less per year as well as an exemption for chemicals whose use will result in little or no human exposure.[83] Manufacturers must apply for these exemptions, which the EPA may deny if it finds that a substance may cause serious health or environmental effects.[84]

Section 6 Authority to Regulate

Section 6 provides another important regulatory tool under the TSCA. This provision authorizes the EPA to regulate the manufacture, processing, distribution, use, or disposal of any chemical substance where there is a "reasonable basis to conclude" that such an activity "presents or will present an unreasonable risk of injury to health or the environment."[85]

The "unreasonable risk of injury" standard requires a factual finding of risk and a normative finding that such risk is unreasonable.[86] In determining whether a risk is unreasonable, the EPA must balance any negative effects to human health and the environment with the benefits derived from use of the substance.[87] Furthermore, under a leading judicial interpretation of Section 6, the EPA must evaluate the availability of substitutes for the substance in question and may apply only the least burdensome regulatory measure that yields an acceptable level of risk.[88] In light of these fairly stringent requirements, it is not surprising that the EPA has not applied Section 6 to any nanomaterial or expressed any plans to do so.

Section 8(e) Notification of Substantial Risk

Finally, Section 8(e) of the TSCA requires manufacturers, processors, and distributors of chemical substances to notify the EPA if they obtain "information which reasonably supports the conclusion that such substance . . . presents a substantial risk of injury to health or the environment."[89] This provision allows companies to exercise their judgment in determining whether to report information and what information to report.[90] Moreover, the provision does not mandate that companies develop health and safety data. Nonetheless, Section 8(e) could serve as an important source of risk information about nanomaterials, given the paucity of such information and the limited resources at the government's disposal for developing it.

The TSCA's Inadequacies

Two critical inadequacies of the TSCA reveal why the statute is not fit to address the potential hazards of nanotechnology, notwithstanding the EPA's incipient initiatives. First, the TSCA places heavy evidentiary burdens on the EPA. Regulation of a chemical substance under Section 6, for example, requires that the EPA demonstrate the existence of unreasonable risk. This standard has been deemed "a failure" because it "has imposed huge information demands, invited contention and judicial intervention, and thwarted regulatory action."[91] Because there is often little information regarding the effects of chemical exposure, the EPA frequently cannot meet this burden of proof.[92] For nanomaterials, the uncertainty is especially great—and the evidentiary standard particularly unattainable—because of the lack of adequate models for predicting toxicity. The variety of nanomaterials and rapid pace of nanotechnology developments exacerbate the EPA's evidentiary difficulties.[93] Even the exercise of Section 4 testing authority may be problematic, since it requires the EPA to demonstrate the existence of potential risk at the same time that testing is necessary precisely because such information is unavailable.[94]

The TSCA's second important inadequacy is its implicit assumption that the absence of information on the risk of a chemical means that no risk exists.[95] As a result of the evidentiary burdens the TSCA imposes on the EPA, substances whose effects are uncertain are treated identically to substances that demonstrably pose no unreasonable risks. Compounding the problem, the TSCA does not require manufacturers to develop health and safety data absent a test rule. Section 8(e), for example, requires a company to notify the EPA only if it obtains information indicating a substantial risk but does not mandate testing that might generate such information. Most nanomaterials are likely to be accompanied by very little toxicity data and will be treated as if they are safe—even though there are serious grounds for believing that some of them are not.

TSCA reform could go a long way toward addressing these concerns by mandating the generation of toxicity data and shifting evidentiary burdens to manufacturers. Recent proposals would require chemical manufacturers to develop and submit a minimum data set for each chemical they produce as well as prove that each chemical used in commerce is safe.[96] The proposals would also explicitly authorize the EPA to declare that a nanomaterial constitutes a new chemical substance that requires a PMN, even where the nanomaterial has the same molecular identity as a macroscale substance listed in the TSCA Inventory.[97] Absent substantial changes in the statute and agency practice under the statute, however, the TSCA remains a poor tool for responding to the potential dangers of nanomaterials.

Other General Environmental Statutes

Environmental statutes other than the TSCA could apply to nanomaterials, but their coverage is too limited to be effective. For example, statutes that govern pollution and waste, including the Clean Air Act and Clean Water Act, have more extensive regulatory structures in place and more successful histories of implementation than the TSCA. These statutes authorize the EPA to define pollutants based on their negative effects and rely on permitting schemes to regulate pollutants at their points of release into the environment.[98] In theory, the EPA could identify nanomaterials as pollutants under these statutes and establish permit requirements and limits on their release. Such limits typically cover pollutants released by production facilities in their waste streams, however, and are poorly adapted to govern substances such as nanomaterials, which are deliberately incorporated into products. Because use and disposal of nanotechnology products are expected to be the greatest source of human and environmental exposure, facility permit limits would be of limited value.[99] Regulation of the release of nanomaterials during or after use does not present an attractive option either. It is hard to imagine, for example, restrictions on use of a sunscreen or socks containing nanomaterials that would be effective in preventing human exposure to nanomaterials or the release of nanomaterials into the environment.

Regulation of the disposal of nanomaterials, perhaps under the Resource Conservation and Recovery Act, would also face serious difficulties. For practical reasons, regulations under the statute currently exempt household waste from the rigorous regulation of hazardous waste.[100] Nanomaterials in consumer products are likely to end up in household waste, commingled with ordinary garbage, and would not be readily separable.[101] The Comprehensive Environmental Response, Compensation, and Liability Act, which focuses on the cleanup of hazardous waste already released into the environment, also is not a promising avenue for addressing nanomaterial hazards. In most instances, it is impractical or impossible to remove nanomaterials from the environment once they are released.[102]

Consumer Product Safety Statutes

Statutes focused on consumer products likewise offer little prospect of effective oversight. The Consumer Products Safety Commission has authority to protect consumers from unreasonable risks of injury associated with consumer products not specifically regulated by certain other federal statutes.[103] According to one estimate, half of all consumer nanotechnology products currently on the market could fall within the commission's jurisdiction.[104] Nonetheless, the commission's regulatory authority is weak, and the agency lacks the expertise and resources to take effective action on nanotechnology products.

The commission generally regulates consumer products within its purview through information disclosure requirements and product safety standards.[105] Under the Consumer Products Safety Act (CPSA), the commission must initially rely on "voluntary standards" to address unreasonable risks of injury associated with a consumer product.[106] If voluntary compliance is inadequate, the commission may establish mandatory standards, but only as "reasonably necessary" to prevent or reduce unreasonable risks.[107] This threshold for regulation is at least as difficult to meet as the standards imposed by the TSCA. Because of the sparse risk data available for nanomaterials, regulating nanotechnology through the CPSA is presently impossible.[108] Moreover, in contrast to the EPA, the commission has no authority to mandate premarket safety testing. Rather than anticipating problems, the commission typically acts only after receiving reports of injuries.[109] The commission also is subject to statutory restrictions on the information it can disclose about a brand or manufacturer.[110] The commission ultimately lacks the resources, experience, and expertise to carry out the research and analysis needed to evaluate and manage the potential dangers of nanomaterials in consumer products.[111]

Product-Specific Statutes

Statutes that govern specific products may offer more effective oversight of nanomaterials in certain contexts. Pesticides, including those incorporating nanomaterials, must be registered with the EPA under the Federal Insecticide, Fungicide, and Rodenticide Act (FIFRA). New drugs, including those containing nanomaterials, are subject to an extensive premarket approval process by the FDA.[112] Applications of nanotechnology in food and cosmetics also fall within the FDA's purview but generally do not require premarket approval.

FIFRA requires that new pesticides be registered with the EPA before they are marketed.[113] The EPA will allow registration only where an applicant demonstrates that the pesticide will perform its intended function without unreasonable harm. Accordingly, the EPA can demand a battery of studies on any pesticide to determine potential impacts on human health and the environment prior to allowing registration.[114] How broadly the EPA will construe its FIFRA authority with respect to nanomaterials remains to be seen, however. In November 2008, a group of nonprofit organizations filed a petition asking the EPA to regulate products containing nanosilver under FIFRA.[115] The group alleges that hundreds of products containing nanosilver have entered the market without the requisite FIFRA registration. These products include a range of items that the average person might not consider to be pesticides, such as food storage containers, pillows, and various types of clothing. As of early 2013, the EPA had not yet acted on the petition. Meanwhile, the EPA's 2011 conditional FIFRA registration of a textile preservative containing nanosilver is the subject of a pending lawsuit filed by environmentalists.[116]

The Federal Food, Drug and Cosmetic Act (FFDCA) requires that new drugs be approved prior to their introduction into commerce.[117] The drug approval process begins with the submission of an Investigational New Drug Application, which must describe specific testing plans and provide the results of pharmacological and toxicological studies on the drug.[118] Based on the application, the FDA decides whether it is reasonably safe to proceed with human clinical trials, which involve three phases of studies to determine effectiveness and toxicity.[119] The drug approval process thus ensures that some health and safety testing is done on new drugs, even when the presence of nanomaterials is unknown to the FDA.[120]

Although the FDA classifies sunscreens as drugs, government oversight of nanomaterials in sunscreens has been minimal. The FDA classifies sunscreens as drugs because they purport to protect the skin against the harmful effects of sun exposure.[121] New drugs require premarket approval, but drugs that contain ingredients generally recognized as safe and effective do not.[122] In 1999, after reviewing only limited toxicity data, the FDA expressed the view that "micronized" titanium dioxide is not a new drug ingredient despite the functional differences between it and larger particles of the substance.[123] This agency pronouncement opened the door to the widespread incorporation of nanomaterials in sunscreens without further safety assessments or oversight. Alarmed by this development, a coalition of nongovernmental organizations petitioned the FDA in 2006 to regulate nanosunscreens and to conduct a programmatic EIS of nanomaterial use in products under the agency's jurisdiction.[124] In a response issued six years later, the FDA declined to prepare a programmatic EIS or to issue nanotechnology-specific regulations but indicated that it was continuing to study the safety of nanomaterials in sunscreens.[125]

Though cosmetics as a category represents one of the most common uses of nanomaterials,[126] its regulation by the FDA is similarly superficial. The FDA does not require premarket approval of cosmetic products and ingredients.[127] The FDA instead places the responsibility to determine the safety of cosmetics on manufacturers. Manufacturers may participate in voluntary programs to file data on ingredients, register manufacturing sites, and report cosmetic-related injuries to the FDA, but there is no requirement that a manufacturer do any of these things.[128] A cosmetic manufacturer essentially may use any ingredient or market any cosmetic unless the FDA proves that it may be harmful to consumers—something that rarely occurs.[129]

Finally, FDA review of nanomaterials in food is currently minimal but appears likely to increase. Here, the FDA is relying on its authority over food additives—the same authority applied to GM foods. As discussed in chapter 2, manufacturers and not the FDA determine whether a substance added to food is GRAS. If a manufacturer makes such a determination, it may market the substance without informing the FDA. Because the FDA leaves it to manufacturers to voluntarily notify it of GRAS determinations and of the presence of nanomaterials in food, the FDA does not accurately know the extent to which nanomaterials are present in the food supply.[130] Nonetheless, a 2012 FDA draft guidance document notes that nanomaterials may raise new risks and require new methods of safety testing.[131] The agency accordingly warns that nanotechnology applications in food may not qualify as GRAS and therefore may require premarket approval.[132]

The Occupational Safety and Health Act

Workers and researchers have perhaps the greatest potential for exposure to nanomaterials, yet the primary statute meant to protect such persons is inadequate. Such workplace exposures fall within the ambit of the Occupational Safety and Health Act, which requires that employers provide workplaces "free from recognized hazards that are causing or are likely to cause death or serious physical harm."[133] This statute gives the Occupational Safety and Health Administration (OSHA) authority to set and enforce standards that require "conditions, or the adoption or use of one or more practices, . . . reasonably necessary or appropriate to provide safe or healthful employment and places of employment."[134] OSHA implements the statute by establishing permissible exposure limits (PELs) for hazardous materials and by mandating measures that help to achieve PELs, such as engineering controls and protective equipment.[135] Substantive, legal, and political constraints, however, have prevented OSHA from issuing health regulations for most of the substances for which regulation has been recommended.[136] Accordingly, various commentators have concluded that the Occupational Safety and Health Act does not protect workers effectively.[137]

With the proliferation of different types of nanomaterials in research and manufacturing settings, OSHA will encounter even greater difficulties in addressing the risks posed by nanotechnology.[138] For example, demonstrating that an employer has violated its general duty to provide a workplace free from recognized hazards is exceptionally difficult in the nanomaterial context because of the uncertainty regarding the hazards of nanomaterial exposure.[139] The same uncertainty also will hamper OSHA's ability to demonstrate the significant risk of harm required for it to issue protective regulations.[140] Moreover, it is not clear how effective the implementation of typical workplace health standards would be. Little information is available regarding the effectiveness of engineering controls and protective equipment in controlling nanomaterial exposure in the workplace.[141]

Summing Up

Existing statutes were not crafted with nanomaterials in mind. These statutes may be of some use after more data is gathered on nanomaterials and specific risks are identified. The main problem, however, is that these statutes lack adequate mechanisms to address the uncertain hazards posed by the increasing use of nanomaterials. Existing laws generally place on regulators the burden of demonstrating harm that is specific to a material. Such a showing is not presently possible. As a result, the statutes offer a relatively languid response when faced with rapid technology development and commercialization. The result—essentially a society-wide experiment on consumers with no controls or systematic follow-up—demands the adoption of a different, more precautionary system.[142]

Tort Law as a Backstop?

Before turning to what such a system might look like, the potential role of tort law in managing nanotechnology's risks merits consideration. That role is likely to be minimal because of the lack of toxicity data on nanomaterials, the lack of disclosure surrounding their use, and the general difficulties of the tort system in redressing toxic injuries.

Tort law deters negligent conduct by threatening exposure to compensatory and punitive liability. In product liability cases, strict liability may apply, meaning that manufacturers and distributors may be held liable regardless of fault.[143] Irrespective of whether negligence or strict liability applies, plaintiffs must prove that the defendant caused their injuries. Plaintiffs in environmental tort cases often face an uphill struggle in demonstrating causation, however.[144] Such plaintiffs must prove both that a defendant's substance is capable of causing the injury (general causation) and that exposure to that substance in fact caused the injury (specific causation). With respect to specific nanomaterials, there is almost no existing data sufficient to establish either type of causation. We know very little about the toxicity of nanomaterials, and what we do know suggests that toxicity will vary widely, depending on size, coating, method of preparation, and other factors. In addition, individual plaintiffs will likely face difficulties in demonstrating that their exposure to a defendant's nanomaterials, as opposed to other substances, caused their injuries.[145] The time gap between initial exposure and injury often hinders the ability of toxic tort plaintiffs with latent injuries to recognize a tortious injury, identify possible causes, and collect evidence.[146]

Nanotechnology manufacturers concerned about liability for acute toxic effects do have some incentive to test their products because these effects are more readily traceable to their source. Testing poses some business risk, however, since adverse results can lead to negative publicity and trigger disclosure and further testing obligations. In addition, incentives to test for chronic or latent effects, as opposed to acute effects, are weak because such tests are expensive and because it is unlikely that a manufacturer will be held liable for these effects.[147]

Tort law consequently is of limited value in persuading manufacturers to learn more about nanotechnology's hazards to human health and in internalizing the costs of those hazards. Tort law is likely to be even less effective in addressing nanotechnology's potential hazards to the environment. As a general matter, liability for natural resource damages resulting from the release of conventional chemicals has been infrequent because of difficulties in determining impacts and establishing causation.[148] Such difficulties are likely to be magnified in the case of damages from exposure to nanomaterials.

Nanotechnology and Public Engagement

Potential health and environmental hazards are not the only aspect of nanotechnology requiring further attention. Greater public involvement in deciding the course of this influential technology is also essential. Reflecting on the missteps made in the development and commercialization of GMOs, the nanotechnology industry openly acknowledges the need to develop public trust in nanotechnology through public engagement.[149] Public engagement should not only foster trust, however. It should also give people a meaningful voice in the kind of society in which they live. In recognition of the importance of public participation, the United States and other countries have made a number of efforts to apply modified technology assessment techniques to nanotechnology.[150] While innovative, these efforts have had little effect on the course of nanotechnology development or policy. As one scholar of public engagement in nanotechnology has remarked, "It has been easier to praise the idea of democratizing science than to achieve it."[151]

The National Nanotechnology Program, an effort to coordinate federal nanotechnology activities pursuant to the Nanotechnology Act, includes as one of its objectives "ensuring that ethical, legal, environmental, and other appropriate societal concerns . . . are considered during the development of nanotechnology."[152] The act directs the program to incorporate public outreach and input "through mechanisms such as citizens' panels, consensus conferences, and educational events."[153] Accordingly, the National Science Foundation has funded Centers for Nanotechnology in Society at Arizona State University (CNS-ASU) and the University of California, Santa Barbara. The mission of these interdisciplinary centers is to study the potential societal impacts of nanotechnology and to engage stakeholders in dialogues about the future of emerging technologies.[154]

The discussion here focuses on CNS-ASU, which has developed a "real-time technology assessment" program that incorporates principles of participatory and constructive technology assessment.[155] The program's components include monitoring of opinions and values among researchers and the public regarding nanotechnology, fostering deliberation and participation involving researchers and the public, and assessing the program's effects on nanotechnology researchers and on nanotechnology in society. These ongoing efforts represent a creative departure from past approaches to technology management. But for various reasons, they are generating little up-front assessment, little public participation, and few effects on nanotechnology policy.

As an example, one important component of CNS-ASU's efforts to engage the public included the organization of the 2008 National Citizens' Technology Forum (NCTF). The NCTF linked six groups of citizens from different parts of the United States in deliberations focused specifically on human enhancement through nanotechnology, biotechnology, information technologies, and cognitive science research.[156] Deliberations were conducted face-to-face and electronically, and each group of citizens drafted a report reflecting the group's consensus. Based on these deliberations and reports, the organizers of the forum found that citizens supported research in this vein if it was coupled with trustworthy oversight. More broadly, the organizers concluded that "average citizens want to be involved in the technological decisions that might end up shaping their lives."

The citizens' reports generated by the NCTF do reflect a slice of public opinion reached after substantial deliberation. Nonetheless, the impact of the NCTF and of other similar participatory exercises has been minimal. Simply publishing citizens' reports hardly ensures any influence on the course of nanotechnology research or regulation.[157] The reports produced by the NCTF had no clear constituency or audience, and efforts to disseminate them were limited.[158] Moreover, the main purpose of the NCTF was to demonstrate the value of deliberative exercises and to investigate ways of structuring consensus conferences rather than to foster widespread public participation or to influence technology development or policy.[159] Indeed, one commentator characterized the NCTF primarily "as a social scientific research instrument" and observed that participants "tended to 'not even bother' to fight for ideas or opinion[s]" because they "'knew that they were part of a research project.'"[160] Although participants in the NCTF undoubtedly learned about and became engaged in the issues, the public at large was not brought into the process. The task of broadening public awareness and engagement has been left largely to the Nano-Scale Informal Science Education Network (NISE Net), which promotes nanotechnology education through science museums.[161] Such relatively narrow efforts have done little to inform the general public about nanotechnology.[162]

CNS-ASU director David Guston, a proponent of public engagement in technology development, concedes that overall nanotechnology research "has grown much larger and faster than the societal implications work that might engage it."[163] CNS-ASU's efforts to encourage nanotechnology researchers to reflect on the broader implications of their research have influenced research focus and design in some individual instances, but such influence has been limited largely to researchers, particularly graduate students, at ASU and two other universities with which it is collaborating.[164] Furthermore, the role of the NCTF and other technology assessment activities within the broader context of nanotechnology research is somewhat unclear. The Nanotechnology Act contains an inherent tension between the goals of promoting rapid nanotechnology development and integrating societal concerns into the research and development process.[165] In sum, ongoing technology assessment efforts with respect to nanotechnology have been small in scale and are unlikely to have a substantial effect on policy. These efforts ultimately may serve as little more than political tools for obtaining public acceptance unless they are broadened and better integrated with the policy-making process.[166]

Future Policies for Nanotechnology

The uncertainty surrounding the health and environmental effects of nanomaterials calls for an approach that promotes the gathering and analysis of risk information. The data generated eventually may inform regulation, but they are unlikely to dispel much of the uncertainty in the short term. In the meantime, nanotechnology development and commercialization will continue. The persistent uncertainty regarding nanotechnology's risks, combined with nanotechnology's spread and its broad and wide-ranging impacts, necessitates greater public awareness and participation. Indeed, neither the need for additional safety research nor the need for public involvement is disputed.[167] There is disagreement, however, regarding how to involve the public and how to proceed in the absence of safety data.

The current state of nanotechnology development offers a narrow window of opportunity to apply a Promethean approach to technology management. Although its presence is expanding rapidly, nanotechnology is at a relatively early stage of commercialization. Neither the technology nor a regulatory approach has become entrenched—at least not yet. In contrast to biotechnology in its early days, nanotechnology researchers and industry profess an openness to and even a desire for public input. The relatively undetermined situation, combined with the expected reach of nanotechnology, calls for a broad examination of policy options to address nanotechnology's potential risks proactively while involving the public in a meaningful way. Such an examination should occur promptly, as the massive investments in nanotechnology and its rapid growth will soon constrain the opportunities for policy change.

Proposals for addressing the potential risks of nanotechnology fall into three categories: (1) regulating nanotechnology under existing law, (2) promoting voluntary initiatives, and (3) enacting reforms through new regulatory authority. Consideration of the strengths and weaknesses of these options underscores the need to depart from past technology management approaches and to develop new regulatory authority.

Reliance on Existing Law

One option for governing nanotechnology would be to employ existing law to address the risks posed by nanomaterials. In a June 2011 policy memorandum, the Obama administration indicated its intent to rely heavily on existing law, and regulators are in the process of developing rules under the TSCA to require notification and testing with respect to some nanomaterials.[168] Proponents of this approach contend that "existing environment[al] laws and their implementing regulations generally are well equipped—in the abstract—to encompass nanomaterials in the context of their respective missions."[169] The use of existing statutes, though requiring modest adjustments, capitalizes on administrative structures already in place and avoids the difficult task of passing new legislation.

The current state of affairs—in which nanotechnology products are entering the marketplace without effective oversight or public awareness—resembles the course followed by GMOs. Using existing law to regulate nanotechnology essentially would mirror the use of the Coordinated Framework to govern biotechnology. As discussed in chapter 2, that approach was problematic in many ways. Nonetheless, Gregory Mandel, a leading critic of the Coordinated Framework, contends that nanotechnology regulation presents a stronger case for using existing law than biotechnology because the "division of authority among agencies for regulating nanotechnology is currently well-aligned with each agency's general mandate and expertise."[170] Granted, the statutes governing some specific classes of products, such as drugs and pesticides, could serve as the basis of adequate oversight, and most other relevant statutes are administered by the EPA, which is presumably qualified to address health and environmental hazards.

However, it would be a mistake to rely on existing regulatory authority alone. Such a course overlooks the unique challenges posed by nanotechnology, ignores the weaknesses of existing law and the importance of public input, and puts society at risk for catastrophic health and environmental consequences. The central premise underlying the Obama administration's approach—that "regulation should be based on risk, not merely hazard, and . . . must be evidence-based"—neglects the difficulties of generating nanotechnology risk information.[171] That approach gives no weight to the uncertain hazards surrounding nanotechnology. Furthermore, the TSCA places such heavy evidentiary burdens on the EPA as to make it almost toothless in the context of nanomaterials. The TSCA also provides little incentive for generating the health and environmental data that are sorely needed. The prospects for identifying and managing the potential hazards of nanomaterials through TSCA are bleak, particularly in light of the rapid rate at which new nanomaterials are being incorporated into commerce.

The role of tort law in governing the risks of nanotechnology will necessarily be limited as well, given the difficulties that plaintiffs are likely to face in establishing liability. Notwithstanding these difficulties, tort concerns have attracted some attention from the insurance industry.[172] One insurer, for example, offers coverage specifically "designed for firms whose principal business is manufacturing nanoparticles or nanomaterials, or using them in their processes."[173] As a general matter, insurers are in the business of assessing risks and can help to spread their economic consequences.[174] Liability insurance is a risk management tool that can shift the deterrent pressure of tort law from potential tortfeasors to insurance companies. Requiring nanotechnology companies to obtain insurance, for example, would shift incentives to develop risk information from those companies to insurers.[175] However, those incentives are relatively weak, and the viability of insurance as a risk management tool ultimately hinges on tort plaintiffs' ability to show causation and on insurers' ability to gauge risks accurately. In the case of nanotechnology, those abilities are extremely limited and will remain so for some time.

Voluntary Measures

Another approach to technology management relies on voluntary environmental programs, which have become an increasingly popular alternative to mandatory regulation in recent years. This growing popularity can be attributed to legislative and regulatory gridlock in Washington as well as decreased societal confidence in government regulation.[176] Voluntary environmental programs may be initiated by private parties or by the government and may involve procedural and substantive measures.[177] Supporters contend that voluntary initiatives are well suited for addressing nanotechnology's uncertain hazards because of their flexibility and relative ease of implementation. In contrast to slower-developing risk-based regulation, voluntary programs arguably allow space for "experimentation, learning[,] and graduated action" by regulators and the regulated community.[178] Voluntary initiatives to govern nanotechnology include the EPA's efforts to solicit data on nanomaterials through the NMSP as well as collaborative work by DuPont and Environmental Defense on a model framework for risk analysis.[179] Future voluntary efforts could include a government-supervised certification scheme for nanotechnology products satisfying certain testing and risk management standards or the cooperative development of guidelines for best nanotechnology management practices.[180]

Empirical analyses have generated mixed results on the question of whether voluntary environmental programs effectively reduce health and environmental risks or produce useful risk information.[181] Participation rates are often low, particularly where participation provides insubstantial benefits to a company and imposes significant costs.[182] Only a handful of companies participated in the NMSP, for example, apparently because most had little to gain from doing so.[183] Under the right circumstances, industry peer pressure, public perception, and the threat of regulation or tort liability may provide companies with sufficient incentives to follow voluntary guidelines or otherwise participate in voluntary programs.[184] With respect to nanotechnology, however, current incentives to develop and participate in effective voluntary programs are weak. First, with nanotechnology research and commercialization occurring in numerous companies spread across a wide range of applications, any peer pressure may be too dispersed to have much effect.[185] Individual companies have little incentive to take part in voluntary efforts, as goodwill is essentially shared among the various companies in the industry. To the extent that reputational benefits may accrue to participants in voluntary programs, companies may gain public recognition from associating with a voluntary program without necessarily producing substantive results.[186] The Nano Risk Framework, developed in 2007 by Environmental Defense, a nonprofit, and the DuPont Corporation, is potentially vulnerable to this problem. The Nano Risk Framework provides a blueprint for companies to identify and assess risks associated with nanomaterials.[187] Where risk data are unavailable, the Framework recommends the use of worst-case default values based on existing assessments of analogous materials. The Framework contains no substantive guidelines regarding what companies should do with the information ultimately generated, however. Although it may be too early to assess the effectiveness of the Nano Risk Framework, the fact that companies can claim adherence to it without undertaking specific risk management measures raises the possibility that the Framework, even if followed, will have little substantive effect.

Additional factors suggest that we cannot rely on voluntary programs alone to govern the risks of nanotechnology. First, there is relatively little consumer pressure to participate in such programs, as many consumers are unaware of the increasingly widespread use of nanotechnology. Moreover, nanotechnology companies face no imminent threat that regulation will be imposed should they fail to participate in the Nano Risk Framework or other voluntary programs. Similarly, the threat of tort liability will hardly prod adherence to voluntary guidelines, since it is improbable at this point that a tort plaintiff will be able to establish liability for any harms caused by nanotechnology.

Voluntary environmental programs also may not achieve the rapid results that are sometimes claimed.[188] The NMSP, touted by the EPA as a means of efficiently gathering nanomaterials risk data that might be used for the basis of regulation if needed, was neither rapid nor effective.[189] The program was officially proposed in 2005 and launched in 2008. The EPA accepted industry submissions for nearly two years but ultimately achieved minimal participation rates and very little useful risk data.[190] Moreover, the meager data that were collected are unlikely to be representative, as voluntary programs tend to attract the participation of the most responsible companies with the least to hide.[191] In the end, reliance on voluntary initiatives may even leave the public and environment worse off by misleading the public into believing that effective oversight is in place while forestalling formal regulation.[192]

Voluntary programs are also criticized for a lack of transparency, accountability, and mechanisms to evaluate and sanction poor compliance.[193] These weaknesses can undermine the efficacy of voluntary efforts, and the NMSP exemplifies some of these weaknesses. Much of the limited information submitted under that program has been concealed from the public under broad assertions that it is confidential business information.[194]

One proposal for voluntary regulation, the "Tested NT" scheme set forth by law professor Gary Marchant and his colleagues, addresses some of the weaknesses commonly found in voluntary schemes but may have limited effect. Under this government-supervised scheme, voluntary certification would be available for products that meet specified safety testing, data disclosure, and risk management standards.[195] The authors of the proposal contend that participation in the "Tested NT" program would be greater than under other voluntary programs because firms would receive something of value—a Tested NT mark that companies could affix to their products. Government supervision and enforcement, combined with full public disclosure of supporting data, would help build public trust in the certification process. As the authors acknowledge, however, the benefits of participating in the program may not be sufficient to outweigh the costs for many companies.[196] Indeed, given the low public visibility of nanotechnology and the inability of consumers to determine whether a particular product contains nanomaterials, the presence of a Tested NT mark on a product might actually heighten concerns. Under such circumstances, companies might prefer to leave products unlabeled than to obtain the certification. Many nanotechnology companies apparently have made a similar calculation already; in an effort to avoid public and regulatory scrutiny, they have deleted claims that their products use nanotechnology from product labels.[197] Without widespread participation, the Tested NT mark will have little salience with consumers, and companies will face little pressure to participate.

Another option to encourage voluntary testing, suggested by law professor David Dana, is to provide nanotechnology companies that engage in health and environmental research with limited protection from tort liability.[198] Under this proposal, Congress would enact legislation to preempt state tort law for companies that perform premarket and postmarket testing. Preemption would apply only to claims that a manufacturer failed to conduct adequate testing or monitoring, however. Claims that a manufacturer failed to respond to actual knowledge of adverse health and environmental effects would not be preempted. Though the proposed liability relief would reduce a significant disincentive to perform testing, the cost of testing and the potential for creating adverse information that could be used later against a manufacturer would likely discourage many companies from participating.[199]

In spite of their shortcomings, voluntary efforts can play a supporting role in the management of nanotechnology's risks. Voluntary testing and monitoring can sometimes address regulators' inability to keep pace by responding more nimbly to a quickly evolving commercial marketplace.[200] For example, the Internet-based GoodNanoGuide, a self-described "collaboration platform designed to enhance the ability of experts to exchange ideas on how best to handle nanomaterials in an occupational setting," could enable the rapid sharing of information on good workplace practices.[201] Voluntary efforts also could lead to the establishment of "green nanotechnology" performance and branding standards that incorporate life-cycle analyses and waste prevention principles.[202] And by requiring adherence to the DuPont–Environmental Defense framework or other voluntary guidelines as a condition for providing coverage, insurance companies can offer incentives for nanotechnology companies to assess and manage risks.[203] The practices and standards developed through such efforts can serve as a starting point or source of information for future regulation. Voluntary efforts ultimately can be an important complement to direct regulation but cannot be an adequate substitute for it. Voluntary initiatives can provide only partial coverage of risks and leave too much control of oversight in the hands of interested parties.

The Case for Nano-Specific Regulation

General Considerations

Nanotechnology demands regulatory attention because there is a reasonable basis for suspecting that some nanomaterials present serious health and environmental hazards. However, in comparison to conventional chemicals, which often pose hazards as well, nanomaterials are surrounded by far greater uncertainty. Little is known about the specific characteristics that are relevant to toxicity, and the models used to predict the toxicity of conventional chemicals are not readily applicable. Under these circumstances, the data-intensive regulatory regimes now in place, which rely heavily on quantitative risk assessment, are a poor fit. Research on material behavior and toxicity, already hampered by limited resources, is unlikely to keep pace with the development of new types of nanomaterials whose characteristics may differ substantially from those of nanomaterials currently being studied. Regulatory efforts based on existing law ultimately are likely to become bogged down.

The lack of risk data and inability to detect nanomaterials complicate the application of any regulatory scheme, new or old, to nanotechnology.[204] Nonetheless, work can begin on designing regulatory regimes that could be quickly enacted should the need arise.[205] Moreover, the absence of data is not an insurmountable barrier to all forms of regulation, and it need not prevent the improvement of existing schemes. At a minimum, regulation should be designed to encourage the generation of risk data. Furthermore, statutes can incorporate standards that require little or no health and safety data as a prerequisite to regulate. Such standards might focus on what is technologically feasible rather than on the extent of risk reduction. In the context of nanomaterials, such standards might require, for example, that companies adopt best management practices to minimize human exposure and environmental release.[206] While technology-based standards are sometimes criticized as rigid and inefficient, they can be implemented more rapidly than standards focused on health risks because of their lesser informational requirements.[207] Technology-based standards also are comparatively simple to define and codify and thus are generally easier to administer and enforce than other types of regulatory standards.

Heightened regulation can impede the pace of innovation or prompt R&D efforts to relocate abroad, of course.[208] It is important not to overstate the opportunity costs of regulation, however. Like the potential harms of nanotechnology, many of the potential benefits of nanotechnology development are surrounded by uncertainty. Indeed, the more dramatic predictions surrounding nanotechnology—suggesting seemingly infinite applications and possibilities—often involve long-term speculation about advances in active nanotechnology and molecular manufacturing.[209] The NNI's 2007 Strategic Plan, for example, proclaims nanotechnology's "potential to transform and revolutionize multiple technology and industry sectors, including aerospace, agriculture, biotechnology, homeland security and national defense, energy, environmental improvement, information technology, medicine, and transportation."[210] Similarly, the National

Science Foundation's senior adviser for nanotechnology declares that the "effects of nanotechnology on the health, wealth, and standard of living for people in this century could be at least as significant as the combined influences of microelectronics, medical imaging, computer-aided engineering, and man-made polymers developed in the past century."[211] If these statements are to be believed, nanotechnology can be all things to all people.

Thanks to these vague promises, the concept of the responsible development of nanotechnology has come to command widespread support from a range of constituencies and interests groups. As Alfred Nordmann and Astrid Schwarz ask rhetorically, "[W]ho could be against the responsible development of nanotechnology?"[212] The breadth and undetermined nature of nanotechnology's potential, in other words, give rise to a seductive power that makes it logically and politically difficult to argue against nanotechnology development.

The projected benefits of emerging technologies are hardly assured, however, as the examples of nuclear energy and biotechnology illustrate. The early promise of nuclear energy was that it would supply a majority of the nation's energy needs in a safe and environmentally sound manner.[213] That promise, however, is far from being realized today, and it may never come to fruition in light of the Fukushima Daiichi disaster. Likewise, in the early days of biotechnology, developers sketched out a vision in which GMOs would cure cancer, address malnutrition, and solve world hunger.[214] Yet four decades after the first genetic engineering experiments, the biotechnology industry has made little progress on these fronts. The benefits of GM crops that have been commercialized, such as herbicide resistance and pesticide resistance, have accrued primarily to farmers and GM developers rather than to consumers. Although nanotechnology encompasses a wider field of research than biotechnology, it resembles biotechnology in the hype that surrounds it.[215] Nanotechnology oversight should consider both promises and pitfalls without succumbing to excessive speculation. Elements of a more effective oversight system should include disclosure of the use of nanomaterials, financial assurance requirements for nanotechnology companies, and more widespread public engagement.

Labeling

Requirements that manufacturers label products containing nanomaterials as such, notify regulators of the use of nanomaterials, and provide workplace warnings regarding potential nanotechnology risks would help the public make informed decisions regarding the use of such products and promote more efficient choices.[216] Consumers could decide whether to purchase conventional products, whose risks may be better known, or "new and improved" products containing nanomaterials. Likewise, better-informed workers could monitor their health more closely and demand greater safety precautions or wage premiums that reflect the uncertain occupational hazards they face. In addition, industry would be motivated to weigh more carefully the competitive advantages of using nanomaterials against the potential for tort liability and other concerns.[217] Labeling and disclosure requirements also would advance important normative goals. Public awareness of nanotechnology would increase, facilitating democratic deliberation and the exercise of personal autonomy.

A labeling requirement need not be onerous. The European Union has instituted just such a requirement with respect to nanomaterials in cosmetic products.[218] Manufacturers must place the word *nano* in brackets after each nanoscale ingredient and provide regulators with information, including safety data, about any nanomaterials used. Such a labeling requirement charts a sensible middle course between sensational language that might trigger an overreaction and excessive detail that might be ignored.[219]

Substantive Regulation

Labeling requirements increase transparency and create better incentives for generating information, but they do not directly address the uncertain hazards of nanotechnology. Among the possible options for achieving the latter, one might envision at one extreme a moratorium or a ban on products containing nanomaterials.[220] The tremendous potential of nanotechnology and the lack of data establishing adverse health effects from all or even most nanomaterials, however, undermine the appeal of this option. Indeed, a complete ban would be politically and practically impossible. Billions of dollars are pouring into nanotechnology research and development, and the public is unlikely to support a ban.[221] Moreover, a U.S. ban would drive nanotechnology research and manufacturing to other countries, with potentially significant impacts on economic and military security.[222]

A more realistic approach should account specifically for the uncertainties and other characteristics of nanotechnology. J. Clarence Davies, a former EPA official, has proposed one such scheme. Observing that remediation of nanotechnology pollution is likely to be difficult and ineffective, Davies argues for preventing pollution by focusing on nanotechnology products.[223] Under his proposal, manufacturers would be required to test nanomaterials, forward test results as well as any reports of adverse effects from exposure to the EPA, and prepare a sustainability plan. The sustainability plan would include a life-cycle analysis and proposed labeling and restrictions. To commercialize a new nanotechnology product, manufacturers would have to demonstrate that the product does not pose unacceptable risks. None of these requirements would apply to nanotechnology products already on the market, although the EPA could regulate such products if they are found to have an adverse effect.

Davies's proposal would create incentives for companies to develop health and environmental risk information on new nanotechnology products. Moreover, this approach would shift the burden of uncertainty away from the public and the environment and to the nanotechnology industry. Whether companies would be able to demonstrate the absence of unacceptable risks is less clear, however. Proving the absence of such risks may be difficult, especially for substances that are as poorly understood as nanomaterials.[224] In addition, present uncertainties may preclude performance of a full life-cycle analysis, although at the least risk assessors should be able to estimate potential exposure levels.[225]

One way to strike a balance between innovation and precaution would be to require nanotechnology companies to post an assurance bond. Under such a proposal, which I have discussed elsewhere in greater detail, companies would have to provide financial assurance in order to introduce nanomaterials or products containing nanomaterials into commerce.[226] The requirement could be waived if a company provides sufficient data to conclude that the manufacture, use, and disposal of the product is safe. A bonding requirement would assure the existence of funds to pay for damages that are subsequently discovered.[227] By shifting the financial burden of uncertainty, bonding also gives nanotechnology companies an incentive to undertake research to demonstrate that their products are safe. Regulators—presumably the EPA—would set the value of the bond at an amount adequate to cover the worst-case scenario, taking into account any existing data on toxicity, routes and levels of exposure, environmental fate and transport, and similarities between the material in question and substances with known toxicology. The value of the bond could be revised periodically to reflect new information, and it would be refundable at the end of a defined period of time if the company could demonstrate lower actual or expected damages than those estimated in setting the bond.

Implementing a bonding system would be preferable to relying on existing statutes and the tort system to manage the uncertain hazards of nanotechnology. For nanotechnology applications now entering the market, costs are so uncertain as to render impossible the risk assessment and cost-benefit analysis inherent in the TSCA. Environmental bonding, in contrast to typical cost-benefit approaches, explicitly "acknowledges uncertainty regarding the value, resilience, and replaceability of biophysical systems by assessing serious ex ante financial responsibility for possible environmental harms."[228] The imposition of responsibility ex ante also distinguishes environmental bonding from tort law, which does little to assure the availability of funds for compensation or cleanup.

Bonding requirements also create stronger incentives than the status quo for toxicity research. The TSCA actually discourages such research by making testing optional while requiring the disclosure of any test results to the EPA. The tort system has a similar effect.[229] A bonding requirement, in contrast, places the responsibility on manufacturers to produce information on health and environmental risks.[230] This approach is fair and efficient because manufacturers will profit directly from nanotechnology and will tend to have the most information about the manufacturing process, their products, and the substances in those products.[231]

Bonding requirements ultimately address critical weaknesses of the status quo while offering a promising middle road for capturing many of the benefits of nanotechnology. Unlike a ban, a bonding system would allow some products containing nanomaterials to go forward into the marketplace.[232] And in contrast to existing law and voluntary approaches, bonding would shift the burden of establishing safety off of regulators and onto the nanotechnology industry.[233] In the words of Doug Kysar, environmental bonding "acknowledges the strength and dynamism of sociolegal systems such as markets by allowing private actors to proceed with potentially beneficial activities despite the existence of a credible risk of harm."[234]

Implementation of an environmental bonding scheme for nanotechnology would face several potential challenges and limitations, however. The most serious of these involve setting bond amounts and addressing liquidity constraints. With respect to setting bond amounts, at least two objections could be made to the use of worst-case analysis. First, requiring companies to post a bond based on such analysis may be inefficient; second, uncertainty makes such an analysis difficult, if not impossible. The former objection assumes that efficiency maximization is the goal of the bond-setting process. Even if such a goal were normatively desirable in the abstract, it makes little sense in the context of nanotechnology, since the lack of data makes any quantitative analysis of costs, including cost-benefit analysis, impossible. Indeed, a bonding requirement would actually create incentives for generating the information that could enable such analyses.[235] The objection to worst-case analysis on the grounds of uncertainty, while having some merit, is hardly fatal. A bonding requirement is a tool for taming the unavoidable uncertainties of nanotechnology: Periodic revision of bond amounts could account for new information, and the refund of excess bond money with interest would mitigate fairness concerns. Given past experience with the use of environmental bonds, the worry is not that bond amounts would be too great but rather that they would be inadequate.[236] The costs of unforeseen effects of a catastrophic nature may exceed bond amounts, or worst-case effects—such as species extinction—simply may not be rectifiable.[237] A bond requirement thus is hardly perfect, but it would internalize costs more effectively than does the status quo.

A bonding requirement also could create liquidity problems for smaller nanotechnology companies and start-ups.[238] Such problems could be alleviated through the use of third-party surety firms or insurance to protect against bond forfeiture, although insurers may be reluctant to enter the market because of difficulties in assessing risks.[239] Liquidity constraints ultimately may favor larger nanotechnology companies with greater access to capital over smaller companies.[240] This will not necessarily quash all innovation, however, or eliminate start-up companies from the industry. The pharmaceutical industry provides a potentially apt comparison: Thanks to the lengthy process and high costs involved in identifying, developing, and seeking FDA approval, smaller companies often partner with larger companies in developing or marketing new drugs. One can envision the formation of similar partnerships in the nanotechnology industry; in particular, large companies that can afford to post bonds may tend to be more involved in the manufacture of goods for the marketplace, while smaller companies may focus on research and development activities not subject to bonding requirements.

Involving the Public

Shaping the future course of nanotechnology should not be left solely to regulators and industry. The time is ripe for actively involving the public in assessing and managing nanotechnology. Although nanotechnology continues to be surrounded by exaggerated claims and vague promises, technology assessment can now move beyond abstractions and consider actual applications. Products on the market and in the development pipeline provide a concrete sense of what nanotechnology can do and of the societal changes it might bring. At the same time, nanotechnology is not yet entrenched within economic and social systems, no regulatory architecture is fixed, and the future of nanotechnology is yet to be determined.

So how can the public play a more active and meaningful role in directing the future of nanotechnology? Labeling and disclosure requirements would lay the foundation for greater public involvement by raising awareness of the growing presence of nanotechnology. Further measures should move beyond providing information and engage the public more directly. First, the public should be incorporated into the process of deciding what research goes forward. Much of the ongoing activity in nanotechnology is occurring in research laboratories, and a great deal of this research—$2 billion worth per year—is federally funded. The awarding of federal grants should explicitly consider the extent to which a grant applicant incorporates public participation and considers public concerns. Review panels that make grant awards might even include laypersons whose input on values-based issues can be particularly useful. Provided that participating laypersons represent diverse views, the grant proposals that emerge from such review are more likely to involve lines of inquiry and applications that are supported by a societal consensus.

Second, technology assessment should be applied more broadly and effectively to nanotechnology. Efforts undertaken by CNS-ASU and others offer encouraging examples of new methods of technology assessment. Many of these efforts, however, have been limited in scope and effect and were designed with the primary purpose of achieving the experimental objectives of the social scientists who carried them out. Adopting various measures could increase the practical impact of nanotechnology assessments. For example, technology assessment organizers could invite policymakers or key support staff to participate. Organizers could also seek out greater media coverage to raise the public profile of nanotechnology issues. Furthermore, the involvement of government and policy organizations in designing and implementing nanotechnology assessment could provide real-world connections and direction that increase the likelihood that such assessments will influence the course of nanotechnology development.[241]

Third, additional methods should be employed to gauge broader public sentiment on specific issues, such as the use of nanotechnology for surveillance or for human enhancement purposes. Public participation is more likely to be constructive if citizens consider specific applications of nanotechnology rather than the field as a whole.[242] Innovative methods of public engagement could include nonbinding referendums or electronic deliberations. Given the vulnerability of electronic consultation processes to mass email campaigns, law professor Oren Perez has proposed greater use of deliberation support systems, such as a Wiki platform, to ensure that contributions promote deliberative discourse and not mere invective.[243]

Perez offers the Environmental Defense–DuPont Nano Risk Framework as one example of a policy document that could be refined through a Wiki platform. The information gathered through such methods may be valuable not only to government policymakers but also to scientists, start-up companies, and investors contemplating further efforts and investments in nanotechnology.

International Governance

Given the increasingly voluminous global trade of goods incorporating nanotechnology as well as the ability of nanotechnology research and manufacturing to relocate, domestic regulation alone will be insufficient. International attention to nanotechnology also is necessary. Although the preceding discussion assumes regulatory implementation by the United States, the impacts of U.S. regulation would not necessarily be confined to our borders. Multinational corporations may choose to follow domestic labeling and disclosure requirements in all markets in which they participate, for example. Moreover, a scheme analogous to that proposed for the United States could be adopted by other nations individually or by international agreement. Harmonizing regulation across nations would simplify compliance for manufacturers, distributors, and processors. It would also help to avoid a race to the bottom in which countries loosen regulatory requirements to seek a competitive advantage.[244] More generally, international cooperation could facilitate the pooling of research efforts and the sharing of data and expertise. Recognizing such benefits, global studies professor Kenneth Abbott and his colleagues have advocated a framework convention for nanotechnology that would provide a flexible institutional structure to respond promptly and in an internationally harmonized manner to emerging nanotechnology risks.[245] Patterned after other framework agreements like the United Nations Framework Convention on Climate Change (FCCC), a framework convention for nanotechnology would initially contain limited substantive commitments yet could evolve as parties engage in dialogue and as risk information develops.

Transnational cooperative efforts thus far have been limited. OECD members are cooperating on several projects pertaining to nanomaterial hazards and regulation, including the previously discussed development of safety data.[246] The joint development of such data could lay the groundwork for the joint development of safety standards. Various government agencies, international organizations, private sector entities, and public interest organizations also have participated in nanotechnology risk and governance initiatives.[247] Unfortunately, international coordination of formal regulation is unlikely at this stage. The negotiation of formal treaties is typically a difficult process that requires the devotion of significant amounts of time and resources. Slow progress in mitigating climate change pursuant to the FCCC has raised doubts about the effectiveness of the framework convention approach to international lawmaking. Furthermore, efforts to regulate technologies typically follow the demonstration of serious and imminent harm, especially for international regulatory efforts, which must overcome states' reluctance to yield even a portion of their sovereign powers.[248] Indeed, many nations, spurred by the fear of falling behind, are racing to develop nanotechnology rather than to regulate it.[249]

A suite of international efforts, including but not limited to a framework convention, ultimately may be necessary to address the challenges posed by nanotechnology. Private or voluntary initiatives to certify nanotechnology products, develop safe practices, or promote information sharing will be helpful and can be adopted across jurisdictions. Official state involvement will likely be needed, however, to provide such efforts with legitimacy and credibility.[250] To allay concerns about the slow pace of formal international action, an international agreement on nanotechnology might begin with those nations having significant involvement in nanotechnology and later expand to include other members of the international community.

Concluding Thoughts

Public consideration of nanotechnology is complicated by its broad and somewhat amorphous boundaries. Nanotechnology encompasses diverse activities, a wide range of substances, and seemingly innumerable applications. Moreover, the financial and institutional momentum behind nanotechnology and the forces of regulatory inertia will hamper efforts to effectively regulate nanotechnology's hazards. As with many other technologies, the benefits of nanotechnology are often immediate and obvious, whereas the harms to health, environment, and society are not. Nanotechnology is not yet fated, however, to follow the same troubled course as GMOs. There is widespread recognition that nanotechnology could involve potentially serious dangers, and stakeholders have stated their willingness to discuss these dangers. In addition, incipient efforts have been made to engage the public, and members of the public who are presented with information on nanotechnology have expressed strong interest in the subject. Although such efforts have so far been inadequate, they suggest the prospect for governments, researchers, industry, nongovernmental organizations, and the public to be involved in shaping the future of nanotechnology. Waiting for more safety data and relying on existing law will not suffice, however. More aggressive efforts must be made to increase public awareness, solicit informed public views, and hold nanotechnology companies responsible for their activities, even in the face of uncertain effects.

CHAPTER 4

Geoengineering

A Technological Solution to Climate Change?

Geoengineering refers to a variety of unconventional proposals for responding to climate change, such as spraying tiny sulfur particles into the atmosphere or fertilizing the ocean with iron to stimulate phytoplankton growth. Climate change undoubtedly poses a critical environmental challenge, as it threatens higher temperatures, rising sea levels, diminished water supplies, impaired food production, and other severe and wide-ranging impacts.[1] Broader consequences of climate change could include threats to public health, ecosystems, and geopolitical stability. Thus far, efforts to address climate change have focused exclusively on mitigation and adaptation as opposed to geoengineering. Mitigation encompasses measures to reduce greenhouse gas (GHG) emissions or to enhance GHG uptake by forests and other carbon sinks, whereas adaptation refers to adjustments in natural or human systems in response to the effects of climate change.[2] On the mitigation front, discussions pursuant to the 1992 Framework Convention on Climate Change (FCCC) yielded the Kyoto Protocol, the only international agreement containing binding limits on GHG emissions.[3] Progress to date in actually reducing GHG emissions has been minimal, however. The emission reductions nominally required by the Kyoto Protocol are insufficient to curb climate change, and recent negotiations, which had been anticipated to yield more drastic and permanent emissions reductions, have accomplished little.[4] Planning of adaptation measures has begun, but it is widely considered to be even more limited than mitigation in its scope and implementation so far.

Against this backdrop, geoengineering has gained visibility as a possible additional weapon against climate change. Geoengineering is a catch-all term for an array of untested and frequently risky climate-manipulation proposals that fall outside the rubrics of mitigation and adaptation. These techniques generally involve the "engineering" of physical or chemical processes at a planetary scale to counter the climate consequences of higher GHG concentrations in the atmosphere.[5] Specific geoengineering proposals include fertilizing the oceans to stimulate phytoplankton growth and thereby store carbon in the oceans, seeding marine clouds to increase their reflectiveness, and deploying a thin layer of sulfur particles in the stratosphere to deflect the sun's radiation. At present, geoengineering technologies are far from mature. No full-scale geoengineering projects have been undertaken, and no geoengineering techniques are ready to be deployed.[6] Research efforts have primarily involved computer modeling rather than field testing, and the climate models in use are admittedly inadequate.[7]

The emerging technology challenge posed by geoengineering is distinct from the others considered in this book in several ways. In contrast to genetic engineering, for example, geoengineering does not involve a set of related scientific techniques. Rather, geoengineering refers to a diverse array of technologies involving differing scientific phenomena and means of deployment. As a result, a tailored approach for analyzing the potential hazards associated with different types of geoengineering may be appropriate. Geoengineering proposals do have one thing in common, however: a shared purpose of countering the elevated carbon levels that cause climate change. The goal-oriented nature of geoengineering argues in favor of coordinated oversight as well as coordination of any geoengineering activities that might take place with climate mitigation and adaptation efforts. Coordination is likely to be most effective through international governance mechanisms. Indeed, because any geoengineering project would be designed to affect global climate—and might have unintended effects on many nations—geoengineering governance is inherently and primarily an international matter.

None of this is meant to suggest that geoengineering research or deployment should necessarily occur. Geoengineering raises a complex suite of policy and ethical concerns. One of the unique concerns raised by geoengineering is that it poses a potential moral hazard. In the context of insurance, moral hazard refers to the reduced incentive of insureds to take reasonable precautions against an accident because they have insurance. Similarly, geoengineering research and development may undermine political, economic, and societal support for the main climate change policy options of mitigation and adaptation. Each of the various geoengineering techniques carries serious and substantial risks, uncertainties, and limitations, however. Accordingly, even strong supporters characterize most geoengineering techniques as no more than a backstop or emergency option. Nonetheless, the moral hazard lies in the possibility that geoengineering could be depicted—and readily misunderstood—as offering a quick and painless solution to the complicated problem of climate change.

The increasing attention to geoengineering raises difficult questions regarding international governance of technology and public participation in technology management. Despite the relatively undeveloped state of geoengineering technologies, national governments, international treaty organizations, and the scientific community are beginning to recognize the need for governance. This early attention offers hope for developing a participatory and Promethean approach to this emerging technology.

Background: Geoengineering Techniques

Proposals to geoengineer the Earth focus on responding to the consequences of higher atmospheric concentrations of GHGs. Such proposals fall into two general categories: carbon dioxide removal (CDR) techniques, which strive to remove carbon dioxide (CO_2) from the atmosphere, and solar radiation management (SRM) techniques, which strive to reflect some of the sun's radiation into space. Atmospheric concentrations of CO_2 are now estimated at 390 parts per million (ppm) and rising, well above preindustrial levels of 280 ppm.[8] While there is much debate regarding the CO_2 levels at which severe or catastrophic consequences might follow, there is international consensus for taking significant mitigating actions of some sort.[9] CDR techniques, by removing CO_2 from the atmosphere, would move GHG concentrations in the atmosphere back toward their natural state. As GHG levels decline, it is expected that the Earth's climate system would return toward earlier conditions. SRM techniques, in contrast, would not lower GHG concentrations but would instead attempt to control climate conditions by reducing the amount of radiation absorbed by the Earth. Because SRM techniques essentially focus on climate change's symptoms rather than its scientific root causes, they tend to involve greater risks and uncertainties. The discussion here briefly considers leading proposals within each category.

Carbon Dioxide Removal

In the global carbon cycle, carbon is exchanged naturally among the atmosphere, the oceans, the earth, and living things. CDR techniques seek to enhance certain parts of this cycle to reduce the amount of carbon in the atmosphere. Examples of CDR techniques include fertilization of the oceans, direct capture of CO_2 from the air, and enhancement of natural chemical processes in which minerals react with CO_2. Land use changes to increase carbon uptake, such as afforestation, also remove CO_2 from the atmosphere, but they typically are not categorized as geoengineering because of their low-tech, more conventional nature.[10]

Ocean fertilization has received the most research attention of all CDR techniques.[11] Ocean fertilization seeks to dramatically accelerate the natural process by which carbon is stored in the deep oceans. Under natural conditions, CO_2 is transferred from the atmosphere to the deep oceans in a process that takes approximately one thousand years. Central to this process are phytoplankton, which live at the ocean surface and convert atmospheric carbon to organic carbon. When the phytoplankton die, much of the carbon they absorbed during their lives returns to the atmosphere through natural decay.[12] The rest of the carbon, however, is transported to the deep oceans as the dead phytoplankton sink. Through ocean fertilization, this process theoretically could take place in years, if not months.[13] The hypothesis underlying ocean fertilization is that the unavailability of various micronutrients limits biological productivity in certain ocean regions. Adding a relatively small amount of these micronutrients in those regions may drastically increase phytoplankton populations and consequently the amount of carbon transported to the deep oceans.

Based on inverse correlations between atmospheric CO_2 concentrations and the amount of iron in atmospheric dust, some scientists have postulated iron to be the most important limiting micronutrient. In theory, the addition of one atom of iron could lead to the sequestration of 100,000 organic carbon atoms. Furthermore, global supplies of iron are believed to be sufficient to support mass fertilization of the oceans at a relatively moderate cost.[14] However, it is not at all certain that ocean fertilization will cause substantial amounts of dead phytoplankton to sink into the deep oceans. In addition, even if ocean fertilization were effective and widely implemented, it could sequester only a modest fraction of the carbon emissions generated by humans each year.[15]

Indeed, the results to date of small-scale ocean fertilization experiments and computer modeling have been unimpressive.[16] In these trials, iron fertilization has demonstrated little ability to promote carbon transfer to the deep oceans. A number of factors appear to be at work, including a scarcity of nutrients other than iron, slower rates of vertical mixing in the oceans than are necessary to remove carbon from the atmosphere, and grazing of phytoplankton by other organisms.[17] Verifying that carbon has actually been sequestered in the deep oceans is difficult, if not impossible.[18] Moreover, iron fertilization appears to stimulate the growth of toxic phytoplankton species in particular, with potentially fatal effects on various marine animals.[19] These findings are troubling, as the small-scale studies that have been carried out can predict only some of the adverse consequences of ocean fertilization. In general, environmental effects are likely to vary with the scale at which ocean fertilization—or any geoengineering—is carried out. Ocean fertilization schemes ultimately risk significant alteration of ocean chemistry and marine ecosystems. Phytoplankton form the foundation of marine food webs, and changes in their populations could lead to the creation of ocean regions low in oxygen, unpredictable ecosystem shifts, and even heightened production of methane and other GHGs.[20]

A second proposed CDR technique, direct capture of CO_2, involves the use of chemical processes to absorb CO_2 from the ambient air. Sometimes referred to as artificial trees, direct capture is akin to existing pollution control technologies for other air pollutants: scrubbers would remove CO_2 from the atmosphere, and the removed CO_2 would then be stored using carbon sequestration techniques.[21] The chemical processes behind direct capture are relatively straightforward, but the fact that CO_2 makes up only 0.04 percent of the atmosphere presents a serious technical challenge to implementation.[22] To get a sense of the technical difficulties, one can compare this figure with the concentration of CO_2 in the exhaust stream of commercial power plants, which exceeds 10 percent.[23] Even at this higher concentration, the expense of carbon capture and sequestration has thus far precluded the existence of a single full-scale commercial power plant that captures CO_2 from all of its exhaust streams.[24] Given the far lower concentration of CO_2 in the atmosphere and the capital and energy requirements of artificial trees, direct capture will surely be less efficient and more costly than postcombustion capture at a power plant.[25] Indeed, the energy required to capture CO_2 from the atmosphere and to store it may generate more than a ton of carbon for every ton of carbon captured if carbon-based fuels are used to power the capture process.[26] Renewable energy could be used to operate direct capture facilities, but the inefficiencies of direct capture suggest that it would be more sensible to dedicate such energy instead to displacing high-carbon energy sources.[27]

Another set of CDR techniques, enhanced weathering, seeks to accelerate natural processes in which CO_2 is removed from the atmosphere through chemical reactions with carbonate or silicate minerals.[28] Enhanced weathering could be implemented on land or in the seas. When added to the oceans, carbonate and silicate minerals not only can sequester carbon but also can counter the acidification of the oceans caused by higher CO_2 concentrations in the atmosphere.[29] A major drawback to enhanced weathering, however, is the immense quantity of minerals that would be needed. To obtain and transport these minerals, large-scale mining and transportation activities, which are costly and damaging to the environment, would have to be carried out.[30]

By reducing GHG levels in the atmosphere, each of these CDR techniques would address the scientific phenomena at the heart of climate change. The use of CDR techniques could complement emissions reduction efforts and could even help drive net carbon emissions negative. Thus far, however, ocean fertilization appears ineffective and problematic, artificial trees inefficient, and enhanced weathering environmentally destructive. Furthermore, because CO_2 remains in the atmosphere for long periods of time, CDR techniques could have only a gradual effect in countering global warming.[31]

Solar Radiation Management

Whereas CDR seeks to reduce GHG concentrations in the atmosphere, SRM seeks to deflect the sun's energy. The Earth's climate remains relatively constant because the Earth radiates heat back into space at approximately the same rate as it absorbs energy from the sun. Greenhouse gases act as a partial insulator in reducing the amount of heat energy radiated back into space.[32] SRM techniques attempt to counter the warming resulting from increased atmospheric GHG concentrations by blocking a fraction of the incoming solar radiation. For example, to counter the warming effect that would result from a doubling of CO_2 concentration, approximately 1.8 percent of incoming solar radiation would need to be blocked.[33] In theory, SRM could occur at the Earth's surface, in different layers of the atmosphere, or in outer space. SRM techniques are particularly controversial, however, because they generally involve even greater risks and uncertainties than those posed by CDR.

For the most part, SRM methods could be deployed more rapidly and inexpensively than CDR techniques. As such, and in light of their risks, SRM techniques have sometimes been characterized as "Plan B"—a potential emergency response to a sudden and catastrophic worsening of climate conditions.[34] Although defining what constitutes a climate emergency is likely to be difficult and controversial, emergency scenarios could include the sudden melting of polar ice sheets, a severe and widespread drought, the shutdown of critical ocean currents that temper regional climates, the release into the atmosphere of vast amounts of methane (a substance with approximately twenty times the warming effect of CO_2), or other unpredictable phenomena.[35] As climate mitigation efforts have continued to stall, however, the temptation to deploy SRM even in the absence of emergency circumstances has grown.[36]

One of the most prominent SRM proposals involves the release of tiny sulfur particles into the stratosphere. Scientists have observed that the unintended addition of sulfur to the stratosphere, whether from volcanic eruptions or industrial activity, produces a cooling effect by causing more sunlight to be reflected into space.[37] To deliberately accomplish a similar result, various proposals suggest the deployment of airplanes, artillery shells, balloons, or even giant towers rigged with hoses to release sulfur aerosols.[38] Such proposals have focused on the use of sulfur aerosols because, when introduced as a gas into the stratosphere, sulfur avoids clumping and other problems that might result if solids were to be used.[39] Delivery of aerosols into the stratosphere—approximately 20 kilometers above the Earth's surface—rather than the lower atmosphere would be necessary to ensure that the particles remain in the atmosphere for more than a few days.[40] In the stratosphere, chemical and microphysical processes would convert the sulfur into particles of suitable sizes to scatter sunlight. These particles would be expected to remain in the stratosphere for one to two years.[41]

The use of stratospheric aerosols is one of the most seriously discussed geoengineering proposals because of its apparent technical and economic feasibility. Initial computer modeling suggests that sulfate aerosols would reduce average global temperature increases and counter changes in precipitation.[42] A scheme to release aerosols could be deployed relatively quickly, proponents contend, and once deployed, would reduce temperatures quite rapidly.[43] The estimated costs of operating such a scheme range from several billion dollars per year (less than 1 percent of annual global military expenditures) to $200 billion per year.[44] Projections of effectiveness and implementation costs are necessarily speculative, however. Scientists have yet to carry out field experiments involving the technique, and maintaining the appropriate size distribution of aerosol droplets will be technically difficult.[45] Indeed, the modeling done thus far rests on grossly simplified assumptions regarding the Earth's climate system.[46] Moreover, the logistics required to support such a scheme are imposing: under one proposed delivery method, a dedicated fleet of specialized aircraft would have to make thousands of flights per day, sustained over hundreds of years, to deliver sufficient sulfur to the stratosphere.[47]

More important, these cost estimates completely disregard the potential for adverse consequences. Even under the limited and idealized modeling that has been carried out, the release of sulfur aerosols would produce significant regional perturbations in climate.[48] One important effect could include modification of the Asian and African summer monsoons, which would have severe ramifications on food supplies for billions of people.[49] The release of stratospheric aerosols could also generate global impacts on the environment and human safety. For example, sulfur aerosols could exacerbate depletion of the stratospheric ozone layer, which provides protection from the sun's ultraviolet rays. While sulfur aerosols themselves do not directly destroy ozone, they can provide a surface for the activation of ozone-destroying chlorine gases already present in the stratosphere.[50] This would magnify the negative effect of these gases on ozone and undermine international progress in phasing out the use of ozone-depleting substances. In addition, the forms of sulfur likely to be used in a stratospheric aerosol scheme—either hydrogen sulfide or sulfur dioxide—are highly toxic and corrosive and thus would necessitate special handling precautions.[51]

Another SRM proposal involves whitening clouds over the ocean. In theory, cloud whitening techniques could significantly increase the Earth's reflectivity, or albedo, and thereby reduce the amount of solar radiation absorbed.[52] Seventy percent of the Earth's surface is covered by the oceans, which have an albedo far less than that of clouds.[53] Just as ship tracks form around the exhaust released by ships traveling across the oceans, clouds could be formed or brightened by seeding them with tiny seawater particles. To generate such particles, aircraft could release a suitable powder, or ships or other sea vessels could produce a mist of sea salt from ocean water.[54] The aerosols that would be used in cloud whitening have a relatively short lifetime (ten days or so), and thus would have to be constantly replenished. The upside of this short lifetime is that any cloud whitening experiment could be stopped rapidly if problems arise.[55] With greater understanding of climate systems, geoengineers might even locate and time cloud whitening to provide localized cooling where needed.[56]

Cloud whitening has been the subject of computer modeling, and researchers are developing plans for field experiments.[57] Because scientists have yet to resolve technical design issues with respect to the spray generator that would do the seeding, however, the costs of carrying out such schemes are unknown.[58] The whitening of marine clouds on a broad scale, moreover, would likely have regional effects on temperature, precipitation, wind, and ocean currents, and these effects require careful consideration.[59]

Yet another SRM approach—a somewhat fanciful one—would involve the deployment of reflectors in outer space. Under such proposals, a shield—comprised of dust particles, dustbin-sized discs, aluminum threads, or large mirrors—would be launched into orbits situating them between the Earth and the sun.[60] The use of space-based deflectors would avoid some of the hazards associated with proposals utilizing aerosol releases, such as ozone depletion. However, there are serious drawbacks. The deployment of space-based deflectors would likely have various regional weather effects.[61] The deflectors would have to be continually replaced at the end of their useful lives, lest rapid climate change occur. In addition, the deflectors would generate debris that could interfere with Earth-orbiting spacecraft.[62] Most important, a space-based approach would be extremely expensive and take decades to implement.[63] These difficulties have led the United Kingdom's Royal Society to conclude that space-based techniques "contain such great uncertainties in costs, effectiveness (including risks) [,] and timescales of implementation that they are not realistic potential contributors to short-term temporary measures for avoiding dangerous climate change."[64]

A major shortcoming of SRM techniques generally is that they have no effect on atmospheric GHG concentrations. As a result, SRM could serve only as a stopgap measure to buy additional time for humankind to reduce GHG emissions or find other means of countering climate change.[65] Through natural processes occurring over the course of centuries, the oceans would absorb the CO_2 already released by humans.[66] But in the absence of dramatic measures to stop the growth of GHG emissions and to reduce atmospheric GHG concentrations, SRM would face what has been called the "termination problem." Once deployed, SRM efforts would have to continue for perhaps several hundred years, for their sudden cessation would result in extremely rapid climate change to which human societies and natural ecosystems would have little time to adapt.[67] Because of the termination problem, SRM techniques require a long-term commitment to their continued operation. As an ethical matter, it is questionable whether current generations should lock future generations into such a commitment; as a practical matter, it is uncertain whether human society could carry it out.[68]

Because SRM techniques do not reduce atmospheric GHG levels, they also would not counter the problem of ocean acidification. The acidity of the ocean is directly correlated to GHG levels in the atmosphere, and increased acidity could lead to the loss of many of the Earth's coral reefs, which serve as important marine habitat.[69] In addition, since certain plant species flourish under high concentrations of atmospheric CO_2, elevated GHG levels would confer a competitive advantage to these species and thus affect terrestrial ecosystems, habitat, and biodiversity as well.[70]

The Need for Governance Now

Though geoengineering proposals are largely theoretical at this time, the preceding discussion demonstrates that there is no shortage of ideas for geoengineering the Earth. Scientists are generally proceeding with caution: Some proposals have been the subject of computer modeling, but in most instances, little or no field research has been done.[71] Full-scale deployment of tested geoengineering systems, if it ever occurs, may be decades away. Consequently, one might question whether any governance of geoengineering is necessary at this time. There are nevertheless several convincing reasons for initiating governance efforts immediately.

Mounting Momentum for Geoengineering Research

First, governance efforts are more likely to be effective if implemented in the early phases of technological development. Although the imposition of burdensome regulatory requirements could slow research and development, past experience with emerging technologies indicates that the opposite danger is more likely. That is, the momentum that builds behind unregulated technologies can overwhelm subsequent efforts to incorporate public input or to impose meaningful oversight. With respect to genetic engineering, for example, the pivotal Asilomar Conference took place just as research efforts were moving forward. Even then, the conference was "too late," as one panelist later remarked. The vast majority of conference participants were already committed to moving forward with recombinant DNA research.[72] Similarly, the momentum to commercialize nanotechnology has swamped efforts thus far to open up the nanotechnology development process to assessment and meaningful public participation.

For geoengineering, now may be the optimum time to discuss and develop governance structures. Research interest is growing among scientists and policymakers, as is commercial interest in geoengineering projects that might eventually generate carbon offsets.[73] These growing interests are demonstrated by a recent and controversial ocean fertilization experiment in which an American businessman scattered 100 tons of iron dust in the Pacific Ocean without government or scientific oversight.[74] Powerful interests that could wield undue influence on policy in favor of deployment have not yet become entrenched, however.[75] Society has not made vast investments of human or financial capital in geoengineering or in a particular geoengineering infrastructure. Nor is society locked into technical commitments by path dependency. Moreover, while climate change is an urgent problem, we are not yet facing the sort of emergency conditions that might preclude careful deliberation, debate, and experimentation before deploying geoengineering. Granted, our ability to develop detailed regulation at this time is constrained by our limited understanding of the concerns posed by various geoengineering techniques. Proactive governance nonetheless can be a means of identifying and analyzing ethical issues, fostering global political discussions, formulating the conditions—if any—that might justify deployment, and guarding against unilateral or hostile deployment.[76] Even a skeletal governance framework could serve as the foundation for articulating more specific rules in the future as circumstances change and more information becomes available.

The need to oversee geoengineering research and not just deployment further necessitates early governance. Research oversight is warranted by the potential for widespread and unintended consequences from field tests and the likely influence of near-term research choices on long-term geoengineering policy. Oversight also is necessary because geoengineering research is directed toward a highly contested goal; such research is not being conducted purely for the sake of discovering new knowledge.[77] Governance of research can determine research priorities, address the permissibility of field tests, ensure the performance of adequate research to identify risks, and set research guidelines.

Research efforts thus far have largely involved theorizing and computer modeling. Theorizing, modeling, and even contained laboratory experiments are necessary to determine whether particular geoengineering techniques might work, and they pose little or no direct risk to the environment. Such processes can yield only so much information, however. Because current climate models rely on gross simplifications, the results generated by such models must be viewed with skepticism.[78] Similarly, contained laboratory experiments might provide support for the principles underlying various geoengineering proposals but can only hint at whether such proposals will work in uncontrolled conditions and on a global scale. A better understanding of the benefits and risks associated with certain techniques requires field tests, and proposals to undertake such tests are growing.[79] Even small-scale field tests can generate their own risks, however. More problematically, small-scale tests are unlikely to dispel much of the uncertainty regarding how the Earth's climate system might react to full-blown geoengineering efforts. As philosopher Martin Bunzl points out, "You cannot encapsulate part of the atmosphere and it is too complex to be able to build a realistic non-virtual model at scale."[80]

An examination of proposals for small-scale testing of stratospheric aerosols reflects some of the limitations at issue. Some observers have suggested that this testing be performed over small areas, but such restrictions would not confine the geographic extent of climate effects. A release of stratospheric aerosols near the North or South Poles, for example, would affect not only climate in the polar regions but also monsoon patterns in the middle latitudes.[81] Another option for conducting small-scale testing could involve the release of very limited quantities of aerosols. Although such testing might avoid drastic negative consequences, the effects of such a release would be virtually impossible to distinguish from the normal variations of weather and climate.[82] Indeed, despite decades of experience with cloud seeding, scientists still cannot determine whether these modest weather modification efforts actually work, even at a local scale.[83] Large-scale field tests that might provide useful and accurate data for assessing a geoengineering proposal essentially would require full-scale implementation of a geoengineering project itself.[84] Large-scale field testing must be subject to oversight, as should any field testing with the potential for adverse effects.

Ease of Unilateral Deployment

The deployment of geoengineering would not require dramatic diplomatic breakthroughs. Moreover, although geoengineering techniques face substantial technical barriers to effective implementation, the principles behind the techniques are not unusually complex. The relative ease with which a single nation might launch geoengineering efforts, however imperfect, provides an added reason for establishing international governance sooner rather than later.

Combating climate change through GHG emissions reductions ultimately will require cooperation among many nations to be effective. Agreeing on and coordinating such an effort is a painstaking process, as such aggregate efforts are particularly susceptible to free riding. Emissions reductions by some countries, in other words, can undermine the incentive for other countries to reduce their own emissions.[85] Geoengineering, in contrast, potentially reframes climate change as a problem for which unilateral rather than aggregate action may provide a solution. If a single nation were to carry out geoengineering, there would be little need for treaty mechanisms to encourage participation or to counter the temptation to backslide on commitments.[86] Nevertheless, such a course would not eliminate the need for international governance to properly authorize geoengineering and to address the potential for adverse effects on millions if not billions of people.[87]

International oversight is important because a number of countries have the economic resources and technical capacity to independently carry out ocean fertilization, stratospheric aerosol release, or other geoengineering projects.[88] Few engineering barriers would prevent a country from releasing sulfur particles into the stratosphere or dumping iron particles into the ocean. However, the potential hazards of geoengineering are poorly understood, and the technical details of *effective* implementation are yet to be worked out. Global governance can safeguard against rogue and potentially reckless efforts to engineer the climate without adequate analysis and international approval. Global governance can also address the potential use of geoengineering as a weapon.

Moral Hazard

A third reason for initiating geoengineering governance now is the potential moral hazard that geoengineering presents. While geoengineering is often characterized as an emergency backstop to mitigation, the potential for its use could undermine public, political, and financial support for the seemingly more difficult task of reducing GHG emissions through development of renewable energy sources, redesign of industrial processes, changes in lifestyle, and other means of mitigation.[89] Nobel laureate Thomas Schelling has argued that "the economics of geoengineering compared with CO_2 abatement . . . transforms the greenhouse issue from an exceedingly complicated regulatory regime to a simple—not necessarily easy, but simple—problem in international cost sharing."[90] Similarly, economist and *Freakonomics* coauthor Steven Levitt contends that geoengineering "could end [the climate] debate" and allow humanity to "move on to problems that are harder to solve."[91] By offering the prospect—or mirage—of a quick, low-cost, and relatively painless solution to climate change, geoengineering could lead to an unwarranted perception that GHG emissions reductions and adaptation measures are unnecessary.[92] A complacent world might even emit more GHGs, generating more climate damage than would have occurred in the absence of any geoengineering efforts.

Citing the limited role envisioned for geoengineering under most proposals, some commentators have expressed doubt that geoengineering would present much of a moral hazard.[93] Under these proposals, geoengineering would simply be "Plan B"—a backup option for rapidly responding to an imminent climate emergency.[94] In addition, proponents of geoengineering research generally emphasize the preliminary nature of their proposals and caution against exclusive reliance on geoengineering techniques. It is unlikely, however, that geoengineering options, once developed, would be confined to a Plan B role.[95] The prospect of spending billions of dollars to develop the technology but not to deploy it is implausible. Already, there are increasing calls for geoengineering to serve as "Part 2 of Plan A"—that is, as a supplement to accompany emission reduction and adaptation measures.[96] Geoengineering options are beginning to attract the support of influential industry groups.[97] Moreover, carbon-intensive industries face strong incentives to persuade the public that geoengineering offers a relatively painless alternative to mitigation. The somewhat tenuous public support for GHG regulation, energy conservation, and other long-term mitigation measures could ultimately erode in the face of geoengineering's bold yet untested promises. Indeed, psychological biases could lead to excessive optimism about geoengineering. For example, overconfidence bias leads people to overvalue the magnitude of a possible outcome and to undervalue the statistical probability associated with that outcome.[98] Consequently, people may give undue emphasis to the dramatic benefits suggested by stratospheric aerosols yet disregard quantitative assessments of risk and uncertainties associated with the technique. Similarly, under the phenomenon of hyperbolic discounting, people tend to give especially little weight to future costs and benefits when comparing them to present ones.[99] This tendency could lead to insufficient consideration of geoengineering's potential downsides, which are not well understood. The slippery slope of technological development combines with the problem of moral hazard to suggest the need for careful framing and oversight of geoengineering research and development.[100]

Complexity of the Problem

Finally, the various ethical, legal, political, and technical issues raised by geoengineering also argue in favor of early governance. Deciding whether to proceed with geoengineering—and if so, how—will be complicated and difficult. Planning and executing any geoengineering scheme on a scale necessary to counter global warming will be logistically complex, and ensuring its reliable operation over decades if not centuries presents daunting challenges of institutional governance.[101] Because geoengineering necessarily has global implications, including potentially catastrophic risks, its governance should be based on international discussions and widespread public input. How that governance might be designed is addressed later in this chapter; for present purposes, the important point is that the process of debating geoengineering and deciding on a governance regime will take time. This process need not and should not await the results of further geoengineering research. As David Keith, a leading proponent of geoengineering research, has recognized, even experiments with minimal environmental impacts can have nontrivial implications because of path dependency and research momentum. Keith accordingly cautions that "[t]aking a few years to have some of the debate happen is healthier than rushing ahead with an experiment."[102] That debate will involve more than a weighing of risks and benefits, for geoengineering raises thorny ethical questions. These questions include whether such techniques are morally permissible, how to define the responsibilities of those countries most at fault in contributing to climate change, what obligations current generations have to future generations, and whether utilitarian concerns justify geoengineering deployment.[103]

In sum, it is not too early to develop governance mechanisms for geoengineering. This conclusion is consistent with the views of many researchers as well as the growing attention that national governments and international forums are devoting to geoengineering. In 2009, for example, the U.S. House Committee on Science and Technology and the United Kingdom's House of Commons Science and Technology Committee began a collaborative inquiry on geoengineering. This inquiry, which included several hearings, resulted in a report on geoengineering "research needs and strategies" authored by the chair of the U.S. House committee as well as a report on the regulation of geoengineering by the House of Commons committee.[104] Similarly, the German Federal Ministry of Education and Research commissioned a 2011 report on geoengineering with the aim of establishing a knowledge base "for public debate and political decision-making" and "stimulating international debate on climate engineering."[105] International treaty organs also have begun to analyze the applicability of existing treaty regimes to certain geoengineering techniques. As our experience with other emerging technologies suggests, oversight of research is critical because the foundation for future developments and for subsequent policy decisions is laid at the research stage.

Governance So Far

There are various options for geoengineering governance, ranging from domestic law to international law and from informal "soft" law to formal regulation. This discussion focuses on international governance options. Although national policies on geoengineering might be more readily established, they are likely to take a backseat to international policies in light of geoengineering's global dimensions.[106] Geoengineering will have global consequences, and domestic regulation by one nation may have little effect on geoengineering activity by other nations. Geoengineering has been analogized to a thermostat for the Earth, and geoengineering governance encompasses questions that must be determined by the global community: Whose hand is on the thermostat, and at what temperature setting?[107] Under one model of international governance, governance may be top-down, formalized in a treaty and in binding legal rules and administered through international organizations and national governments. Alternatively, governance may be bottom-up, expressed through research guidelines or informal norms developed by the scientific community or a broader set of stakeholders. Incipient efforts at geoengineering governance reflect both approaches. Given the limitations of each, it is likely that both approaches will be necessary—though perhaps not sufficient—to govern geoengineering.

Formal Governance

At present, no international treaty agreements directly address geoengineering. Treaty-based international oversight of scientific research activities is similarly rare. Nonetheless, several treaties could be interpreted or extended to apply generally to geoengineering or geoengineering research. In addition, other, narrower treaties arguably could govern certain specific types of geoengineering.

A leading example of a treaty that might generally apply to geoengineering is the United Nations Framework Convention on Climate Change (FCCC).[108] Signed in 1992, the FCCC is the framework agreement under which international negotiations to address climate change have taken place for two decades. The commitments made in the FCCC are general in nature, focus primarily on encouraging mitigation, and create no clear obligations with respect to geoengineering. Nonetheless, as a framework convention, the FCCC contemplates the formation of subsequent protocol agreements on specific matters as further information develops and international support builds. The FCCC may be a useful mechanism for addressing geoengineering because almost all nations of the world are parties to the FCCC, there are already well-established institutions for administering and implementing the treaty, and these institutions could coordinate geoengineering efforts with mitigation and adaptation strategies to combat climate change.[109] Although the parties to the FCCC have yet to address geoengineering, the Intergovernmental Panel on Climate Change, a scientific body that supports the FCCC by assessing information on climate change, is now convening expert meetings on the subject.[110]

The Convention on Biological Diversity (CBD), which has broad membership and focuses on the conservation of biodiversity, is another treaty that could be applied to geoengineering.[111] The CBD does not directly address geoengineering or climate change, but its members have an interest in these topics insofar as they affect biodiversity. Ocean fertilization experiments have been of particular concern to the CBD regime. In 2008, the parties to the CBD issued a decision "requesting" that member states ensure that ocean fertilization projects do not occur absent "an adequate scientific basis on which to justify such activities" and "a global, transparent and effective control and regulatory mechanism."[112] The decision allows "small-scale scientific research studies within coastal waters" to proceed, subject to several conditions. Such projects must be (1) "justified by the need to gather specific scientific data"; (2) "subject to a thorough prior assessment of the potential impacts . . . on the marine environment"; (3) "strictly controlled"; and (4) "not . . . used for generating and selling carbon offsets or any other commercial purposes." Advocates of further research have attacked this exception for small-scale studies in coastal waters as arbitrary and overly restrictive. Specifically, they argue that experiments must take place in the open ocean and over large areas to generate useful information regarding the effectiveness of ocean fertilization.[113] In a subsequent decision, CBD members nonetheless extended to all geoengineering activities the approach they had adopted for ocean fertilization. This more recent decision urges that no geoengineering activities take place unless "science based, global, transparent and effective control and regulatory mechanisms" are in place.[114] The decision, which contains an exception for "small scale scientific research studies . . . in a controlled setting," also demands an adequate scientific basis for geoengineering and consideration of environmental, social, economic, and cultural impacts. The legal impact of the decision has been questioned, however, because of its nonbinding nature and because of the nonclimate focus of the CBD regime.[115]

Parties to the London Convention and London Protocol (LC/LP), a marine pollution treaty regime that regulates the dumping of waste into the sea, have also considered the permissibility of ocean fertilization experiments.[116] Each of these treaties prohibits ocean dumping, which is defined to include "any deliberate disposal into the sea of wastes or other matter from vessels, aircraft, platforms or other man-made structures at sea."[117] "Placement of matter for a purpose other than the mere disposal thereof," in contrast, is allowed as long as such placement is not contrary to the purposes of the LC/LP.[118] In 2007, the parties to the LC/LP agreed that ocean fertilization falls within the jurisdiction of the LC/LP and that "given the present state of knowledge regarding ocean fertilization, . . . large-scale operations [are] currently not justified."[119] The parties subsequently adopted a resolution distinguishing between "legitimate scientific research"—which would be regarded as placement for a purpose other than disposal and therefore permissible—and other ocean fertilization activities (such as full-scale deployment), which "should not be allowed."[120] Individual nations are to apply an assessment framework to determine "with utmost caution, whether a proposed ocean fertilization activity constitutes legitimate scientific research."[121]

Given the number of ocean fertilization experiments thus far, it is not surprising that formal international attention has concentrated on this type of geoengineering. Indeed, existing treaty regimes are arguably better suited to address ocean fertilization than other geoengineering techniques such as the release of stratospheric aerosols. Whereas the oceans are subject to the governance regimes of the LC/LP and the more general United Nations Convention on the Law of the Sea, there is no instrument designed to provide overall protection for the global atmosphere.[122] By and large, states have sovereignty over the airspace above their territories, subject to regional treaties and international norms regarding transboundary harm.[123] If stratospheric aerosol experiments were to be seriously considered, one international regime that could come into play is the Montreal Protocol.[124] As mentioned earlier, sulfur aerosols intensify the ozone-depleting effect of chlorine gases already present in the stratosphere. The Montreal Protocol restricts the consumption and production of ozone-depleting substances and thus could be amended to include sulfur aerosols as regulated substances.[125] Similarly, the Convention on Long-Range Transboundary Air

Pollution, a regional framework agreement that provides support for protocols addressing specific classes of pollutants, could serve as a forum for considering the governance of stratospheric aerosol release.[126]

Informal Governance: Asilomar 2.0

Thus far, formal governance of geoengineering has largely involved ad hoc responses to proposals for field research on ocean fertilization. Informal governance efforts, which are also emerging, have attempted to address the subject of geoengineering research more broadly. One of the most prominent of these efforts to date took place in March 2010 at a conference organized by the Climate Response Fund, a nonprofit foundation. The stated purpose of the conference was "to discuss and develop a set of voluntary guidelines, or best practices, for the least harmful and lowest risk conduct of research and testing of proposed climate intervention/geoengineering techniques."[127] To that end, the organizers invited geoengineering scientists, social scientists, government employees, representatives of nongovernmental organizations, and members of the media.[128] Modeling the conference after the 1975 Asilomar meeting on recombinant DNA, the organizers praised the earlier meeting "as a landmark effort in self-regulation by the scientific community."[129]

The 2010 Asilomar Conference (informally dubbed Asilomar 2.0) did not take place without controversy. The Climate Response Fund faced conflict-of-interest allegations based on financial and familial ties between the organization and a private geoengineering company.[130] More significantly, the conference drew criticism that it was "moving us down the wrong road too soon and without any speed limit."[131] In a letter to the organizers, the ETC Group and other organizations attacked the legitimacy and transparency of the conference proceedings. These critics declared that decisions regarding geoengineering belong to "the community of nations and peoples," not to an elite group invited by a committee of "almost exclusively white male scientists from industrialized countries." The letter further asserted that the development of any research guidelines would be premature absent an initial determination that geoengineering is "technically, legally, socially, environmentally and economically acceptable."[132]

Notwithstanding the critics' charges, conference organizers had made some effort to address social and ethical issues, as a comparison of Asilomar 2.0 with the 1975 Asilomar proceedings on recombinant DNA demonstrates. In the original Asilomar meetings, for example, participation was heavily skewed in favor of hard scientists, and the proceedings were structured to sidestep ethical and social issues.[133] With the exception of a luncheon speaker and a concluding panel, discussion was restricted to technical issues, such as how recombinant DNA experiments could be done more safely. Asilomar 2.0 organizers viewed the original Asilomar proceedings as a success, and invitees to Asilomar 2.0 were limited in geographic diversity.[134] Asilomar 2.0 nonetheless involved a wider range of participants, including social scientists, ethicists, and individuals from nongovernmental organizations. Furthermore, the conference discussions explicitly considered questions of public participation and informed consent.[135]

Asilomar 2.0 ultimately fell short of its stated purpose. The conference's steering committee produced a general statement endorsing further research on geoengineering's efficacy and risks as well as broader public participation in decisions concerning research and implementation.[136] The committee also issued a report containing various recommendations, including open and cooperative research within an internationally supported framework, iterative and independent assessments of research progress, and public participation and consultation in research planning and oversight.[137] However, conference participants were unable to generate and agree on a set of voluntary guidelines for geoengineering research and testing, as had been anticipated.[138]

Moreover, figuring out how to operationalize public participation and informed consent proved to be a particularly difficult challenge. Despite frequent references to involving the public and obtaining consent from those potentially affected, conferees made little progress in determining how to achieve those objectives.[139] As one panelist noted, meaningful public participation in geoengineering policy poses a seemingly intractable challenge when the majority of the population in some developing countries has never even heard of global warming.[140] The Asilomar steering committee recognized the difficulty but ultimately left it unresolved:

> As the scale of climate engineering research expands beyond national borders, . . . the identification and responsibilities of potential decision-making institutions becomes more problematic, especially given the widening range of social and cultural perspectives deserving consideration. For governance of such research, new or modified roles for existing institutions or new governance mechanisms may be needed.[141]

The establishment and design of international institutions that are participatory and accountable is a critical task highlighted by Asilomar 2.0's shortcomings.

Governance Design

Left unregulated, geoengineering efforts by researchers, private parties, and individual nations could entrench particular geoengineering methods that society might find objectionable, heighten international tensions, and precipitate climate or other ecological disasters. As interest in geoengineering rises, robust governance mechanisms are needed to ensure that any activities carried out in this field are done so responsibly and under international oversight. Individual decisions should be made cautiously and with an eye toward potential long-term ramifications. Moreover, those who may be affected by field experiments or deployment should be informed and provided with some opportunity for input. This section offers observations and recommendations for promoting governance of geoengineering along these lines.

Formally Attending to Geoengineering

First, present mechanisms for geoengineering governance are inadequate. Though a completely new treaty regime may not be necessary, the international community has substantial work to do. Simply relying on a patchwork of preexisting treaties crafted to address other problems will likely result in gaps and misapplications analogous to those observed in applying the Coordinated Framework to biotechnology in the United States. For example, use of stratospheric aerosols—a technique likely to attract attention because of its seemingly low cost—might go virtually unregulated unless new treaty arrangements are made. Yet international agreement on how to govern the proposed use of stratospheric aerosols is particularly necessary because of its likely widespread and severe adverse consequences.

Even for methods of geoengineering that are becoming subject to ad hoc oversight under individual treaty regimes, such as ocean fertilization, exclusive reliance on such regimes may be problematic. Existing treaties were not designed to address geoengineering and thus may prove to be a poor regulatory fit. Other than the FCCC, existing treaties are not focused on the problem of climate change and may give insufficient attention to the issue. In addition, piecemeal application of existing treaty regimes to geoengineering proposals could result in potentially inconsistent policies and a lack of coordination with other climate change measures.

Absent binding treaty obligations, opponents of geoengineering may invoke international environmental norms such as the precautionary principle or the obligation to avoid transboundary harm to halt deployment or large-scale experimentation. The application of these norms, however, will be open to debate. For example, although the obligation to avoid transboundary harm is widely accepted, nations may dispute whether a specific climate effect constitutes significant harm, whether the effect was caused by geoengineering, and whether the effect is justified by national or global benefits. With respect to the precautionary principle, the status and meaning of the principle are contested as a matter of international law.[142] Even if this were not so, it is far from clear what the precautionary principle might require or forbid with respect to geoengineering in specific contexts. In a climate emergency, for example, the principle even might support geoengineering deployment as a precautionary hedge against catastrophic risk.

Despite its limitations, existing international law can provide a starting point for the development of future governance. Specifically, geoengineering should be addressed within the structure of the FCCC, notwithstanding that regime's general emphasis on mitigation. A primary principle of the FCCC is that parties "should protect the climate system for the benefit of present and future generations of humankind."[143] That premise arguably encompasses the governance of geoengineering. Moreover, the FCCC already has established a regular forum for considering climate-related issues and has at its disposal technical bodies that can facilitate research, peer review, and development of consensus in this area.[144] Given that some types of geoengineering projects might be substituted for emissions reductions, it makes little sense to develop an entirely separate international regime to address geoengineering.[145] In addition, while actions taken under existing treaty regimes other than the FCCC can provide useful contributions to geoengineering governance, the FCCC should serve as the overarching structure and coordinating mechanism for formal geoengineering regulation. The framework for assessing ocean fertilization research proposals under the LC/LP, for example, could provide a template for more comprehensive guidelines under the FCCC for evaluating geoengineering research in general.

Should the FCCC or other existing treaty mechanisms prove too unwieldy to address the need for oversight, individual nations nevertheless should push for international dialogue. A task force of the Bipartisan Policy Center, for example, advocates that U.S. government agencies communicate with counterpart research organizations and government institutions to initiate international cooperation on geoengineering.[146] Such efforts could begin to forge norms to govern geoengineering research and lay the groundwork for subsequent, more formalized oversight.

Developing Top-Down and Bottom-Up Governance

A second observation regarding geoengineering governance is that both top-down and bottom-up approaches to governance are needed. Formal lawmaking processes, particularly at the international level, can be slow, difficult, and potentially ineffectual. For an example of the challenges of making or amending treaty regimes, one need only look to the negotiations aimed at limiting global GHG emissions. The Kyoto Protocol, negotiated in 1997 as a follow-up to the FCCC, contains binding emission limits applicable to certain industrialized countries from 2008 to 2012.[147] Negotiations to develop a successor to the Kyoto Protocol have been under way for several years and were slated to culminate in a final agreement at the 2009 Climate Change Conference in Copenhagen. That conference, however, yielded only a nonbinding accord under which parties agreed to submit voluntary, self-determined national emissions reduction targets, and successive meetings have produced little more than an agreement to negotiate a binding mitigation regime by 2015.[148]

Of course, achieving agreement on binding emissions reductions is particularly challenging because such efforts are susceptible to free riding by nonparticipating countries.[149] Reaching a consensus on geoengineering rules might be easier because doing so would not require commitments to vast changes in energy production, economic structures, or individual behavior. In addition, the negotiation process could be streamlined by designing such an agreement as a protocol to the FCCC and by holding discussions between key nations representing different viewpoints on geoengineering. The potential for values conflict nonetheless would make such negotiations very challenging. Nations are likely to disagree on what risks are acceptable, how to cope with uncertainty, how to compensate those harmed by experiments or deployment, and whether geoengineering is ethically appropriate. Moreover, such negotiations would take place in the shadow of broader discussions on climate change and could not be divorced from them. Some countries might perceive any geoengineering efforts as a dangerous attempt by industrialized nations to avoid emissions reductions and evade responsibility for past pollution.[150] Others might view geoengineering as a troubling but nevertheless essential option for warding off national or regional climate disaster.[151] Whether the nations of the world can agree on rules for geoengineering may depend on how restrictive such an agreement might be. At the least, however, an international agreement can establish ground rules for research and provide a forum for resolving conflicts.[152]

The obstacles to making and enforcing formal legal arrangements suggest a need for less formal, bottom-up approaches as well. International cooperation and dialogue among scientists on geoengineering can generate research norms and informal oversight mechanisms. A voluntary code of conduct for geoengineering researchers combined with an informal research governance framework, for example, would offer a potentially more immediate and flexible governance option than more formalized arrangements. These sorts of bottom-up efforts can raise awareness of normative concerns, begin to address the current lack of systematic oversight, and lay the groundwork for more formal regulation.[153] In 2009, a group of British academics took an initial step in this direction by drafting the Oxford Principles. Presented as a proposal to the British Parliament, the Oxford Principles call for (1) geoengineering to be regulated as a public good; (2) public participation in geoengineering decision making; (3) public disclosure of geoengineering research and open publication of results; (4) independent assessment of the impacts of geoengineering research; and (5) establishment of governance structures prior to deployment.[154] The principles are quite general in nature and could serve as the basis for top-down regulation or bottom-up norms development.[155] In either case, however, more work will be necessary if the principles are to serve as an effective guide to research conduct. A British parliamentary committee has suggested, for example, that the second principle, public participation, "needs to spell out . . . what consultation means and whether, and how, those affected can veto or alter proposed geoengineering tests."[156] The principle, moreover, does not specify who the public is or how conflicting preferences might be resolved.

Unfortunately, bottom-up development of norms and norm internalization do not necessarily occur in timely or predictable ways. Moreover, it is doubtful that a bottom-up approach alone will produce sufficiently effective, transparent, and accountable mechanisms for assessing and controlling research efforts. Voluntary cooperation is exactly that—voluntary—and various actors, whether nations, corporations, or individuals, might decline to participate to seek military, commercial, or personal advantage. Just as important, such an approach, if unaccompanied by formal international approval, would lack political legitimacy. The Asilomar 2.0 meeting was caricatured as an elitist gathering to decide the world's future, and other informal efforts to develop a voluntary code of conduct likely would be subject to similar attacks. In addition, as the Oxford Principles illustrate, voluntary codes tend to be highly general and thus of limited value. Furthermore, those codes that do contain more detail often suffer from weak implementation.[157]

Given the controversy likely to accompany field tests that could have transboundary effects, such tests require some sort of formal sanction that can confer political legitimacy. Geoengineering efforts that could undermine international cooperation on GHG emissions reductions or commit humankind further on a path toward eventual deployment also should be subject to a similar approval process.[158] At least one commentator has proposed that there be no constraints on "modest low-level field testing,"[159] but the difficulty of defining such a concept and the need to establish trust and transparency suggest that even low-level field tests should be subject to disclosure and formal oversight. Field tests—if they take place at all—should occur under the aegis of an international political process. Voluntary codes of conduct can buttress international oversight by encouraging compliance with formal regulatory requirements. Standing alone, however, a bottom-up approach will not be viewed as legitimate even if it incorporates public input. Exclusive reliance on self-regulation risks a repeat of mistakes made in the development of biotechnology—mistakes that in this instance could precipitate international conflict or lead to a backlash against future geoengineering.

Tailoring Governance to Different Types of Geoengineering

A third important point is that geoengineering governance cannot rely on a one-size-fits-all approach. Rather, governance should be tailored to reflect the distinct risks and uncertainties of different geoengineering methods and different categories of geoengineering research. CDR techniques treat the scientific root cause of climate change—that is, elevated GHG concentrations in the atmosphere. To the extent that techniques such as artificial trees can serve as substitutes for mitigation and are not likely to pose significant new risks, existing governance mechanisms may suffice or be readily adapted.[160] Although ocean fertilization is classified as a CDR technique, however, it will require closer oversight because of its serious potential to affect ocean ecosystems in adverse or unexpected ways.

SRM techniques require closer scrutiny than CDR techniques because the former generally present greater uncertainties as well as higher probabilities of wreaking havoc with the Earth's climate. SRM techniques themselves would generate climate risks, and these techniques do nothing to lower GHG concentrations, which would reach levels unprecedented in recent geological history. SRM techniques also require special attention because their relatively low cost and rapid effect—at least in theory—make it tempting to deploy them, perhaps unilaterally, in a climate emergency. Governance of SRM thus will need to address issues such as monitoring of unauthorized SRM projects.[161] More important, the uncertainties surrounding SRM techniques, combined with our limited ability to reduce that uncertainty through field tests, demand particular caution in deciding how to handle SRM. Renunciation of SRM options might even be warranted by the termination problem: It may simply be impossible to create institutions that reasonably assure the continued operation of SRM techniques over centuries.

Geoengineering governance should also reflect distinctions between different categories of research. The Solar Radiation Management Governance Initiative, an ongoing effort to explore governance issues raised by SRM research, has suggested that possible research activities be categorized according to increasing levels of apparent risk: nonhazardous studies such as computer modeling, contained laboratory studies, or passive observations of natural phenomena; small field trials; medium field trials; and large field trials.[162] Research posing lesser hazards presumably would require less oversight than research posing greater hazards, although the difficulty of distinguishing between the risks posed by field trials of different scales suggests that oversight efforts should not place much if any weight on such distinctions.

Addressing Liability

A fourth observation regarding geoengineering governance pertains to liability for harms caused by geoengineering. Liability provisions should be included in any geoengineering governance regime. Monetary damages should be available to compensate for adverse effects of geoengineering and to restore damaged environments to the extent possible. However, any liability provisions should play only a secondary role in the management of geoengineering risks. Demonstrating causation and determining which injuries merit compensation would be complicated. Attributing specific droughts, floods, or other climate phenomena to geoengineering would be difficult if not impossible in light of the complexity and variability of the climate system.[163] And determining a baseline against which to measure damages would be controversial, as suggested by the possibility of compensation claims by countries that would benefit from warmer temperatures and other characteristics of a world without geoengineering. A fund to cover damages and clear guidelines for its use should be established prior to any significant geoengineering field tests.

Monetary damages ultimately are likely to be a poor remedy for many of the harms that may result from geoengineering. Pecuniary damages will be inadequate to compensate for loss of life. Damaged ecological resources, such as extinct species or degraded ecosystems, may be irreplaceable. And given the complexity and poorly understood nature of the global climate system, it may be impossible to reverse processes that geoengineering sets in motion. As a result, any liability scheme is unlikely to make whole those nations and individuals harmed by geoengineering. For many of the same reasons, an environmental assurance bond requirement similar to that proposed for nanotechnology would not be a suitable primary mechanism for governing geoengineering.[164] The potential harms are simply too irreversible, irremediable, and catastrophic for monetary damages to suffice. Just as common law tort provides for injunctive relief in situations where damages are inadequate,[165] the difficulty of establishing, measuring, and making up for adverse consequences calls for a cautious approach to geoengineering.

Incorporating the Public and Public Values

Finally, meaningful public participation and some form of consent are necessary to legitimate any significant geoengineering research or implementation.[166] As already discussed, public participation and consent on an issue such as geoengineering are extremely difficult to operationalize, especially on a global basis. Not surprisingly, concrete suggestions on exactly how to bring the public into geoengineering governance are few and far between.[167] Truly democratic control of geoengineering may be impossible, as there is no global public that shares a collective identity, nor are there plausible political structures for democratically controlling international institutions.[168]

Although the ideal of global democratic control may be unachievable, the option of leaving geoengineering policy to individual nations or to the scientific community is unacceptable. Such a course would give no voice to those most affected and risk global catastrophe and conflict.[169] As a practical matter, then, the question is how to foster three essential characteristics of global governance: openness, so that decision-making processes are fair and transparent; responsiveness, so that decisions reflect public attitudes and values; and responsibility, so that governance accounts for effects on various communities and on present and future generations. At the very least, valid consent to geoengineering activity requires formal international consultation and approval. Generally speaking, matters affecting the global community are proper subjects of international consideration, whether through the United Nations or through other international political forums. With respect to geoengineering, the Conference of the Parties of the

FCCC—the formal annual gathering of the international community to assess climate policy—would be a logical place to begin. In light of the difficulties encountered in negotiating reductions in GHG emissions, securing consensus on geoengineering policy and management in that setting would be challenging, to say the least. Such a task is not impossible, however, as suggested by the consensual processes in which the parties to the LC/LP and CBD have begun to develop rudimentary positions on certain aspects of geoengineering. In addition, nonconsensus processes, in which decisions are made by a supermajority, offer a less cumbersome management alternative that might be particularly suited for technical decisions.[170]

Discussions between official state representatives in such forums provide only a starting point for deliberation.[171] The widespread and potentially life-changing effects of geoengineering demand a more open and participatory process.[172] Steps should be taken to improve the representativeness and deliberative quality of international bodies and organizations that might craft geoengineering policy. These steps would not only promote legitimacy but also lead to better-informed decisions.

Transparency in state-to-state deliberations and decision making promotes accountability and is a necessary condition for informed public debate.[173] Official meetings as well as informal discussions at which important decisions are made should be open to outside observers. Such meetings could even be made available to the public via webcast or other means. Transparency should also extend to any research projects that are undertaken and to resulting data and analysis. Public access does not guarantee that citizens will be able to fully understand or assess proceedings and their ramifications. Nevertheless, interested individuals or groups should have an opportunity to gather information, formulate views, and raise concerns with political officials.[174]

As with other emerging technologies, public awareness campaigns, national referendums, opinion polls, and efforts to promote discussion in concrete and electronic public forums can all be used to engage the general public in the geoengineering debate. Not all nations will choose to implement these tools or can afford to do so, however. Many people, particularly those who struggle for daily subsistence, have little interest or time to devote to climate change, let alone geoengineering. Even if significant numbers of people do become engaged in the issue, the prospect of involving hundreds of millions of people worldwide in a participatory exercise to determine geoengineering policy is simply not practical. Such difficulties have raised doubts about the democratic legitimacy of international governance institutions generally.[175] Democratic governance is not merely concerned with the aggregation of individual preferences, however.

Democratic governance is also concerned with reasoned and informed deliberation, and global institutions of governance at the least should strive to promote such deliberation.[176] Ensuring that alternative views and concerns are represented in geoengineering policy-making processes, whether top-down or bottom-up, will require the development and use of creative mechanisms.

Nongovernmental organizations, community organizations, labor unions, and other civil society organizations have a critical role to play in this process. Thousands of such organizations have participated in plenary sessions and committee meetings at recent international environmental conferences, including conferences to address climate change.[177] These organizations typically do not directly craft the text of international agreements. Nevertheless, they have helped to frame issues, set agendas, develop policy options, shape official state positions, and monitor state commitments.[178] As a number of global governance scholars have recognized, civil society organizations can function as pluralistic intermediaries between international legal regimes and the publics the regimes ultimately govern. These organizations "can give voice to citizens' concerns and channel them into the deliberative processes of international organizations."[179] They also may "make the internal decision-making processes of international organizations more transparent to the wider public and formulate technical issues in accessible terms."[180]

With respect to future international discussions on geoengineering, civil society organizations can provide alternative views, raise overlooked concerns, and promote transparency. As in other international environmental forums, these organizations can provide policy information and should have access to plenary as well as committee meetings. These organizations must be allowed a more active role, however, if they are to function effectively as agents of accountability. Acting like ombudsmen, civil society organizations should be permitted to raise concerns directly to policymakers and to demand public responses to those concerns. For example, if the approval of geoengineering research projects is ultimately entrusted to a review board, the board should be required to respond to questions from civil society organizations and to publicly explain its decisions.[181] Requiring decision makers to provide public justification for their actions not only facilitates the external evaluation of specific actions but also fosters institutional legitimacy.[182]

Civil society organizations are not elected, of course, and thus should not be characterized as democratic representatives or as aggregators of individual preferences.[183] Nor should it be assumed that such organizations necessarily put the general public interest ahead of their own institutional interests; they are voluntary bodies that articulate the views of particular subsets of society.[184] Nonetheless, in demanding public justifications for decisions and providing critical perspectives, these organizations can contribute to democratic legitimacy by fostering more accountable and deliberative global governance.[185] To do so effectively, organizations engaging in the policy-making process should themselves be transparent and independent; they should not serve merely as a front for governments or special interests.[186] Moreover, these organizations should be especially solicitous of marginalized viewpoints and the concerns of those most vulnerable to the effects of geoengineering experiments or projects.[187]

With respect to complex global issues such as geoengineering, there is a serious risk that the views of citizens in developing countries will be underrepresented because of lesser capacities to gather information and formulate concerns.[188] Partnerships between NGOs in the North and South could serve as important means of transferring technical information and facilitating broader participation.[189] Policy-making processes can be designed to give special weight or attention to contributions by organizations that speak for the underrepresented or those most affected. Furthermore, just as formal policy-making bodies must become more inclusive, scientific organizations involved in developing codes of conduct must reach beyond the scientific community and the developed world to engage and respond to people whose interests and concerns might otherwise be ignored.

Concluding Thoughts

Although climate change poses grave threats, geoengineering must not be viewed as a simple technological solution to the climate change dilemma. Geoengineering itself undoubtedly would create its own problems, and the subject presents difficult political, ethical, and technical challenges for humankind. In light of the global and consequential nature of these challenges, research and development into geoengineering should not proceed without external oversight and public engagement. Geoengineering is at a relatively early stage of technological development and thus offers a promising opportunity to apply a Promethean approach to technology management. Establishing governance structures to oversee any geoengineering field research or deployment is essential. Just as important will be a willingness on the part of scientists and the international community to consider their obligations to each other, to the public, and to future generations.

CHAPTER 5

Synthetic Biology

The New Biotechnology

Mary Shelley's *Frankenstein* is a tale about life creation gone awry and more broadly a warning about the dangers of unreflective scientific ambition. The desire to create life nonetheless persists and has found expression in the emerging technology of synthetic biology. Sometimes described as "genetic engineering on steroids," synthetic biology has the aim of creating new forms of life.[1] Researchers in this field hope to construct synthetic organisms that will serve as the basis for producing renewable fuels, developing personalized medical treatments, enhancing crop production, and remediating environmental pollution. Like the emerging technologies already considered, however, synthetic biology holds both promises and perils, demanding careful attention and public deliberation.

Introduction

What Is Synthetic Biology?

Building on the techniques of conventional genetic engineering, synthetic biology promises an even wider array of innovations and applications than its predecessor. Like conventional genetic engineering, synthetic biology aims to produce new or improved traits by tinkering with the genetic code. But unlike conventional genetic engineering, synthetic biology is not confined to manipulating genetic material that can already be found in existing organisms, nor is it practiced solely by molecular biologists. At the heart of synthetic biology is the design of novel genetic sequences and novel organic molecules. The persons engaged in this design process include not only biologists but also chemists, engineers, physicists, computer scientists, and even persons lacking formal scientific training. If successful, synthetic biology would offer a far more powerful and efficient way to engineer desired traits than conventional genetic engineering. Synthetic biologists ultimately seek to assemble entire genomes from a library of interchangeable genetic sequences and thereby to create artificial living systems.[2] In short, synthetic biology "aims to merge engineering approaches with biology" and to use engineering design and development techniques to create new forms of life.[3]

Synthetic biology techniques are already showing promise in various fields, including improving the yield of algae that produce biofuels and altering bacteria so that they can target and invade cancer cells.[4] Within the next decade or two, the ability to design new biological materials and chemicals using synthetic biology could lead to the development of new drugs and drug delivery systems, biosensors, bio-based manufacturing, and microbes that can digest environmental toxins.[5] Through synthetic biology, "living factories" filled with synthetic microbes may produce needed quantities of industrial chemicals, drugs, and biofuels. Further into the future, synthetic biologists may create entirely new organisms that bear little resemblance to species currently in existence.[6] With respect to this last goal, however, synthetic biology is at an early stage of development roughly comparable to the state of recombinant DNA research in the 1970s.[7]

In current synthetic biology experiments, scientists are just learning to manipulate biological materials that they have isolated from cells.[8] One of the leading breakthroughs to date, for example, involved the creation of a synthetic bacterial genome.[9] In this work, researchers constructed a synthetic chromosome matching the genome of an existing bacterium. The researchers then used the synthetic chromosome to replace the DNA of cells from a second bacterium species, and the resulting cells functioned in a manner identical to cells of the first bacterium species.[10] Synthetic biologists' goal of designing and constructing truly novel organisms from standardized genetic parts alone, however, is far from being achieved.[11]

Concerns Raised by Synthetic Biology

Notwithstanding the modest progress to date, commentators predict that "relatively untrained people using commonly available equipment and materials" could one day undertake the synthesis of new organisms.[12] At present, one can easily order customized DNA sequences over the Internet and set up an amateur do-it-yourself (DIY) biology lab for a few thousand dollars. Community labs are springing up, enabling amateur scientists to dabble in DIY biology with minimal investment. Indeed, college and high school students are conducting rudimentary experiments in synthetic biology, their exploits highlighted at the annual International Genetically Engineered Machine (iGEM) competition.[13]

The projected hazards of synthetic biology in some ways resemble those posed by GMOs. Like conventional genetic engineering, synthetic biology would create living organisms that are capable of replicating, mutating, and evolving. The development and release of such organisms could lead to biosafety hazards—unintended adverse effects on people and the environment. The hazards posed by synthetic organisms, however, would be far more uncertain than those associated with GMOs, particularly as scientists create novel organisms that are increasingly dissimilar to already existing organisms.

In conventional genetic engineering, genes are inserted into a host organism whose traits are generally well understood. Although the gene insertion may have unpredictable effects, much of the host's genome remains intact, thus providing a baseline for predicting the traits of the engineered organism. A synthetic organism whose genome is assembled from the ground up, however, would possess no such baseline. Such an organism may have emergent and unexpected properties that cannot be predicted simply from the individual genes used to construct the organism's genome.[14] Specifically, the biological environment of a gene can be as important as the gene sequence itself in determining the function and expression of the gene. Temperature, light, and the presence of drugs or chemicals, for example, can influence whether a specific gene is expressed.[15] Thus, even if a synthetic organism's genome is derived primarily from an existing "parent" organism, the complexity of the synthetic organism may make any estimate of hazards based on the parent organism's risk profile erroneous.[16] Furthermore, synthetic biology may pose especially uncertain dangers because the first organisms likely to be synthesized, bacteria and other microorganisms, have the ability to mutate and evolve rapidly. In contrast to more complex organisms (including genetically engineered food crops), microorganisms can pass through multiple generations within a matter of hours and readily exchange genetic material with other organisms, and thus can undergo unpredictable evolutionary changes very quickly.[17] The release of synthetic organisms, whether deliberate or not, ultimately challenges traditional risk assessment methods and poses serious and potentially catastrophic health and environmental hazards.[18]

These biosafety concerns are matched by concerns about biosecurity—

that is, concerns focusing on the deliberate misuse of the knowledge, techniques, and products of synthetic biology. Persons who might engage in misuse include bioterrorists as well as "bio-hackers"—amateur scientists who would fashion experimental organisms to gain attention or generate mischief. The potential for misuse already exists, as DNA synthesis techniques can be used to assemble the genomes of the flu virus and other known pathogens.[19] Knowledge of a viral genome alone is not sufficient to create a disease-forming pathogen, but it is an important step. As synthetic biology techniques advance and become more accessible, bioterrorists and bio-hackers may someday be able to create synthetic organisms with relative ease.[20] The potential for misuse, theft, or deliberate release of harmful agents has generated comparisons of synthetic biology to nuclear technology and led to proposed limits on the production and dissemination of research data.[21]

Synthetic Biology Regulation: Existing Law

As is the case with most other emerging technologies, there are no regulatory regimes specifically directed toward governing synthetic biology. Laws developed in other factual contexts may be relevant, but the "cross-borderness" of synthetic biology will make the application of these laws particularly challenging. Specifically, because synthetic biology crosses the borders of scientific disciplines, industrial sectors, and geopolitical units, its governance will require bridging cultural divides and building international structures and norms.[22]

Relevant Domestic Law

From a policy perspective, synthetic biology is likely to be framed as an extension of conventional genetic engineering, just as conventional genetic engineering was framed as an extension of conventional plant breeding. Synthetic biology poses a far greater danger of "dual use," however, in that it may be applied toward both beneficial and harmful ends. Nonetheless, unless new laws or regulations are put in place, domestic oversight of synthetic biology will consist largely of the same haphazard scheme that applies to conventional genetic engineering.[23] That is, the NIH guidelines for GMO research would govern some synthetic biology research, and the Coordinated Framework would govern the production and use of synthetic organisms.

The NIH guidelines specify safety practices and containment procedures for research involving recombinant DNA molecules. Experiments involving greater levels of estimated risk receive correspondingly higher levels of scrutiny.[24] Recent revisions to the guidelines make explicit their applicability to synthetic nucleic acids placed in cells, organisms, or other biological systems.[25] The guidelines apply only to research funded by the NIH, however. They do not apply to research funded by other sources, except to the extent that other sources require compliance or researchers voluntarily comply. The guidelines' limited applicability will be of growing concern as more and more synthetic biology experiments take place in community and private labs. Moreover, the NIH has no direct regulatory authority, as the guidelines are effectuated through contractual provisions. In addition, specific application of the guidelines occurs primarily through institutional biosafety committees established by research institutions receiving funding from the NIH.[26] These committees are local institutions, and the oversight they provide varies widely. More critically, the risk-based precautionary measures the guidelines prescribe are especially challenging to apply to synthetic biology experiments because the guidelines assume the possession of basic knowledge about the risks involved. Synthetic biology experiments will raise risks and uncertainties that are more difficult to gauge ex ante than is the case with conventional genetic engineering research because they involve novel organisms or organisms generated from the genomes of multiple other organisms.

Production and use of synthetic organisms would be governed largely under the Coordinated Framework already developed to oversee GMOs.[27] Under the Framework, regulatory oversight is divided among multiple agencies, including the EPA, APHIS, and the FDA. In addition to the regulatory programs already considered in chapter 2, the EPA's use of the Toxic Substances Control Act (TSCA) to regulate genetically engineered microorganisms would also be of interest because many synthetic biology products will involve engineered microorganisms. Rules issued pursuant to Section 5 of the TSCA require companies to give the EPA notice before testing genetically engineered microorganisms outside a noncontained facility and before producing genetically engineered microorganisms for a commercial purpose.[28] Based on these notices, the EPA may impose controls to protect against unreasonable risks, but the data that companies submit may be inadequate for the EPA to make properly informed judgments of risk. In addition, the use of specific engineered microorganisms identified by the EPA as involving lesser risks are potentially exempt from EPA review altogether.[29] To govern the novel circumstances and biosafety concerns associated with synthetic biology, regulators ultimately will have to stretch existing health and safety laws further than ever before. These laws rely heavily on risk assessment to generate the information that regulators use to determine appropriate levels of containment and other biosafety measures. Given the tremendous uncertainties surrounding the biosafety risks of synthetic organisms, such information often will be unavailable or too unreliable to provide adequate protection.

Biosecurity regulations, such as the Federal Select Agent Program, are essential for addressing dual use potential and would apply to some synthetic biology activities. The Federal Select Agent Program limits the possession, use, and transfer of biological agents and toxins that could threaten public health and safety or animal and plant health and safety.[30] The program requirements apply not only to listed agents and toxins but also to the nucleic acids and genetic elements that encode for them.[31] Administered by the Centers for Disease Control and Prevention (CDC) and APHIS, the program tracks the use of specified biological agents and requires registration of facilities that handle them. An individual who submits the genome sequence and requests the synthesized DNA for the Ebola virus (a select agent), for example, is subject to the program's registration requirements.[32] These requirements apply only to agents and toxins that regulators have specifically listed, however. Novel agents produced through synthetic biology would not be subject to Select Agent Program oversight unless they are identified and added to the program list.[33] Synthetic gene fragments that bioterrorists could assemble to create listed biological agents or toxins might also escape review.

Other existing biosecurity measures also may be relevant. These measures include export regulations, screening of gene synthesis orders, and outreach to those engaged in synthetic biology experiments. U.S. Department of Commerce export regulations, for example, limit the export of materials having both civilian and military applications.[34] A license may be required to export these dual use materials, and product recipients are screened against lists of proscribed users. In practice, however, these regulations generally have not been applied to orders for synthetic DNA, and the volume of such orders would make the regulations cumbersome to apply.[35] Another measure that addresses biosecurity risks involves the voluntary screening of synthetic DNA orders by DNA providers. Guidance issued by the U.S. Department of Health and Human Services encourages screening of orders to determine if select agents are being sought as well as screening of customers to determine if they are legitimate.[36] At present, such screening is voluntary, however, and the overall ease of obtaining materials and supplies for synthetic biology poses a challenge to these and other means of oversight. Finally, in an effort to foster responsible conduct among DIY biologists, the FBI has developed an education and outreach program. By establishing cooperative relationships between DIY biologists and local law enforcement, this program and similar efforts seek to improve information flow, increase awareness of how research findings might be misused, and reduce DIY labs' vulnerability to exploitation by would-be bioterrorists.[37]

International Law

At the international level, there are likewise existing but limited governance regimes that could serve as a basis for beginning to address synthetic biology's biosafety and biosecurity concerns. For example, the Convention on Biological Diversity (CBD) requires parties to establish means of managing biosafety risks associated with the use and release of living modified organisms resulting from biotechnology "as far as possible and as appropriate."[38] This language hardly creates any specific or binding commitments, however, and the treaty has not been ratified by the United States. More concrete obligations relevant to biotechnology are found in the Cartagena Protocol on Biosafety, which was negotiated under the aegis of the CBD. The Cartagena Protocol requires that an exporting country give notice to the importing country prior to the transboundary movement of living modified organisms intended for release into the environment.[39] The importing country has a right of informed consent and must ensure that a risk assessment is performed before deciding to allow or prohibit such transboundary movement. That decision may take into account adverse effects that are uncertain. If living modified organisms are being exported for use as food or feed, the protocol requires only labeling, and it imposes no obligations where modified organisms are exported for contained use.

The protocol's definition of "living modified organism"—"any living organism that possesses a novel combination of genetic material obtained through the use of modern biotechnology"—appears sufficiently broad to encompass synthetic organisms.[40] Accordingly, the protocol's notice and consent provisions would likely apply to the transboundary movement of synthetic organisms intended for environmental release. Many synthetic biology applications, however, would not be subject to these provisions because they would not be intended for environmental release. More generally, the protocol's scope is relatively limited: It recognizes states' ability to restrict the importation of potentially harmful modified organisms intended for release and provides a mechanism for doing so, but it does not create a general regime for overseeing the hazards of synthetic biology.[41]

Notwithstanding the limits of the CBD and the protocol, synthetic biology has attracted the attention of these treaty regimes. A 2010 decision issued by the Conference of the Parties to the CBD urges members to take a precautionary approach to the use of modified organisms for biofuel production and to the field release of synthetic life, cells, or genomes.[42] These sorts of precautionary measures, however, may lie in tension with World Trade Organization (WTO) rules if they affect international trade. The WTO's Agreement on the Application of Sanitary and Phytosanitary Measures, for example, generally requires that health and safety measures affecting international trade be based on a risk assessment and/or existing international standards.[43] Although it is uncertain how such apparent conflicts will be resolved, parties to both treaty regimes may be inclined to follow issue-specific directives, notwithstanding any restrictive effects on trade, than general free trade rules.[44]

Instruments relevant to addressing the biosecurity concerns of synthetic biology include the 1925 Geneva Protocol, the Biological Weapons Convention (BWC), and the Australia Group Guidelines. The Geneva Protocol generally prohibits the use of chemical and biological weapons.[45] The BWC goes beyond the Geneva Protocol by prohibiting the development, production, stockpiling, acquisition, and retention of biological agents or toxins "that have no justification for prophylactic, protective, or other peaceful purposes."[46] Although the agreement requires parties to take necessary measures to prevent the misuse of biological agents and toxins within their territories, this specific requirement has not been widely implemented.[47] The BWC does not restrict research on biological weapons or govern the development, production, or use of biological agents that have peaceful purposes. Accordingly, the regime has struggled to address the potential for dual use, as might be manifest in the development of facilities that appear civilian in nature but are intended for military purposes. The misuse of agents and equipment normally used for peaceful purposes and the misuse of information generated for scientific advancement have similarly confounded implementation of the BWC.[48] Finally, the Australia Group Guidelines reflect an effort by various BWC members to harmonize their export controls in furtherance of their obligation under the BWC not to transfer biological agents or technologies likely to be used as weapons.[49] These guidelines apply to listed biological agents and equipment. The Australia Group set up a synthetic biology advisory body in 2008, and the guidelines cover organisms containing nucleic acid sequences coding for listed microorganisms or toxins, including those whose genetic material has been "produced artificially in whole or in part."[50] However, the guidelines are voluntary and apply only to the exportation of listed agents and equipment. Such oversight will prove increasingly inadequate as synthetic biologists develop novel organisms that fall outside those lists.[51] The Australia Group Guidelines and similar formalized arrangements could serve as a foundation for managing the biosecurity risks of synthetic biology, but more oversight ultimately will be needed.

In addition to these arrangements, general principles of international law also may apply to synthetic biology. These include the principle of transboundary harm, which obligates states not to cause harm to other states, and the precautionary principle, which allows states to adopt measures responding to uncertain threats of serious or irreversible environmental harm.[52] These principles reflect widely held expectations regarding state responsibilities and are likely to inform discussions on international oversight. They are general in nature, however, and are not consistently applied. Because these principles do not offer clear and specific guidance on how to manage synthetic biology's risks, further explication will be necessary if they are to serve as useful elements in the governance of synthetic biology.

Proposals for Regulating Synthetic Biology

Critical governance challenges posed by synthetic biology include the uncertain hazards of synthetic organisms; the potential for their evolution, escape, or misuse; the ease of obtaining synthesized DNA and conducting uncontrolled experiments; and the morality of synthesizing new forms of life. In response, commentators have offered an array of policy proposals, many of which resemble those suggested for governing other emerging technologies.[53] At one end of the spectrum, a coalition of civil society groups has called for a moratorium on the release and commercial use of synthetic organisms and their products until comprehensive and precautionary oversight mechanisms are in place.[54] At the other end of the spectrum, the biotechnology industry has rejected calls for new oversight, arguing instead for the continued application of existing law—including in particular the Coordinated Framework—possibly supplemented with voluntary guidelines.[55]

Potential governance measures can be categorized according to whether they would address biosecurity or biosafety concerns. Proposed biosecurity measures include educating researchers regarding biosecurity risks, expanding the role of institutional biosafety committees to include biosecurity concerns, stepping up government surveillance or oversight of the synthetic biology community, requiring gene synthesis firms to screen DNA orders, requiring owners of DNA synthesizing equipment to obtain licenses, and restricting publication of information that could be used for malicious purposes.[56] While these measures are not mutually exclusive, they do vary in terms of their ability to be administered, acceptability, and likely efficacy. Surveillance of synthetic biology research taking place outside government or university laboratories, for example, would be difficult, intrusive, and resource intensive. These challenges will only be magnified as research efforts advance and as DIY biology expands. In contrast, focusing on a relatively limited class of regulated entities through a mandate that gene synthesis firms screen orders would be less intrusive and easier to administer. Screening efforts may not identify all gene sequences of concern, however, and stringent screening standards could inadvertently thwart legitimate research.[57] In addition, would-be bad actors might circumvent mandatory screening by placing orders with overseas or black-market gene synthesizers. The gaps in each of these approaches have led to growing support for enhancing awareness within and outreach to the synthetic biology community as a means of encouraging responsible conduct. Efforts might include the development of a code of ethics applicable to all synthetic biology researchers.[58] While such a code might curb reckless experimentation, it would have limited effect in countering deliberate misuse by bioterrorists.

Many of the methods proposed to increase biosecurity, particularly those involving education and outreach, would also help reduce biosafety risks. These methods are unlikely to address the uncertainty surrounding synthetic biology's hazards, however. Risk research is essential but can resolve only some uncertainties and will surely reveal new ones. To cabin potential hazards, the government could make the containment measures found in the NIH guidelines mandatory and apply them to all synthetic biology research. In the absence of reliable methods for assessing potential hazards, regulators could even designate maximum levels of containment as the default standard for synthetic biology experiments. Such requirements presumably would increase the cost and difficulty of research, however, and might even increase biosafety and biosecurity risks by driving DIY biologists underground.

Indeed, recent experiences at the Synthetic Biology Engineering Research Center illustrate the difficulty of integrating biosafety and biosecurity concerns as well as broader values and approaches into the research process. The Center, a joint project of five research universities, developed a research program to bring together hard scientists pursuing specific engineering goals in collaboration with social scientists promoting social and ethical reflection.[59] The social scientists were frustrated in their efforts to engage with synthetic biology scientists and engineers, however. According to the project's lead social scientists, the hard scientists were disinclined to reorient their research objectives or daily practices, and they viewed biosafety and biosecurity issues largely as "the work and responsibility of others—industrial partners, government agencies, or unspecified others."[60]

Beyond the laboratory environment, regulation could include a prohibition on the release of synthetic organisms into the open environment or a requirement that such organisms be studied in a simulated environment and determined to be ecologically safe prior to release.[61] Such restrictions seem sensible at this stage, given the vast uncertainty that exists and the catastrophic effects that otherwise might result. Finally, the engineering of safety features into synthetic organisms themselves could serve as an additional means of managing biosafety risks. It may one day be possible, for example, to incorporate genetic code into a synthetic organism's genome to limit its life span, its ability to transfer genes to other organisms, or its capacity to survive or reproduce outside of a controlled environment.[62] Caution is warranted against exclusive reliance on such measures to manage risk, however. Genetic mutations may negate any limitations that are engineered into a synthetic organism, and evolutionary pressures would favor the survival and spread of organisms with such mutations.

At the international level, alternative governance mechanisms may prove especially helpful in addressing synthetic biology's risks. Here, formal regulation occurs slowly, and treaty arrangements are generally weak. International efforts nonetheless may help to facilitate the development of national regulations.[63] Specifically, the Cartagena Protocol could be modified to require that notice be given to importing countries prior to any transboundary movement of synthetic organisms. This or other treaty regimes could also clarify the ability of states to regulate synthetic biology in a precautionary manner. In addition, international coordination can promote consistent regulation across different jurisdictions and thereby help to avoid a regulatory "race to the bottom." Informal governance mechanisms could also play an important role in ensuring that synthetic biology activities and accompanying hazards do not simply move to jurisdictions lacking any oversight. Such informal mechanisms may include entities that look quite different from the professional societies or industry codes of conduct that typically come to mind. A London School of Economics paper suggests, for example, that the iGEM competition has had a "central role . . . over the formation of international research culture" in synthetic biology.[64] The competition attracts young synthetic biology researchers from many countries, and it has facilitated global discussion of risks and ethical issues.[65] Recognizing the potentially broad influence of entities such as iGEM while ensuring that they are also attuned to public concerns is essential.

Further Concerns

Existing law potentially applicable to synthetic biology focuses on risks of physical harm. For the most part, proposed biosafety and biosecurity measures are similarly narrow in scope. But synthetic biology raises further concerns that do not directly involve physical harm. These concerns are mainly ethical in nature, drawing on philosophical and religious beliefs. Although ethical matters lie outside the boundaries of conventional risk assessment, they merit serious consideration in light of the potentially broad and pervasive role of synthetic biology. These concerns lie at the heart of what synthetic biology does and will surely inform the public's nascent views regarding synthetic biology.[66] Some ethical concerns are relatively unique to synthetic biology: The potential to synthesize life according to human design "establishes a new concept of life" and brings to the fore questions regarding the role of human beings in the universe and the value we place on living things.[67] Other ethical concerns, such as increased economic disparity and inequitable access to technological benefits, raise distributional issues common to all emerging technologies.

The pursuit of the design and creation of life from nonliving matter fuels the objection sometimes made that synthetic biologists are "playing God."[68] Literally understood as an objection based on religious beliefs, the contention is that humans are intruding on a realm of activity—creating life—that properly belongs to a higher being. In its secular form, the contention is that the synthesis of new organisms crosses inherent moral boundaries in the operation of the universe or the natural environment.[69] This argument, while essentially deontological in nature, has a consequentialist aspect as well. The hubris reflected in undertaking activities not fully understood and best left alone can lead to ultimately catastrophic results.[70]

A related ethical concern charges synthetic biology with blurring the distinction between living organisms and machines. Such blurring could foster moral confusion and diminish the value that humans accord to living things. Characterized by human design and control, machines lack any moral status. Synthetic organisms, however, would have characteristics of both machines and living organisms. On the one hand, synthetic organisms would be designed by humans with specific applications in mind. On the other hand, they would be composed of cells comparable to those found in living things. They would also possess other characteristics of living things, including self-regulating and self-organizing capacities and the ability to reproduce.[71] The ambiguous moral status of synthetic organisms raises difficult questions regarding humans' duties toward those organisms. Furthermore, the creation of artificial life forms threatens to dispel the sense of mystery that life holds and reduce appreciation of natural life forms, biodiversity, and human life itself.[72] Philosopher Bryan Norton fears, for example, that synthetic biology will promote "a static conception of biodiversity as a stockpile of parts" and undermine support for ecosystem protection.[73] Why protect nature, one might ask, once we have learned to engineer and improve on its parts? Although conventional genetic engineering also arguably conceives of living things as objects to be manipulated, synthetic biology takes this instrumental approach to an extreme. Dominated by engineering principles and industrial metaphors, synthetic biology seeks to produce "living machines" in the pursuit of narrow anthropocentric goals.[74]

Supporters of synthetic biology offer several responses to these ethical concerns. First, they contend that the field represents just another step in a continuum of human activity.[75] Humans have a long history of intervening in natural processes and of employing living organisms to suit their needs. Under this view, synthetic biology is the latest rung on a ladder that includes animal domestication, plant hybridization, genetic engineering, and cloning. Proponents of this viewpoint admit that synthetic biology involves a different kind of modification than previous or existing practices—synthetic organisms would be created from nonliving matter and would not be part of an evolutionary line of development.[76] Whether the products of evolutionary processes have particular ethical value as such, however, is open to debate. This is not to suggest that the creation of synthetic organisms possessing certain qualities, such as consciousness, poses no ethical problems. The ethical quandary would arise primarily as a result of the organisms' possession of those qualities, however, and less from the specific technique leading to creation of the organisms.

Synthetic biology proponents note further that present research efforts are directed primarily at creating microorganisms and contend that these relatively simple organisms merit no substantial moral consideration.[77] There is little reason to expect synthetic biology efforts to remain so confined, however. The aims of synthetic biologists include the eventual creation of higher life forms. Harvard molecular geneticist George Church, for one, envisions using synthetic biology to create parasite-resistant crops and to resurrect extinct species.[78] Another leading synthetic biology researcher, Steven Benner, has the avowed goal of creating organisms that are capable of evolving in a manner unimpeded by the "baggage" of natural organisms.[79] Synthetic biology could be applied to humans as well in the form of programmable personal stem cells or human embryos rewired to have viral immunity. These applications clearly raise serious ethical concerns (examined in greater detail in chapter 6) regarding the manipulation of higher life forms, including human enhancement.[80]

Finally, synthetic biology proponents assert that the "playing God" conceit is misleading. Synthetic biology involves the formation of life from existing matter and thus does not conflict with common accounts of divine creation, in which a divine being creates matter out of nothingness.[81] Indeed, a presidential commission examining ethical issues associated with synthetic biology encountered no "specific objections to current research efforts in synthetic biology based on the [official] views of organized religions."[82] The concern that synthetic biologists are "playing God" nonetheless resonates with a significant portion of the general public, particularly those holding strong religious beliefs.[83] Such concerns are likely to increase as synthetic biologists seek to create more sophisticated organisms.

Where might these ethical concerns fit in the governance of synthetic biology? Present initiatives to address the biosafety and biosecurity hazards of synthetic biology are necessary, but these narrowly focused efforts miss the larger picture. While current applications involving synthetic biology techniques may appear incremental in nature, synthetic biology is not simply an extension of conventional genetic engineering. Synthetic biology raises new ethical dilemmas and renews old ones. Moreover, these dilemmas lie at the core of synthetic biology's goals and activities. Consideration of these ethical issues should not simply be tacked on at the end of the day, after the technology has already been developed. Rather, they must be incorporated into incipient and future governance efforts.

Increasing public awareness of synthetic biology is essential to fostering deliberative processes that include ethical concerns. The presidential commission's report on the ethics of synthetic biology made several constructive recommendations in this regard, such as developing public and private initiatives in science education, creating independent mechanisms for fact-checking claims about synthetic biology, and soliciting public input in the making of policies regarding synthetic biology.[84] Such efforts would constitute only a first step in establishing a public discussion that actually will influence the course that synthetic biology takes.

Conclusion

Synthetic biology's promise of wide accessibility to relatively untrained persons outside of typical research environments is generating both excitement and worry. While the claims surrounding synthetic biology surely involve some exaggeration, the advances made to date, combined with our experience with conventional genetic engineering, suggest that synthetic biology will have the power to alter radically our economic processes and our environment. We must ultimately choose how to exercise that power and in so doing wrestle with fundamental ethical questions concerning our understanding of life and the rights and responsibilities that attend its manipulation.

CHAPTER 6

Human Enhancement and General Reflections on Managing Emerging Technologies

The technologies considered in the preceding chapters are directed outwardly at transforming the environment around us. But might we also turn our technological power inwardly to transform ourselves? The so-called convergence of technologies toward human enhancement raises this question, and behind these technologies lies an agenda that presents ethical concerns at least as troubling as those surrounding synthetic biology. This final chapter begins by considering human enhancement technologies and the related philosophy of transhumanism, allowing further exploration of the ethical dilemmas generated by the development and use of emerging technologies.

This discussion also offers a starting point for reflecting more generally on emerging technologies and considering broader lessons for applying a Promethean approach. Ethical issues in particular must play a central role in future debates regarding emerging technologies. There must be ethical reflection by scientists, developers, and users as well as broader public participation in a technology management informed by ethics. For such reflection and participation to be meaningful, decision-making processes must be truly open to a full spectrum of outcomes ranging from adoption to rejection. Policymakers and the public must have access to respected and unbiased expertise to foster informed discussion and evaluation of conflicting claims. Establishing sources of such expertise and strengthening regulatory institutions to act on that expertise in conjunction with public input are essential. Finally, we must develop new institutions and approaches to respond dynamically to rapid and unpredictable technological change.

Technological Convergence for Human Enhancement

New technologies build on previous advances and can integrate multidisciplinary innovations. Synthetic biology, for example, is sometimes described as a product of the convergence of genetic engineering, nanotechnology, and information technology. That is, synthetic biology involves the manipulation of genetic material, DNA; DNA is a 2.5 nanometer-wide molecule; and DNA encodes information in a manner akin to digital computing.[1] More generally, the term *technological convergence* refers to collaborative and synergistic efforts in various technological fields. Such efforts may generate benefits in health care, energy, education, and infrastructure as well as dramatic economic and social changes.[2]

In the United States, the concept of technological convergence has become associated with a particular purpose: "enhancing" human beings. Most prominently, a 2002 National Science Foundation–sponsored report predicted a convergence of four technological fields—nanotechnology, biotechnology, information and communication technology, and cognitive science (NBIC)—for improving human performance. The authors of that report, Mihail Roco and William Bainbridge, contend that NBIC convergence will enable not only a "unified understanding of the physical world from the nanoscale to the planetary scale" but also seemingly fantastical changes in the mental and physical capacities of human beings.[3] The specific applications they imagine in the not-too-distant future include direct connections between the human brain and machines, human bodies engineered to be more durable and resistant to aging processes, and human genomes altered to enable "widespread consensus about ethical, legal, and moral issues."[4] Through such changes, Roco and Bainbridge assert, "humanity would become like a single, distributed and interconnected 'brain,'" and "world peace, universal prosperity, and evolution to a higher level of compassion and accomplishment" can be achieved.[5] Futurist Ray Kurzweil predicts a similar series of developments, which he terms the "singularity"—a rapid expansion in intelligence resulting from the continuing exponential development of technology that "will allow us to transcend [the] limitations of our biological bodies and brains."[6]

These startling visions of convergence do not reflect ongoing technological developments as much as they embody a particular ideological perspective on technological change: transhumanism. The transhumanist movement does not merely predict radical technological change, however; it embraces technology as a means of "redesigning the human condition, including such parameters as the inevitability of aging, limitations on human and artificial intellects, unchosen psychology, suffering, and our confinement to the planet earth."[7] Technology development has historically focused on manipulating the external environment. Transhumanism, in contrast, emphasizes the application of technology to modify humans themselves. Some enhancement technologies, such as wearable sensors, would be external to the body and temporary, while other technologies, such as implants or new sensory organs, would be internal and permanent.[8] Rejecting ethical objections to human enhancement, transhumanists argue for an individual right to use technologies freely to improve human capabilities.[9]

Much of the transhumanist vision is speculative and far from realization. At present, enhancement technologies internal to the human body are relatively mundane. Examples, including cochlear implants and prosthetic limbs, are directed primarily at addressing specific deficiencies and are generally inferior to the normal functions they are intended to replace.[10] Biomedical enhancement techniques, such as medications to enhance memory and alertness, generally offer only modest improvements in performance.[11] Projections of more dramatic change, such as extended life spans and superintelligent mergers of humans and machines, often rest on dubious assumptions that technologies will develop at exponential rates.[12]

Though largely hypothetical at this point, human enhancement technologies nonetheless merit society's attention. U.S. government agencies, including the Department of Defense, Department of Commerce, and National Aeronautics and Space Administration, have actively supported the NBIC initiative.[13] Moreover, ongoing research suggests that some technological changes projected by supporters may be feasible. For example, scientists are conducting rudimentary experiments to enhance sensory perception using subdermal nonoparticles and to connect the human brain with a computer via implanted microchips.[14] Likewise, the Department of Defense's Advanced Research Projects Agency (DARPA) is sponsoring research to create soldiers who are more metabolically efficient. Possessing enhanced strength and endurance, these soldiers would be able to function for a week without sleep.[15] In light of the substantial resources that the military, researchers, and other interests are devoting toward human enhancement technologies, broader social engagement with the risks and ethical concerns surrounding these technologies is essential.

Human enhancement technologies undoubtedly have the potential to pose serious risks to health, society, and the environment. Fatalities that have occurred in gene therapy trials suggest that the experimentation necessary to develop enhancement technologies may involve tragic consequences.[16] Technologies to enhance physical or mental performance might cause serious health problems, such as those observed with steroid use.[17] Enhancements to increase human longevity may have not only side effects on recipients but also broader consequences for society. These include heightened economic strains, changed social dynamics, and exacerbated environmental stresses.[18] Similarly, enhancements conferring individual competitive advantages may generate widespread economic and social disruption.[19] The effects of human enhancement technologies ultimately will be impossible to fully predict or control.[20]

Given the undeveloped state of human enhancement technologies, it is unsurprising that relatively little law on the subject exists. Research trials in this area will be subject to laws governing experimentation with human subjects. These laws aim to reduce risks to human subjects and ensure that subjects provide voluntary, informed consent. Specifically, regulations governing federally funded research require that institutional review boards review, approve, and monitor research involving human participants.[21] Nonbinding authorities, including the Belmont Report and the Nuremberg Code, articulate ethical principles that are likewise designed to protect the autonomy and well-being of human subjects.[22] These authorities do not apply outside the research context, however, and thus are of limited relevance to the overall governance of human enhancement technologies. The application and use of enhancement technologies ultimately could fall within the FDA's jurisdiction over drugs and medical devices.[23] Like its present regulation of drugs and medical devices, FDA oversight under existing authorities would likely focus on the safety and effectiveness of enhancement technologies.[24]

While serious, worries about safety, effectiveness, or even societal effects pale in the face of the ethical dilemmas that human enhancement technologies raise. For many people, human enhancement technologies trigger moral unease or even outrage. Amid such concern, invocation of conventional policy tools such as risk assessment, cost-benefit analysis, and even the precautionary principle seems almost beside the point.[25] The ethical concerns of human enhancement technologies demand close attention and social deliberation.

Although justifications offered for human enhancement sometimes include collective goals such as promoting economic growth,[26] proponents emphasize the potential for human enhancement technologies to further personal autonomy, happiness, and self-fulfillment.[27] The argument, in a nutshell, is that individuals should be free to choose the improved physical, mental, and other capabilities that enhancement technologies will make available. Supporters of enhancement technologies further contend that therapeutic and enhancement applications of a technology will be indistinguishable. Following this line of thought, technology appropriate for correcting widely acknowledged deficiencies should likewise be accepted for enhancing persons who have no such deficiencies.[28] Some commentators even suggest that members of the present generation may someday be obliged to use enhancements for the benefit of future generations, just as parents presently have an obligation to seek the best medical treatment for the health of their children.[29]

This last argument hints at the potential for human enhancement technologies to decrease rather than to increase personal autonomy. In the face of competitive forces and social pressures favoring human enhancement, individuals will not be truly free to reject it.[30] Skeptical of claims of heightened autonomy, opponents assert that human enhancement technologies will impinge on human autonomy, human dignity, and human nature. To skeptics, human enhancement is less likely to lead to a transhumanist paradise than a dystopian Brave New World in which natural reproduction is obsolete and inner peace is available through pill or procedure.[31] Though nominally subject to personal choice, human enhancement decisions may simply further the economic, social, and political agendas of powerful interests. An additional concern is that by reducing the need for effort or struggle, human enhancements arguably would hinder moral development and deprive life experiences of interest or meaning.[32] Critics offer steroid use among athletes as an example of how resulting achievements are "less real, less one's own, [and] less worthy of our admiration."[33] Human enhancements similarly may undermine the worth of human accomplishments and life experiences.

Human enhancement technologies also raise concerns of fairness and equity. Enhanced persons presumably will gain advantages in aspects of life ranging from education to employment.[34] Disparities of access to enhancement technologies will arise not only between the well-to-do and the poor but also between countries that are technologically advanced and those that are not.[35] Differences resulting from enhancement technologies could serve as the basis for discrimination, social stratification, and class conflict.[36] The relative permanence of anticipated enhancements within humans means that resulting inequities are likely to persist and grow over time.

Contemporary thinkers have articulated in various ways their concerns that human enhancement technologies may undermine our basic humanity. Political philosopher Michael Sandel, for example, argues that these technologies "represent a kind of hyperagency—a Promethean aspiration to remake nature, including human nature, to serve our purposes and satisfy our desires."[37] Such an aspiration, he contends, promotes an illusory sense of control and undermines our sympathy and appreciation for others and the world around us—our home and source of sustenance. Such a view stands in contrast to the transhumanists' rejection of "our confinement to the planet earth" and their perception of the natural world as a limitation to be overcome. Political scientist Francis Fukuyama similarly contends that human enhancements put at stake human nature itself, which he views as the basis for our moral sensibilities and the foundation for human rights.[38] Perceiving a resemblance between human enhancement efforts and past eugenics campaigns, other critics question the assumption underlying such pursuits that "undesirable characteristics can be identified definitively and easily eliminated."[39] The concept of enhancement necessarily involves normative judgments that specific characteristics—and people with those characteristics—need enhancing. Even if there existed a consensus regarding the desirability of certain characteristics, there is no assurance that recipients of enhanced characteristics would apply them toward positive ends.[40] Contextual factors influence human personality and experience, and the specific enhancements contemplated by transhumanists are not likely to correlate directly with the ultimate outcomes they might desire.[41] Underlying the enhancement project is ultimately a paradigm in which humans are treated "not as ends in themselves but as means for the production of benefits" for society.[42]

Defenders of human enhancement respond that the development of new technologies to improve ourselves is a basic aspect of human nature.[43] Bioethicist Arthur Caplan contends further that neither individual happiness nor the cultivation of virtue necessarily depends on the struggle or suffering that human enhancement might eliminate.[44] Other observers question the concept of human nature itself, arguing that it is ill defined and has little value as a normative guide. Transhumanist James Hughes asserts that human enhancement technologies ought to be evaluated according to whether they encourage "our capacities for consciousness, feeling, reason, communication, growth and empathy" and discourage "greed, hatred, ignorance, violence, sickness and death."[45] Whether these characteristics are part of human nature, he suggests, is irrelevant. As Hughes's comments reflect, the persuasiveness of the ethical arguments surrounding human enhancement may depend in part on the purposes and effects of enhancement.

General Lessons

There are no easy answers to the difficult ethical questions raised by human enhancement technologies. Nor are there obvious governance arrangements that will ensure a full airing of these questions or adequate oversight of these technologies. As with other emerging technologies, the adoption of a more proactive and deliberative approach by governments, stakeholders, civil society organizations, and the public is essential. Because they raise issues common to emerging technologies in a manner readily grasped, human enhancement technologies can serve as a useful starting point for considering general lessons for managing emerging technologies.

The Importance of Ethics and Values

First, ethical concerns often will be central to debates and policies regarding emerging technologies. The potential health and environmental hazards associated with such technologies unquestionably must be studied and considered. Governments and technology developers have tended to give inadequate attention to the downsides of emerging technologies; to the extent that they have contemplated adverse effects, they generally have focused on physical harms as opposed to ethical concerns. The neglect of ethics results in part from the unquantifiable nature of ethical issues, which the risk assessments and cost-benefit analyses that dominate public policy tend to disregard. American democratic traditions demanding that policy justifications be based on public reason are also to blame. Such demands may lead policy discussions to shy away from ethical concerns that involve contested values.[46]

For many emerging technologies, however, physical harms are only part of the picture. Emerging technologies frequently have critical ethical dimensions as well. Public unease with genetic engineering and especially GM animals, for example, is rooted not only in worries about health effects from consuming GM foods but also in deep moral objections. Those moral objections are reflected in the notion of GM "contamination" and in allegations that genetic engineers are "tinkering with Nature" or "playing God." Moreover, studies consistently find that cultural values play a critical role in informing public perceptions of emerging technologies. Religious beliefs are one factor that strongly influences popular views regarding synthetic biology, with highly religious persons expressing greater anxiety regarding synthetic biology's risks.[47] Similarly, public perceptions of the risks and benefits of nanotechnology depend largely on cultural and political dispositions.[48]

Values affect not only how emerging technologies are perceived but also how emerging technologies are framed. For example, the concept of "converging technologies for human enhancement" is a value-laden expression, not a neutral description of objective scientific inquiry. The phrase assumes a goal—human enhancement—that is surely contested.[49] Furthermore, the phrase treats an uncertain phenomenon, convergence, as inevitable. Technologies do not converge on their own, nor must the NBIC technologies inexorably lead toward human enhancement projects. Just as specific technology policy measures should be open to societal debate and determination, so should the technologies themselves and how they are framed.

The prominent ethical dimensions of emerging technologies underscore two critical needs: increased ethical reflection among scientists, and public participation in technology management. Several steps may be taken to address these needs. With respect to ethical reflection among scientists, the training of graduate students in the sciences should include ethical deliberation and analysis.[50] Because formal ethics instruction alone may have limited effect, the incentives and culture of work and academic settings must also reinforce ethical considerations.[51] Professionalization—the use of professional associations as a tool for governance—offers one option for ingraining values into scientific practice. One proposed oversight strategy for synthetic biology, for example, envisions a professional organization that would control entry to the profession, set ethical standards of practice, and hold members responsible for adherence to those standards.[52] Professionalization potentially offers the benefits of oversight by officially sanctioned bodies and the flexibility associated with codes of conduct. Professionalization ultimately can encourage scientists to "think and act like doctors" in developing an innate sense of moral obligation and fiduciary duty.[53] Professionalization alone, however, does not ensure political accountability or adequate supervision. The values of the professional organization may not reflect the values held more generally by society, and the effectiveness of the organization's efforts may depend on its ability and willingness to enforce sanctions against violators of its standards. Furthermore, in the case of synthetic biology specifically, it may be especially difficult to operationalize governance through a professional organization because researchers come from different disciplinary backgrounds with distinct cultures and codes of conduct.[54]

Ethical deliberation by scientists alone, moreover, will not be enough. Emerging technology decisions must involve the public and reflect public values. Whether emerging technologies ought to be directed toward human enhancement, for example, is not a question that should be left exclusively to scientists, venture capitalists, grant-makers, or military officials. Nor do marketplace decisions by consumers suffice to give the public an adequate voice in ethical discussions on technology, even when consumers are fully informed. The observation that "all customers are human beings, . . . but . . . not all human beings are customers" captures the notion that the publics affected by emerging technologies extend far beyond a technology's direct consumers.[55] Questions of how to govern transformative technologies are for society as a whole to decide through broad and inclusive debate. That debate, facilitated by various TA techniques, should inform players involved in developing and commercializing emerging technologies as well as government officials responsible for funding research and making technology-related policy decisions.

Science and technology studies scholar Sheila Jasanoff uses the term *bioconstitutionalism* to describe how biotechnology and other modern technologies are changing our basic understandings of rights: who holds them, what they protect, and who is responsible for taking care of technologically manipulable life.[56] One theme of Jasanoff's work is that technology, ethics, and law are intertwined more than ever, a point underscored by the examples of synthetic biology and human enhancement. It will not be enough to refer to static constitutional guarantees or natural law concepts to resolve the questions that will confront us. Emerging technologies challenge us to reconsider fundamental social commitments, ethical boundaries, and the entities to which we have ethical obligations, a collective task that requires generous opportunities for meaningful input.

Preserving Decision Space and Enabling Meaningful Participation

Unfortunately, pronouncements regarding the need to consider ethical concerns raised by emerging technologies often assume a predominantly instrumental purpose. Philosopher George Khushf, for example, calls for ethical reflection on converging human enhancement technologies as a means to "remove an intrinsic barrier to the rate of development of science and technology."[57] Indeed, public engagement efforts to date with respect to emerging technologies have tended to serve as a means of facilitating technology acceptance rather than as a means of understanding and addressing public concerns. Outreach by the biotechnology industry, for example, sought to "educate" the public about the benefits of GMOs, not to engage in discourse that might affect how genetic engineering technology would be used. More extensive efforts have been made to engage the public with respect to nanotechnology, but such efforts have primarily advanced the interests of the nanotechnology industry as well as the interests of the social scientists carrying out participatory TA exercises. Thus far, these efforts have had no discernible effect on the course of nanotechnology development or on public awareness of nanotechnology.

If public engagement is to influence how technologies develop and are managed, it must expand beyond publicity campaigns and perfunctory participatory exercises. To enhance the effect of public engagement, public input should be early, ongoing, and solicited through diverse and numerous channels. These channels might include technology referendums, participation in research funding decisions, and expanded participatory TA efforts. Furthermore, engagement processes should be open, allowing participants to voice a wide range of concerns, including social and ethical concerns. The criteria for technology management decisions must incorporate these concerns and not be confined to a narrow weighing of costs and benefits. In other words, when it comes to emerging technologies, we should be open to the possibility of the ethical "no."

Ensuring that decision-making processes truly embrace the possibility of technology rejection is difficult. Rarely has humankind chosen to forgo a technology based on ethical grounds alone. There are innumerable instances where we have replaced obsolete technologies with more advanced ones. And as exemplified by the U.S. ban on DDT and Germany's decision to phase out nuclear power, there are occasions where we have rejected a technology after deploying it and then deciding that the hazards and uncertainties associated with its use are too great.[58] Only rarely, however, have ethical objections alone blocked the development of a technology ex ante.

Those rare instances in which ethical concerns have prompted limitations on research and development tend to involve technologies of war or technologies affecting the integrity of the human body or human life. International restrictions on biological and chemical weapons exemplify the former; limits on stem cell research, reproductive human cloning, and eugenics exemplify the latter.[59] These examples suggest a starting point for identifying technologies that may be most ethically problematic. Many of the technologies on the horizon implicate comparable concerns. Human enhancement technologies call into question physical integrity and human identity, synthetic biology tinkers with life-forming processes, and developments in both areas as well as nanotechnology and even geoengineering could give rise to military applications. These technologies therefore warrant careful ethical deliberation and forbearance.

A simple yet important measure for preserving decision-making space is to affirm the existence of that space.[60] Technology development and dissemination reflect choices by governments, corporate entities, and individuals. Though technological change may be inevitable, the employment of a particular technology is not. Accordingly, the use of rhetoric that rings of technological determinism, such as the phrase *converging technologies for human enhancement*, should be avoided. Official pronouncements that technology decision processes are open to a wide range of outcomes, coupled with active outreach to those holding divergent views, can lay the foundation for open and vigorous debate. As supporters of particular technologies may have little incentive to minimize technologically determinate rhetoric, listeners who encounter this rhetoric should take a critical and questioning approach. Governments and NGOs can cultivate such an approach by promoting informed public awareness of emerging technologies and their ramifications and by highlighting the socially constructed nature of technology. The public must step up as well. Citizens must become informed about emerging technologies and must be willing to express their views in participatory forums and to their leaders.

Another mechanism for creating or preserving decision-making space, particularly in circumstances of uncertainty and limited information, involves the use of firm timetables to consider or reconsider technology policy decisions. Participatory TA meetings and technology referendums should be scheduled to occur periodically, and they should be better linked with policy-making processes. Likewise, laws governing emerging technologies can be slated for review after defined time periods or triggering events. Such an approach can direct legislative and public attention to emerging technologies, encourage reconsideration of existing policies in light of updated information, and thus promote policy experimentation.[61] There is unfortunately no guarantee that adaptive governance will result, however; empirical analyses of sunset laws indicate that it is difficult to compel legislatures to significantly alter laws already passed or to eliminate agencies already established.[62] But requiring periodic review can at least ensure a future hearing and may leave stakeholders open to a wider range of policy options by putting less at stake in any particular decision. For example, a moratorium on further research in a controversial area, combined with a firm commitment to revisit the issue at a fixed time in the future, may prove acceptable under circumstances in which an absolute ban may not. In short, an approach that contemplates future reconsideration of issues may promote ongoing deliberations on difficult ethical matters and ameliorate the tendency of parties with differing views to assume entrenched positions.

The Need for Trustworthy Expertise

A third lesson for emerging technologies highlights the importance of developing trustworthy sources of expertise. Dramatically conflicting claims often surround the prospects of emerging technologies. Technology forecasting often has an imaginative quality that makes it more akin to science fiction than to objective scientific analysis. Proponents may envision material plenty, miracle drugs, and novel abilities as the consequences of a particular technology, whereas opponents may predict dehumanization, environmental degradation, and uncontrollable Frankensteinian creations. These divergent claims were made in the early days of recombinant DNA research, and they are being made today with respect to nanotechnology and synthetic biology. A few of the predictions made by either side may turn out to be reasonably accurate, but most will prove completely wrong. Though technology forecasts frequently are inaccurate, they often have very real consequences. Visions of the future may evolve into research goals and social agendas.[63] Funding for the NBIC initiative, for example, has surely benefited from the ambitious projections of its promoters. Moreover, technology forecasts can serve as the foundation on which actors form expectations and build plans and can influence public deliberations regarding the acceptability of research and development efforts.

Given the significant role of technological predictions, the public and policymakers would benefit from assistance in winnowing through and evaluating those predictions. That assistance should be based on unbiased expertise. Expertise not only must be unbiased, but it also must be *perceived* as unbiased if it is to influence public deliberations in an effective and appropriate manner. This is not to assert that experts should have determinative roles in technology management or that scientific input will eliminate the need for value judgments. Credible scientific expertise is nevertheless essential to the public's evaluation of emerging technologies.[64]

Establishing impartial and trustworthy expertise presents a difficult challenge. Scientific inquiry is increasingly subject to private influence and direction. Private funding, which is responsible for two-thirds of scientific research in the United States, plays a growing role in supporting academic research.[65] Unfortunately, private funding can lead to skewed research, suppressed results, and undisclosed conflicts of interest.[66] The politicization of science further threatens to undermine expert impartiality and authority. With increasing frequency, industry and special interests seek to influence political outcomes and stymie regulation by manipulating the results of scientific inquiry or attacking the scientific record.[67] The 2010 "Climategate" controversy, in which leaked emails of climate scientists were portrayed as evidence of deliberate data manipulation, illustrates the political controversy that can attend scientific findings. Although the scientists in question largely were cleared of wrongdoing and the science behind climate change remains solid, public trust in scientists reportedly suffered as a result.[68]

Notwithstanding these developments, scientists generally continue to command public confidence.[69] This reservoir of confidence can serve as a foundation for establishing credible institutions and processes to analyze claims about emerging technologies. One possible source of scientific and technical expertise could be a reconstituted OTA, as discussed in chapter 1. An OTA designed to respond to Congress's requests, however, would have a relatively restricted ambit and might be perceived as overly political. Expanding the mission of the National Academy of Sciences or establishing a new, independent government agency charged with sifting through conflicting technology claims and disseminating its analysis to the public could offer attractive alternatives.[70] Perhaps the greatest challenge for a new agency of this sort would be to earn public trust and respect for unbiased analyses. Governance of the agency by an independent board with the authority to establish committees to study emerging technologies is one possible means of fostering the agency's political independence. To promote widespread consideration of the analyses produced by the agency, diverse and active public outreach would be necessary.

Credible expertise might also be provided by nongovernmental entities. Public universities, whose core mission includes the generation of knowledge for the betterment of society, must recognize their obligation to contribute to public awareness and debate on emerging technologies. Given the knowledge to which they have access and the respect they generally command, public universities are well situated to serve as a source of trusted expertise. Although declining public financial support threatens to undermine the independence of these institutions, they are finding novel ways to contribute to the public good. For example, a growing number of universities participate in science cafés, public gatherings in which scientists present short lectures and then engage in informal discussion with attendees.[71] Another mechanism for sharing university expertise with the public is a "science shop," in which a university solicits questions from interest groups as potential subjects for research projects.[72] Beyond the initiatives of individual institutions, the creation of a network of independent and nonpartisan organizations, as proposed by technology scholar Richard Sclove, also could complement any governmental technology assessment function.[73] Titled the "Expert and Citizen Assessment of Science and Technology Network," this network of universities, science museums, policy organizations, and other nonprofits could engage citizens on technology issues in broad and extensive ways. Functioning without congressional approval or appropriation, the network would be relatively immune to political whims and political pressure. If such a network coexisted with a government TA agency, moreover, the network could independently verify or critique the agency's analyses. To gain credibility, however, the network's processes and analyses would have to be well executed, nonpartisan, and transparent. Participation of nonprofit organizations in such a network, for example, might leave the network vulnerable to charges of bias. Full disclosure of funding sources and private affiliations would be critical to establishing such a network's credibility, as would participation by well-regarded institutions.

Independent agents with scientific expertise also may arise to help sort through competing claims. These agents might function in a manner akin to FactCheck.org or snopes.com, websites that provide relatively trusted means for the public to determine the accuracy of factual claims. FactCheck.org, a project of the Annenberg Public Policy Center at the University of Pennsylvania, evaluates the accuracy of politicians' claims.[74] Snopes.com, run by two private individuals, investigates the facts behind popular rumors.[75] Of course, the tasks these sites perform are far simpler than technology assessment: They evaluate relatively straightforward factual assertions rather than complex scientific claims and predictions. Nevertheless, the attention and credibility gained by these websites suggests the potential for universities, other organizations, and even individuals to establish themselves as impartial and respected sources of expert information.

A Vigorous Role for Regulators and International Oversight

Experience from numerous technologies—and technological failures—demonstrates that technology matters cannot be left to the free market. Various features of emerging technologies render them especially susceptible to market failure, including information asymmetries between manufacturers and consumers, externalities from technology adoption and use, and uncertainty regarding the consequences of technology. Just as important, while properly functioning markets can generate economically efficient outcomes, they do not account well for equity and other ethical concerns that emerging technologies often raise.

Even strong proponents of nanotechnology, synthetic biology, and other emerging technologies considered in this book generally acknowledge the need for some external oversight, although the desired form and extent of that oversight is subject to disagreement. Taken together, the case studies reveal a systematic tendency to underregulate known risks and to overlook uncertain hazards. The GMO case study illustrates the importance of early involvement as well as the dangers—and temptation—of relying on existing laws. Nanotechnology so far has followed the developmental path taken by GMOs, notwithstanding assurances from nanotechnology companies and developers that things will be different this time. Other emerging technologies may likewise escape adequate oversight unless governments and other stakeholders take prompt action. At the least, regulators must assess whether existing laws can adequately govern an emerging technology's risks and consequences or whether additional statutory authority is needed. That assessment should not be limited to a narrow calculation of quantifiable hazards or even those hazards subject to qualitative description. Rather, that assessment must incorporate public values, the expression of which is likely to vary among technologies and their applications. Beyond assessment, government must set boundaries for permissible action and hold technology developers and companies responsible for the consequences of their actions, whether through direct regulation, bonding schemes, liability regimes, or other means.

Although any regulation of emerging technologies will likely occur primarily at the domestic level, the transboundary nature of modern technologies and their effects demands international oversight as well. For all emerging technologies, this transboundary nature demands at a minimum that greater efforts be made to study potential hazards and to coordinate regulatory standards. For geoengineering and other emerging technologies intended to be global in scope and effect, international deliberation and governance will be essential. Unanimous agreement on the course to be followed is improbable given the values at stake, but careful collective attention can reduce the likelihood of international conflict or ecological disaster.

The ultimate purpose of government technology policy should not be the mere maximization of economic or technological efficiency. Rather, technology policy should encompass protection of all members of the community and promotion of conditions that reflect the visions and values of that community.

New Tools for Dealing with Uncertainty and Rapid Change

A Promethean approach to emerging technologies demands that we develop and utilize new tools for dealing with uncertainty and rapid change in a global context. Current approaches to emerging technologies, whether relying on self-regulation or government oversight, are inadequate. Self-regulation too often equates to nonregulation, and government oversight is frequently overdue, fragmented, and ineffective.

Instituting Futuring Analyses

Particularly pressing is the need for information that would better equip societies to deal with uncertainty about future impacts of emerging technologies. Risk assessments are critical but cannot eliminate uncertainty or surprise even if supported with ample resources. Techniques for analyzing the future, such as horizon scanning and scenario analysis, can complement risk assessment by generating information that will enhance social resilience in the face of uncertainty. Horizon scanning involves the systematic collection of early warning signals of changes in an operating environment.[76] Activities of the Health and Safety Executive (HSE), the British government agency responsible for workplace safety, provide one example. The HSE reviews developments in technology, industry, relevant socioeconomic trends, and other areas for workplace health and safety implications. This ongoing review has led to the identification of synthetic biology, molecular manufacturing, and human performance enhancement, among others, as topics warranting further monitoring.[77] A complement to horizon scanning, the technique of scenario analysis entails the imagining of plausible future scenarios based on an exploration of current conditions, processes driving changes in those conditions, and critical uncertainties and assumptions regarding future developments.[78] While predicting the long-range future of socioecological systems is impossible, scenario analysis can facilitate exploration of policy options and preparation for contingencies by sketching out possible futures and developmental paths that might lead to those futures. Tools such as horizon scanning and scenario analysis cannot eliminate uncertainty but can promote greater preparedness and counter the complacency that risk assessments may foster.

A number of U.S. institutions could carry out futuring analyses specific to emerging technologies. The Council on Environmental Quality, for example, already possesses statutory authority to study emerging technologies and their environmental impacts. However, the CEQ historically has had limited manpower, resources, and executive support. Perhaps a more promising locus for futuring analyses would be regulatory agencies such as the EPA. One proposal would create within these agencies the position of early warning officers, who would provide strategic reconnaissance on emerging phenomena with potentially adverse implications for human health and the environment.[79] Recent developments that might have attracted the attention of such officials and prompted proactive responses include the rapid expansion of fracking activity in various regions of the United States as well as the explosive growth in deepwater offshore drilling. Futuring analyses can directly support agencies' performance of their regulatory functions by identifying potential activities to monitor and study and by informing potential regulatory standards. Making public the reports and findings generated through such analyses would promote public awareness of emerging technology issues and increase the pressure on agency heads and other policymakers to respond appropriately.

Futuring analyses can also be incorporated across the federal government by modifying how agencies carry out their NEPA obligations. Scenario analysis can enrich EISs by facilitating exploration of the risks and sensitivities that may affect future outcomes.[80] Similarly, worst-case analysis of environmental impacts, once mandated under NEPA regulations, could be reinstituted and paired with best-case analysis.[81] Though sometimes neglected, analysis of catastrophic consequences is arguably required by NEPA as long as the probability of those consequences is not insignificant.[82] If incorporated into decision-making processes, futuring analyses would provide agencies with a better appreciation of large-magnitude impacts that might otherwise be ignored because of their low or uncertain probabilities.

Increasing Agency Responsiveness

Fostering institutions capable of responding more nimbly to new developments and information should be another priority. Generally speaking, agency rulemaking processes are slow and difficult to navigate. Agencies must be provided with greater flexibility if they are to keep pace with changing technologies, but accountability and public input should not be sacrificed in the process. Commentators have suggested various modifications to conventional rulemaking that may enhance agency agility. The use of electronic communications to conduct rulemaking, for example, can streamline participatory processes.[83] Furthermore, incorporating a menu of options within a promulgated rule may allow an agency leeway to respond to successive developments without having to institute further rulemaking processes.[84] This latter approach carries with it a greater risk that the agency might abuse its discretion, however, and potentially obscures from the public the policy decisions being made. Indeed, courts may be skeptical of such an approach and deem the agency's selection of a previously promulgated option to be a further rulemaking, thus requiring compliance with the rulemaking provisions of the Administrative Procedure Act.

More drastic changes will be needed to address the lack of coordination among regulators and the generally inadequate authority of agencies to oversee emerging technologies. Consolidating the functions of multiple regulatory agencies could reduce regulatory complexity, overlap, and fragmentation. Political scientist and former EPA official J. Clarence Davies has proposed a merger of the EPA and five other existing regulatory agencies into a new entity, the Department of Environmental and Consumer Protection.[85] In contrast to existing agencies, this new agency would be primarily "a scientific agency with an oversight component." The agency ideally would have an extensive capacity to collect scientific information with the goal of becoming aware of new problems as they arise. This information may also serve as the basis for risk assessments and technology assessments. In addition, the agency would conduct a range of functions currently carried out by its predecessor agencies, including product regulation, pollution control, and health and environmental monitoring.

Although the proposed Department of Environmental and Consumer Protection likely would benefit from various synergies, it also would face substantial challenges. For one, integrating the diverse functions and cultures of different agencies would not be easy, as suggested by difficulties encountered in the creation of the Department of Homeland Security.[86] Davies' proposed consolidation would involve fewer existing agencies than the 22 that were merged into Homeland Security, but the agencies to be merged—including the EPA, U.S. Geological Survey, and Consumer Product Safety Commission—are rather distinct. Moreover, centralization does not guarantee greater effectiveness and may instead create a more sprawling bureaucracy for department leaders to master and reduce potentially beneficial competition between regulatory agencies.[87] Large organizations can be slow to act and reluctant to innovate.[88] Finally, Davies's emphasis on the proposed agency's scientific mission could come at the expense of effective regulation. Scientific research and monitoring would be essential functions, but these activities should be directed at supporting the agency's oversight responsibilities. Legal authorities governing the agency would ultimately have to be crafted carefully to ensure that its scientific functions do not lead to neglect of its regulatory functions.

Bolstering the general oversight authority of regulators is critical but not sufficient. Thanks to the breadth, power, and unpredictability of emerging technologies, we must increasingly be concerned with the possibility of catastrophic risks. Disaster relief measures, tort remedies, and other after-the-fact responses certainly are essential in recovering from catastrophes. The adequacy of these responses is challenged by catastrophic events, however, because of the widespread, severe, and often irreparable harms that result.[89] Further, an overly heavy emphasis on after-the-fact responses fails to acknowledge the role of humans in causing or contributing to many disasters. Accordingly, a statute equipping regulators with the authority to take preemptive action to address potential catastrophic hazards is needed. Unlike existing laws that expand government authority under ongoing emergency conditions,[90] the proposed statute would focus on enhancing the government's ability to address ex ante the catastrophic risks of emerging technologies.

Defining what hazards might qualify as catastrophic would be critical to ensuring proper application of such authority. Factors that Congress might consider or specify in its delegation of authority include the number of people potentially affected, the extent of possible environmental harm, the seriousness of harm at issue, the permanence or irreversibility of any effects, and the human contribution to the hazard. The new authority, which could be delegated to existing regulatory agencies or to a consolidated agency like that envisioned by Davies, should include the ability to study and monitor for catastrophic hazards. The authority should also include substantive power to mandate that appropriate parties assess such hazards and take action to prevent them from occurring. The agency could identify developments of concern on its own or with the assistance of an advisory committee comprised of representatives of various stakeholder interests. The agency then might require parties engaged in developing an emerging technology with catastrophic potential to take actions in response. These actions might include studying catastrophic risks carefully before proceeding with further development, conducting ongoing monitoring for harm, establishing redundant safety mechanisms, or avoiding activities or uses that could be particularly destructive. Because the probability of a catastrophic occurrence may be low or unquantifiable, the threshold required for regulatory action should be minimal: The regulatory agency should be expected only to demonstrate a reasonable basis for finding that catastrophic risk may be present.[91] To guard against the potential abuse of its authority, the agency could be required to revisit its orders periodically and to consider and respond to public comments on the orders as part of its review. In addition, the agency's actions should be subject to judicial review, albeit under a relatively deferential standard equivalent or comparable to the "arbitrary or capricious" standard that courts typically apply to informal agency action.

Encouraging Norms Development

Finally, although self-regulation cannot serve as a substitute for independent and democratically accountable oversight, industry and stakeholder involvement in creating norms for technology development is also important. The ignorance, uncertainty, and rapid change associated with emerging technologies present tremendous challenges for even well-equipped and capable regulators seeking to establish suitable norms and regulatory regimes. Under such circumstances, bottom-up, stakeholder-driven efforts to develop norms can be of some assistance. Examples of such efforts include the Nano Risk Framework, the iGEM competition, and the Asilomar conferences addressing recombinant DNA research and geoengineering. None of these efforts involved significant government or public involvement, however. "Contextualizing regimes" that engage both private and public actors in the iterative development of norms offer a potentially more open and accountable alternative. In these hybrid regimes, associations or other private actors participate in the elaboration and continuing review of norms, subject to government coproduction and oversight. One example of a contextualizing regime is the California Leafy Greens Products Handler Marketing Agreement, a state-supervised, trade association–led effort to respond to the problem of fresh produce contamination by certifying compliance with safety standards and best practices.[92]

Like bottom-up norm development processes and in contrast to ordinary regulation, contextualizing regimes can better facilitate policy experimentation and adaptation. Such regimes, law professors Charles Sabel and William Simon argue, can be especially useful where specific "correct" norms are not yet known, where norms are likely to change, or where specific applications of norms vary widely across a range of contexts—conditions that frequently characterize emerging technologies.[93] Contextualizing regimes can also enable swifter responses to acknowledged problems, particularly where government action is stymied by political resistance, cumbersome administrative processes, or inadequate resources. Compared to purely private initiatives, contextualizing regimes can offer greater transparency and accountability to the extent that they are subject to government oversight and public participation. A government role may be especially important where stakeholders face little incentive to generate voluntary norms or to follow them; under such circumstances, regulators might prompt stakeholder action by encouraging stakeholder meetings, promulgating recommended guidelines, or threatening direct regulation. Contextualizing regimes are akin to other increasingly common collaborative approaches, such as negotiated rulemaking, which conceive of law more as shared-problem solving than as an ordering activity.[94]

Emerging technologies raise new issues and circumstances to which the application of existing norms is often unclear. Accordingly, transparent and participatory processes for developing norms to govern emerging technologies generally should be encouraged. These processes offer additional tools for reducing risks, tackling uncertainty, and engaging stakeholders. However, such processes ultimately lack the accountability and legitimacy of democratic governance. They cannot replace active government oversight, which at a minimum forms the backdrop against which such processes take place. Responsive government ideally does much more than this, of course: It must look out for the people it represents and account for the interests of the underrepresented and future generations.

Concluding Thoughts

Emerging technologies offer humans both promise and peril. There is no shortage of glowing accounts regarding what these technologies might do *for* us; there is a pressing need, however, for closer attention to what these technologies might do *to* us. In the absence of careful societal oversight, technology tends to develop pursuant to narrow visions of interest. This is not to say that scientists or companies engaged in technology development are purely self-interested. To the contrary, they frequently act in accordance with their conceptions of the social good. Nonetheless, such conceptions often neglect vital societal interests and ethical concerns as well as the potential for adverse consequences. The tools and proposals considered in this book represent a more Promethean approach that can help us to avoid past mistakes in technology management. Although we cannot fully resolve the uncertainty that underlies the dilemma of technology control, we can build more robust and participatory institutions that promote more conscious and informed decisions about our technologies and reduce our vulnerabilities to technological surprises.

Conclusion

The belief that technology can solve humanity's problems is popular and deeply entrenched. Such faith in technology is not irrational. Technology has raised living standards, enabled longer lives, and brought numerous other benefits. Further innovation is indispensable to improving human health and combating environmental degradation. Faith in technology, however, should not be blind. A more informed understanding of technology must acknowledge its influence on all aspects of the world around us. Technologies shape not only our natural environment but also social organization and the distribution of economic and political power. The breadth and depth of technology's influence will only increase in the future, as emerging technologies hold unprecedented power to remake the world in positive, negative, and unexpected ways.

The first step in a Promethean approach to emerging technologies is to appreciate technology's pervasive and varied reach. Technologies often have unforeseen effects, and they rarely remain confined to the boxes we might assign to them. We must also acknowledge our inability to wall off science and technology from society's values. Scientific inquiry and technological development are not purely objective pursuits nor do they occur independently of private agendas. Technological development is spurred on by a mixture of the desire for short-term private profit and considerations of long-term public good. Systemic biases are also at work, as professional norms and government funding promote science in the service of innovation and production in preference to science directed at identifying adverse effects.[1]

The second step in a Promethean approach is to reconfigure our laws and policies on emerging technologies. Today's predominant approach to regulating technological hazards is far too narrow. It focuses on harm after the fact, relies too exclusively on technical knowledge and risk assessment, and pays little heed to uncertain hazards. As a result, it has undermined valued practices and put at risk human lives. Indeed, the introduction of technologies without adequate oversight effectively has conscripted each of us into society-wide experiments. This experimentation has occurred without consent, proper controls, or meaningful external review.

This "old way" of technology management will prove especially inadequate in the face of swiftly evolving technologies and uncertain yet potentially catastrophic hazards. As exemplified by the Coordinated Framework governing GMOs and as suggested by the oversight of nanotechnology to date, the old way leans heavily on existing law, which is often ill-equipped to address uncertain risks, let alone new risks. Indeed, reliance on existing law to govern emerging technologies may be worse than having no applicable law at all. A facade of oversight tends to foster complacency and prevent effective reform.

Risk assessment and other elements of the old way do have important roles to play. Technical analyses of emerging technologies should be thorough and ongoing. Because of the potential influence of such assessments and the path-dependent nature of technology development, even early assessments should involve the public. At the same time, we must recognize the limitations of technical assessments and accept the fact that there will always be uncertainty as we make decisions on technology. Uncertainty may reflect the indeterminacy of future events or the inevitable limits of human inquiry. Such uncertainty may reasonably give us pause, whether in moving forward with a technology or in regulating it. We must also be aware, however, that uncertainty can be socially constructed. As climate change skeptics and the tobacco industry have demonstrated, uncertainty can be a potent tool to block or dilute regulation.[2] In light of the incentives powerful economic interests possess to manufacture uncertainty, we should be leery of it as an excuse for regulatory inaction.

For emerging technologies, what we don't know about adverse consequences will frequently overshadow what we do know. Yet it is often not possible to wait for uncertainty to be resolved before acting. Society continuously makes decisions on technology, whether by providing financial support, sitting back as research and development proceed, or imposing regulation. Uncertainty matters, as the precautionary principle reminds us, and must be accounted for rather than ignored. Legal and policy tools for responding to uncertainty, including insurance, environmental assurance bonding, adaptive management, and expanded tort liability, are important to consider but alone are not enough. Enhanced public outreach and engagement are essential to inform choices regarding how to proceed amid substantial uncertainty and to resolve technological choices that are often ethical choices.

Is a Promethean approach to emerging technologies feasible in the face of international competition and constrained resources? Globally uniform regulatory standards are generally lacking, and global consensus on technology management is not likely in the near future. Accordingly, it is reasonable to be concerned that technology development efforts could move to countries with relatively lax regulation. Furthermore, the ongoing economic crisis arguably has increased pressures to weaken regulatory oversight. To policymakers, a Promethean approach simply may appear too ponderous and costly.

Careful reflection on the events that precipitated the crisis, however, underscores the value of a Promethean approach to problems of risk and uncertainty. Specifically, the crash of the U.S. housing market followed years of loose lending and minimal oversight. The ensuing global financial crisis has been attributed to the combination of "high risk, complex financial products; undisclosed conflicts of interest; and the failure of regulators, the credit rating agencies, and the market itself to rein in the excesses of Wall Street."[3] Lenders injected new and greater risks into the U.S. financial system through complex new financial technologies and sloppy lending practices. Credit rating agencies facilitated the securitization of risky loans by issuing inflated ratings. Investment banks spread the risks to the global financial system and magnified them by promoting products based on these loans. Finally, regulators failed to ensure sound lending practices and ignored growing signs of trouble.[4] The moral of the story should be clear: We need stronger oversight as well as greater global cooperation to make that oversight effective.

Signs indicate that world leaders and national governments are beginning to appreciate the need for global cooperation and regulatory harmonization. In response to the financial crisis, for example, the 2008 G-20 Washington Summit issued a declaration regarding common principles for financial market reform.[5] The declaration specifically acknowledges the need for "intensified international cooperation among regulators and strengthening of international standards . . . to protect against adverse cross-border, regional and global developments affecting international financial stability."[6] This conclusion is just one of several insights from the financial crisis that may be generalized to the management of risk and uncertainty. Not surprisingly, these insights have much in common with the themes of this book.

First, catastrophic risk can fester in unregulated and underregulated markets. Private actors face strong incentives to act in their own interests, and their actions can affect entire economic and social systems. Government oversight is necessary to keep interested parties reasonably honest. More important, such oversight also must protect the public interest. Analyzing and addressing long-term, societal, and systemic risks, whether generated by financial markets or by emerging technologies, is essential.

Second, low-probability, high-impact events—so-called black swans—do happen.[7] Economists, investors, and policymakers discounted or disregarded the prospects of a financial meltdown.[8] Although lack of information did play a role, numerous warning signs went unheeded. Perhaps the events they portended seemed improbable. Perhaps key players had a strong interest in ignoring warnings. Or perhaps psychological biases distorted the judgments of those players. Whatever the causes, similar neglect of possible adverse consequences from synthetic biology experiments or geoengineering projects could prove catastrophic. Although black swan events are unlikely, their disproportionate impact demands heightened attention and forbearance.

Third, we ignore uncertainties at our peril. The complex financial instruments that contributed to the financial crisis and their accompanying risks were often poorly understood.[9] The idealized economic models under which many actors were operating compounded that lack of understanding by generating overly optimistic projections and fostering a misleading sense of control.[10] Emerging technologies raise even greater uncertainties than do complex financial instruments. By definition, we have little or no experience with emerging technologies. Accordingly, as in the case of nanotechnology's potential hazards, we often lack reasonable models for estimating their effects. Even when relevant models do exist, the lack of empirical verification for those models cautions against heavy reliance on them. In the end, emerging technologies present inevitable uncertainty we must acknowledge and in some cases regulate.

Fourth, risk can spread and expand rapidly in our interconnected, technology-enabled world. Financial risk was not confined to the housing market, contrary to some people's expectations. Nor was the resulting crisis limited to the United States. The hazards posed by emerging technologies likewise will resist ready containment thanks to the power and global reach of those technologies. Some emerging technologies, such as synthetic biology, will create living things with the ability to replicate and mutate, making them especially unpredictable and difficult to manage. Developing physical, regulatory, and other mechanisms to restrain or avoid hazards is critical.

Finally, people matter and want to be heard. Public activism by both Tea Party supporters and Occupy Wall Street demonstrators counters any notion that the public is disinterested in policy matters. Although these movements have different goals and ideologies, they share a common frustration with the status quo. Distrust of government is at record levels, as is distrust of corporations, the media, and other social institutions.[11] Some skepticism toward powerful institutions is healthy. But widespread distrust of government suggests that it is failing to reflect the popular will and act in the public interest. In this environment, public participation and careful oversight of emerging technologies are hardly luxuries to be abandoned in difficult economic times. Rather, they are necessary elements for advancing democratic goals, achieving social progress, and forestalling or preparing for future crises. Even if the details of gene synthesis or nanomanufacturing lie beyond ordinary understanding, citizens can become sufficiently informed on emerging technologies to hold and express views regarding cost-benefit trade-offs, tolerance for risk, and the ethics of manipulating life.

If mechanisms of public participation are incorporated into the assessment and management of emerging technologies, one might nonetheless question their practical effect. On the one hand, participatory exercises without a clear constituency or audience, such as the National Citizens' Technology Forum, have had experimental value but little measurable impact on policy. On the other hand, the ongoing societal debate in Europe regarding GMOs, fostered through various participatory efforts, has contributed to GMO policies that account for health, environmental, and ethical concerns far more extensively than has been the case in the United States. These contrasting experiences suggest that engagement of the public must be intentional, multipronged, and directed specifically at generating meaningful policy payoffs.

Notwithstanding the polarization of U.S. politics, we can have a civic debate on our technological and environmental future with vigorous public participation. The land use planning process provides an example where active public involvement is widely accepted and sometimes influential. That process may give persons who own property near a proposed project—presumably, the persons most likely to be affected—special protection or participatory privileges.[12] Participation in land use matters is not limited to such persons, however, but is open to the broader public, which is invited into the process through notices or active outreach. The general public typically can access public records, attend meetings at which decisions are made, and participate in public hearings. These participatory avenues reflect a shared understanding that land use decisions may affect the entire community and that the community consequently should have a voice in those decisions. The land use planning process is sometimes cumbersome, and effective public participation may require the assistance of attorneys and experts to navigate procedural and technical obstacles. Nonetheless, there is little doubt that public participation frequently influences land use decisions.[13]

The Promethean approach would expand this participatory norm beyond the land use planning context. Just as land use decisions influence our built environment, technology decisions shape our overall human environment, though often in more subtle, pervasive, and unpredicted ways. Each of the emerging technologies considered in this book has the power to fundamentally reshape our world and our lives. Genetic engineering is altering what we eat, introducing new organisms into the environment, and transforming agricultural systems and their associated economies. Nanotechnology's eventual impacts may be greater and more widespread, as it generates a myriad of new materials, products, and processes—as well as potential new sources of pollution. Geoengineering is perhaps the most obvious example of a technology directly aimed at changing our environment. And synthetic biology promises more radical powers of life creation than conventional genetic engineering. The sweeping changes that emerging technologies entail demand more careful oversight and public involvement. We owe the environment—and each other—no less.

Notes

Preface

1. See Steven E. Barkan, *Law and Society: An Introduction* (Upper Saddle River, NJ: Pearson Prentice Hall, 2008), 23–24.

2. See Lawrence M. Friedman, *American Law* (New York: Norton, 1984), 4–11; William N. Eskridge Jr., and Gary Peller, "The New Public Law Movement: Moderation as a Postmodern Cultural Form," *Michigan Law Review* 89 (1991): 718 ("For legal process, . . . enactment of a statute was merely the beginning, and not the end, of lawmaking.").

3. Henry M. Hart Jr. and Albert M. Sacks, *The Legal Process: Basic Problems in the Making and Application of Law* (Westbury, NY: Foundation, 1994), 148.

4. Friedman, *American Law*, 10.

5. See Jasanoff, "Introduction."

Introduction

1. See Westrum, *Technologies and Society*, 7 (defining *technology* as "those material objects, techniques, and knowledge that allow human beings to transform and control the inanimate world"); Pitt, *Thinking about Technology*, 30–31 (linking technology to work, defined as "the deliberate design and manufacture of the means to manipulate the environment to meet humanity's changing needs and goals"); Schnaiberg, *Environment*, 115 ("technology, by definition, always involves some processes of ecosystem withdrawals and additions").

2. See Kline, "What Is Technology," 211 (describing usage of the term technology to describe a system that "extend[s] human capacities").

3. Allenby, *Reconstructing Earth*, 9.

4. See Uppenbrink, "Arrhenius and Global Warming," 1122. See also Fleming, *Historical Perspectives*.

5. For a history of the development and introduction of DDT, see Russell, "Strange Career," 770.

6. EPA banned DDT in 1972 and classifies DDT as a probable human carcinogen based on animal testing. See "Integrated Risk Information System: p,p'-

Dichlorodiphenyltrichloroethane (DDT) (CASRN 50-29-3)," EPA, http://www.epa.gov/ncea/iris/subst/0147.htm. The Agency for Toxic Substances and Disease Registry describes DDT as a neurotoxicant and potential endocrine disruptor (U.S. Department of Health and Human Services, *Toxicological Profile for DDT, DDE, and DDD* [2002], http://www.atsdr.cdc.gov/toxprofiles/tp35.pdf). Nonetheless, the human health effects of DDT exposure remain subject to some debate. See, e.g., Smith, "How Toxic Is DDT?," 268 ("DDT can cause many toxicological effects but the effects on human beings at likely exposure levels seem to be very slight."); David J. Hunter et al., "Plasma Organochlorine Levels," 1256 (observing no evidence of increased risk of breast cancer among women with relatively high levels of DDE, the principal metabolite of DDT).

7. Marla Cone, "DDT Use Should Be Curtailed, Left Only as a 'Last Resort' in Some Malaria-Plagued Areas, Scientists Say," *Environmental Health News*, May 4, 2009, http://www.environmentalhealthnews.org/ehs/news/ddt-only-as-last-resort.

8. Glendenning, "Notes," 603–4.

9. See Pitt, *Thinking about Technology*, xi ("[T]here is no single thing called 'Technology.'").

10. Carol Dougherty, *Prometheus* (New York: Routledge, 2006), 3; Thomas Bulfinch, *Mythology: The Age of Fable* (Garden City, NY: Doubleday, 1968), 14.

11. Dougherty, *Prometheus*, 3.

12. Aeschylus, *Prometheus Bound*, trans. Henry Howard Molyneux (London: Murray, 1892), 8–9.

13. Dougherty, *Prometheus*, 19.

14. Ibid., 4. Although the ancient Greeks understood *Prometheus* to mean "forethinker," an alternative etymology links the *meth* component of the name to the Sanskrit root *math*, meaning "to steal."

15. Ibid., 42–43; Hesiod, "Works and Days," in *Theogony; Works and Days; Shield*, trans. Apostolos N. Athanassakis, 2nd ed. (Baltimore: Johns Hopkins University Press, 2004), lines 81–96.

16. Schnaiberg, *Environment*, 138.

17. See, e.g., Grushcow, "Measuring Secrecy," 61–62.

18. Jasanoff, *Fifth Branch*, 77 (contrasting research science with regulatory science, which "is rarely innovative and may never be submitted to the discipline of peer review and publication").

19. See Weber, "Experience-Based and Description-Based Perceptions," 109 (describing phenomenon of hyperbolic discounting); Lazarus, "Environmental Law after Katrina," 24.

20. See EPA, *Life Cycle Assessment: Principles and Practice*, 2006, http://www.epa.gov/nrmrl/lcaccess/pdfs/600r06060.pdf.

21. See Albert C. Lin, "Unifying Role," 898–99.

22. World Commission, *Precautionary Principle*, 14.

23. See, e.g., Scheuerman, "Constitutionalism," 353 (quoting John Dewey's observations regarding challenges posed to liberalism by rapid technological and social changes).

24. Cf. Pitt, *Thinking about Technology*, 87–99 (contending that claims for the autonomy of technology "remov[e] the responsibility from human shoulders for the way in which we make our way around in the world").

25. According to the National Science Foundation, industry was the source of approximately two-thirds of R&D expenditures in 2007. See National Science Foundation, *National Patterns of R&D Resources: 2007 Data Update*, 2008, Table 5, NSF 08-318, http://www.nsf.gov/statistics/nsf08318/.

26. See Bryner, "Science, Technology, and Policy Analysis," 9; Schnaiberg, *Environment*, 129.

27. See Bryner, "Economic Policy," 181.

28. Davies, *Nanotechnology Oversight*, 7.

29. In *Collapse*, Jared Diamond has documented the collapse of ancient societies as a result of their exceeding their ecological limits on a relatively small scale. Turning his attention to the present, Diamond warns that "today's larger population and more potent destructive technology, and today's interconnectedness pos[e] the risk of a global rather than a local collapse" (521).

30. See, e.g., Intergovernmental Panel, *Fourth Assessment Report*, 8.

31. Barnosky et al., "Has the Earth's Sixth Mass Extinction Already Arrived?," 51.

32. See Intergovernmental Panel, *Climate Change 2007: The Physical Science Basis*, 96–97

33. T. J. Blasing, "Carbon Dioxide Information Analysis Center, Recent Greenhouse Gas Concentrations," http://cdiac.ornl.gov/pns/current_ghg.html.

34. Intergovernmental Panel, *Fourth Assessment Report*, 2–4, 7–11; Intergovernmental Panel on Climate Change, *Climate Change 2007: Impacts, Adaptation, and Vulnerability*, 11–18; CNA Corporation, *National Security*, 10; Stern, *Economics of Climate Change*, 37; California Climate Change Center, *Our Changing Climate;* Jeffrey Kluger, "By Any Measure, Earth Is at the Tipping Point," *Time*, April 3, 2006, 30.

35. See Speth, "Creating a Sustainable Future," 11–13; Rockström et al., "Safe Operating Space," 472 (proposing boundaries of unacceptable environmental change for nine Earth processes).

36. See World Wildlife Fund, *Living Planet Report*, 2.

37. See Vivienne Walt, "The World's Growing Food-Price Crisis," *Time*, February 27, 2008, http://www.time.com/time/world/article/0,8599,1717572,00.html; Lester Brown, "Eradicating Hunger," 43, 46; Julian Borger, "Feed the World? We Are Fighting a Losing Battle, UN Admits," *The Guardian*, February 26, 2008, International sec., 18. Per capita grain production has declined since 1984, and there were an estimated 854 million undernourished people worldwide in 2006 (United Nations Food and Agriculture Organization, *State of Food Insecurity*, 8).

38. Worm et al., "Rebuilding Global Fisheries," 578, 584; Worm et al., "Impacts of Biodiversity Loss," 787–90.

39. United Nations Environment Programme, *Global Environmental Outlook*, 129. See also Postel, *Last Oasis.*

40. See Hunter et al., *International Environmental Law and Policy*, 74–75.

41. Most famously, Thomas Malthus postulated that "population, when unchecked, increase[s] in a geometrical ratio, and subsistence for man in an arithmetical ratio" (*Essay*, 21). Although subsequent advances in agricultural and industrial technology undermined Malthus's assumption that food production could only increase arithmetically, those advances have been detrimental to our resource base

and to the environment as a whole. See Philip Appleman, introduction to Malthus, *Essay*, xviii. For more recent views echoing Malthus, see, e.g., Ehrlich, *Population Bomb*, 44 ("There is not enough food today. . . . If the pessimists are correct, massive famines will occur soon."); Meadows et al., *Limits to Growth*, 22 (predicting "sudden and uncontrollable decline in both population and industrial capacity").

42. Diamond, *Collapse*, 505 ("The rapid advances in technology during the 20th century have been creating difficult new problems faster than they have been solving old problems.").

43. Worldwatch Institute, *State of the World*, 5–6.

44. Jared Diamond, "What's Your Consumption Factor?," *New York Times*, January 2, 2008, A17.

45. See Adler, "Are We on Our Way," 26 ("Most futurologists say technology is developing at exponential rates."); Smart, "Discussion," 993 ("technological capacity and technological innovation have always accelerated since the birth of human civilization"); Modis, "Forecasting," 378 ("The accelerating amount of change in technology, medicine, information exchange, and other social aspects of our life, is familiar to everyone."); Linstone, "Technological Slowdown," 195. Technological acceleration is one aspect of the more general phenomenon of social acceleration, which encompasses faster social change and a speedier pace of life in addition to technological acceleration. See Rosa, "Social Acceleration," 82.

46. See Rosa, "Social Acceleration," 82; Davies, *Oversight*, 15; Rejeski, "Public Policy," 48. Whether the pace of technological development is accelerating is subject to some debate. A leading proponent of technological change, Ray Kurzweil, points to growth in data storage capacity, increases in the number of patents, and other indices to argue that a law of accelerating returns applies to technology. Indeed, Kurzweil contends that technological changes are progressing toward a "singularity"—a point in the future when technological advances begin to happen so rapidly that normal humans will be unable to keep pace with artificial intelligence and cybernetically augmented humans. See Kurzweil, *Singularity Is Near*, ch. 2. Other observers, however, conclude that innovation rates have peaked and are now decreasing. See Huebner, "Possible Declining Trend," 980–81 (citing calculations of per capita patenting frequency and "important events in the history of science and technology"); Ayres, "Turning Point," 1188 (focusing on relative lack of change in "ordinary everyday activities"); Modis, "Discussion," 987.

47. Roco and Bainbridge, *Converging Technologies*, 1.

48. Davies, *Oversight*, 15.

49. Guston and Sarewitz, "Real-Time Technology Assessment," 96.

50. Guston, "Insights," 78–80; Guston and Sarewitz, "Real-Time Technology Assessment," 93; Genus, "Rethinking Constructive Technology Assessment," 13.

51. John P. Holdren, Cass R. Sunstein, and Islam A. Siddiqui, *Principles for Regulation and Oversight of Emerging Technologies*, March 11, 2011.

Chapter 1

1. See, e.g., Baumol, *Free-Market Innovation Machine*.

2. See Bijker, *Of Bicycles, Bakelites, and Bulbs*, 3–4; Andy Stirling, "'Opening Up'

and 'Closing Down': Power, Participation, and Pluralism in the Social Appraisal of Technology," *Science, Technology, and Human Values* 33 (2008): 263.

3. See Collingridge, *Social Control*, 19 (discussing "the dilemma of control" of technology).

4. See Albert C. Lin, "Deciphering the Chemical Soup," 958–71.

5. See Tomain and Burton, "Nuclear Transition," 363–66; Yellin, "High Technology and the Courts," 498.

6. See Katie Howell and Patrick Reis, "Interior Issues New Offshore Drilling Rules, Holds Firm on Moratorium," *New York Times*, September 30, 2010, http://www.nytimes.com/gwire/2010/09/30/30greenwire-interior-issues-new-offshore-drilling-rules-ho-82704.html?pagewanted=1.

7. See Tribe, "Technology Assessment," 621–22; Westrum, *Technologies and Society*, 325.

8. See Vig and Paschen, "Technology Assessment," 8.

9. Kysar, "Climate Change," 8–9 n. 42. See also Tribe, "Technology Assessment," 622.

10. See Mark B. Brown, *Science in Democracy*, 86 (discussing approaches "in which experts determine the means of politics and citizens choose the ends" or in which "a parallel distinction [is made] between technical knowledge and political judgment").

11. Collingridge, *Social Control*, 19; Winner, *Autonomous Technology*.

12. See Frank Fischer, "Are Scientists Irrational?," 54. In addition to risk assessment, technology assessment can also involve other types of technical analyses such as mathematical modeling, technological forecasting, and scenario building. See Westrum, *Technologies and Society*, 328–29.

13. See Doremus et al., *Environmental Policy Law*, 395–98.

14. See Woodhouse, "Toward More Usable Technology Policy Analyses," 16; National Research Council, *Science and Judgment*, 165–66 (criticizing overreliance on "artificially precise single estimates of risk").

15. National Research Council, *Science and Judgment*, 165 (discussing sources of uncertainty); Albert C. Lin, "Unifying Role," 968–69.

16. Elizabeth Fisher, Pascual, and Wagner, "Understanding Environmental Models," 271–72; Valve and Kauppila, "Enacting Closure," 353.

17. See Beck, *Risk Society*, 71 (criticizing assumption "that so long as risks are not recognized scientifically, *they do not exist*—at least not legally, medically, technologically, or socially, and they are thus not prevented, treated or compensated for"); Jasanoff, *Designs on Nature*, 265 ("In the United States, a preferred method for displaying objectivity in public decisions has been to clothe the reasons for allocative choices as far as possible in the language of numbers.").

18. Jeroen C. J. M. van den Bergh, "Optimal Climate Policy Is a Utopia: From Quantitative to Qualitative Cost-Benefit Analysis," *Ecological Economics* 48 (2004): 385–93; Farber, "Uncertainty," 901.

19. See Mary Douglas and Wildavsky, *Risk and Culture*, 7–9; Frank Fischer, "Are Scientists Irrational?," 55; Plough and Krimsky, "Emergence," 8–9.

20. Sarewitz, "Anticipatory Governance," 98.

21. See Stirling, "Opening Up or Closing Down?," 218, 220–23.

22. See ibid., 220; Wickson, Delgado, and Kjølberg, "Who or What Is 'The Public'?," 757.

23. Stirling, "Opening Up or Closing Down?," 220. See also Habermas, *Postnational Constellation,* 62–69, 76, 110–18 (contrasting liberal and conservative notions of democratic politics and the effects of globalization and providing normative justifications for public participation in decision making).

24. Shrader-Frechette, "Evaluating the Expertise of Experts," 117 (arguing for the right of public participation in risk assessments because they have consequences for public welfare). As Stirling has pointed out, the narrow involvement and "opaque technical procedures associated with expert analysis . . . conflict with Habermasian principles of 'ideal speech' [and] with Rawlsian notions of 'public reason'" ("Opening Up or Closing Down?," 221). See also Habermas, *Theory of Communicative Action,* 1:25; Rawls, "Idea of Public Reason Revisited," 765; Locke, *Two Treatises of Government,* 374–81 (discussing how assertion of negative rights brings with it protection from government intervention and an opportunity for individuals to assert private interests).

25. See Habermas, *Theory of Communicative Action,* 2:86, 98–101, 363–73 (discussing an "ideal speech situation" in which every competent actor has a right to participate in societal discourse); Rawls, *Political Liberalism,* 7, 9, 214–20 (propounding "public reason" as an ideal mode of deliberation for issues of public concern in a pluralist society and contending that an "overlapping consensus" can be reached through recognition of a core set of substantive moral principles common to a "reasonable" fragment of society); Habermas, *Legitimation Crisis,* 17–24, 95–111 (advocating that a universal pragmatism requiring preacknowledgment among participants that the existence of an ideal speech situation is possible and that participants engaged in competent discourse can reach a consensus that is representative of the general will of the public).

26. Dewey, *Public and Its Problems,* 15–16. See also Mark B. Brown, *Science in Democracy,* 141.

27. Mark B. Brown, *Science in Democracy,* 141–43.

28. Shrader-Frechette, "Evaluating the Expertise of Experts," 116; Stirling, "Opening Up or Closing Down?," 275. See also Beck, *Risk Society,* 29 ("[O]ne must assume an *ethical point of view* in order to discuss risks meaningfully at all."); Slovic, "Risk Game," 19–22. In addition, chemical risk assessments generally rest on the dubious assumption that people are exposed only to the chemical being tested and not also to other chemicals that may have synergistic effects (Jasanoff, "Technologies of Humility," 239).

29. Shrader-Frechette, "Evaluating the Expertise of Experts," 117. See also Stirling, "Opening Up or Closing Down?," 224 (noting that "probability theory underlies the entire activity of risk assessment, yet its applicability is seriously constrained by recognition of intractable states of uncertainty, indeterminacy and ignorance"); Klüver et al., *European Participatory Technology Assessment,* 173; Frank Fischer, "Are Scientists Irrational?," 59 ("As the move to basic cultural orientations is in significant part a response to the fact that science cannot supply the needed answers, it is thus anything but irrational."); Collingridge, *Social Control,* 16–18 (discussing the broad and unexpected consequences of popularization of the automobile).

30. See Harry Collins and Evans, *Rethinking Expertise,* 19–35 (distinguishing popular understandings of science from knowledge and expertise of specialists).

31. See Stirling, "Opening Up or Closing Down?," 222; Harry Collins and Ev-

ans, *Rethinking Expertise*, 48–49, 136–37; Mark B. Brown, *Science in Democracy*, 233–34; Kleinman, "Beyond the Science Wars," 139 (discussing examples of informed participation by laypersons in scientific and technical dialogues).

32. See Habermas, *Theory of Communicative Action*, 1:25; Leach and Scoones, "Science and Citizenship," 24.

33. See Mark B. Brown, *Science in Democracy*, 78–85; James Madison, *The Federalist No. 63*, in *Federalist Papers*, ed. Shapiro, 320 (noting "how salutary will be the interference of some temperate and respectable body of citizens, in order to check" the potential for misguided and rash decisions by the general public).

34. See Morgan and Peha, "Analysis, Governance, and the Need," 3, 11; Rodemeyer, "Back to the Future," 3; Bryner, "Science, Technology, and Policy Analysis," 6; Plein and Webber, "Role of Technology Assessment," 147.

35. See Albert C. Lin, "Unifying Role," 898–99.

36. Farina and Rachlinski, "Foreword," 268.

37. See Schnaiberg, *Environment*, 131 (contending that "[i]n the absence of a public sector debate over 'science policy,' piecemeal decisions continuously supported an exponential increase in energy-intensive production" and chemical production in the United States).

38. See Frank Fischer, "Are Scientists Irrational?," 54; Wynne, "Risk and Environment," 460 ("Risk has become the form of public discourse through which public meaning is given to technology and innovation.").

39. See Doremus et al., *Environmental Policy Law*, 395–99.

40. See Leach and Scoones, "Science and Citizenship," 15, 22 ("Liberal understandings of citizenship . . . hold faith in the modern state's expertise [and] defer decisions to elected elites, who historically have been highly reliant on accredited scientific and technocratic expertise.").

41. See Wynne, "Risk and Environment," 468–69; Jasanoff, "Technologies of Humility," 239. See also Fischhoff, "Public Values," 77 (describing origins of risk research in industry efforts to manage internal affairs and suggesting that risk analysis continues to address "the problems and [speak] the language of industry").

42. National Research Council, *Understanding Risk*, 23–24; Slovic, "Risk Game," 23.

43. Andrews, *Humble Analysis*, 48.

44. See Margolis and Guston, "Origins, Accomplishments, and Demise," 66; Paul Recer, "Office of Technology Assessment Killed," *St. Paul* (MN) *Pioneer Press*, September 27, 1995, 2C.

45. 2 U.S.C. § 471(a) (2006). See Margolis and Guston, "Origins, Accomplishments, and Demise," 54–57.

46. 2 U.S.C. § 472(c) (2006).

47. See Margolis and Guston, "Origins, Accomplishments, and Demise," 59; Vig and Paschen, "Technology Assessment," 9.

48. Andrews, *Humble Analysis*, 181; Carson, "Process, Prescience, and Pragmatism," 249; Sarewitz, "This Won't Hurt a Bit"; Sclove, *Reinventing Technology Assessment*, 10–11.

49. See Margolis and Guston, "Origins, Accomplishments, and Demise," 61–62, 71–72; Vig and Paschen, "Technology Assessment," 9–10.

50. See Margolis and Guston, "Origins, Accomplishments, and Demise," 70–71.

Congress eliminated funding for the OTA but did not formally abolish the agency (Legislative Branch Appropriations Act, 1996, Pub. L. No. 104-53, 109 Stat. 468 [1995]).

51. Peha, "Science and Technology Advice," 20–21.

52. Knezo, *Technology Assessment*, 4–6. See also, e.g., GAO, *Technology Assessment: Cybersecurity for Critical Infrastructure Protection* (2004), http://www.gao.gov/new.items/d04321.pdf; GAO, *Technology Assessment: Using Biometrics for Border Security* (2002), http://www.gao.gov/new.items/d03174.pdf; GAO, *Technology Assessment: Climate Engineering: Technical Status, Future Directions, and Potential Responses* (2011), http://www.gao.gov/new.items/d1171.pdf.

53. See Hill, "Expanded Analytical Capability," 115–16; Francesca T. Grifo, written testimony, in *Hearing on 2011 Appropriations before the Subcommittee on Legislative Branch Appropriations of the House Committee on Appropriations*, 111th Cong. (2010), 4, http://www.ucsusa.org/scientific_integrity/solutions/big_picture_solutions/restoring-the-ota.html.

54. 42 U.S.C. § 6614(a)(5) (2006). OSTP's statutory mission is to serve "as a source of scientific and technological analysis and judgment for the President with respect to major policies, plans, and programs of the Federal Government" (§ 6614((a)).

55. Stine, *President's Office*, 5–7, 24–25. OSTP receives external advice from the President's Council of Advisors on Science and Technology (PCAST), whose members are selected from industry, education, research, and other nongovernmental institutions. PCAST conducts workshops and convenes technical advisory groups that could carry out technology assessment functions (9, 24–25).

56. See Stine, *Science and Technology Policymaking*, 29; Rodemeyer, "Back to the Future," 3–4; Ahearne and Blair, "Expanded Use," 118–23. The NRC is the operating arm of the National Academies, which include the National Academy of Sciences, a private, nonprofit organization established by congressional charter (118).

57. Sclove, *Reinventing Technology Assessment*, 19; Rodemeyer, "Back to the Future," 4. In addition, the National Academy of Sciences, and thus the NRC, can reject congressional requests (*Hearing on 2011 Appropriations*, 4).

58. 42 U.S.C. § 4332(2)(C) (2006).

59. Ibid.; 40 C.F.R. § 1506.6 (2010).

60. The 2009 federal budget, for example, provided $1.5 billion to support nanotechnology research and development through 13 federal agencies (Office of the U.S. President, National Nanotechnology Initiative, *FY 2009 Budget and Highlights* [2008], http://www.nano.gov/NNI_FY09_budget_summary.pdf.

61. Mandelker, *NEPA Law and Litigation*, § 1:1. See also 40 C.F.R. § 1500.1(a) (2010) (characterizing NEPA as "our basic national charter for protection of the environment").

62. 42 U.S.C. § 4332(2)(C).

63. *Robertson v. Methow Valley Citizens Council*, 490 U.S. 332, 349 (1989).

64. 40 C.F.R. §§ 1501.4, 1508.9 (2010).

65. Lindstrom and Smith, *National Environmental Policy Act*, 8. Senator Henry Jackson, NEPA's main sponsor, touted NEPA as "the most important and far-reaching environmental and conservation measure ever enacted by the Congress" (115 Cong. Rec. 40,416 [1969]).

66. 42 U.S.C. § 4332(2)(C) (2006). Although the statute requires an EIS for agency "proposals for legislation," NEPA regulations define legislation narrowly to include only legislative proposals "developed by or with the significant cooperation and support of a Federal agency" and to exclude request for appropriations (40 C.F.R. § 1508.17 [2010]).

67. *Robertson*, 490 U.S. at 353 (noting that NEPA relies "on procedural mechanisms—as opposed to substantive, result-based standards"); *Vt. Yankee Nuclear Power Corp. v. Natural Res. Def. Council, Inc.*, 435 U.S. 519, 558 (1978) (stating that NEPA's "mandate to the agencies is essentially procedural"). See also Binder, "NEPA, NIMBYs, and New Technology," 11 (arguing that "those seeking to obstruct the application of new technology and processes [through NEPA] will have to find a new legal tool to rely on" because agencies need only make requisite disclosures).

68. Council on Environmental Quality, *National Environmental Policy Act*, iii ("Some [agencies] act as if the [EIS] is an end in itself, rather than a tool to enhance and improve decision-making."); Karkkainen, "Toward a Smarter NEPA," 922–23 (contending that NEPA documentation is often "overstuff[ed] . . . with information from every available source, regardless of its quality" but that little evidence suggests that the information discussed in such documentation actually influences agency decision making).

69. *Aberdeen & Rockfish R.R. v. Students Challenging Regulatory Agency Procedures*, 422 U.S. 289, 320–22 (1975).

70. *Kleppe v. Sierra Club*, 427 U.S. 390, 406 (1976). As various critics have pointed out, the preparation of environmental documentation in practice often remains separated from much of the decision-making process. See Farber, "Adaptation Planning," 10,609; Houck, "How'd We Get Divorced?," 10,648–49.

71. *Vt. Yankee Nuclear Power*, 435 U.S. 519.

72. Ibid., 557–58.

73. Cooper, "Broad Programmatic, Policy, and Planning Assessments," 117–18.

74. Houck, "How'd We Get Divorced?," 10,648.

75. See Erik Fisher and Miller, "Collaborative Practices," 376; Newberry, "Are Engineers Instrumentalists?" 109, 112–13. See also Holbrook, "Assessing the Science-Society Relation," 438 (describing institutionalization of Vannevar Bush's views regarding the degree of autonomy necessary for basic scientific research).

76. See Jasanoff, *Designs on Nature*, 235–36 (describing effect of industrial funding of university research and of the Bayh-Dole Act, which allowed recipients of federal grants to patent discoveries funded by federal money).

77. See Esty and Winston, *Green to Gold*, 12.

78. The concept of corporate social responsibility (CSR), broadly defined, refers to socially beneficial decisions and actions by firms that go beyond the legal minimum (Portney, "Corporate Social Responsibility," 108). A narrower definition of CSR focuses on profit-sacrificing behavior by businesses acting under moral or social obligations ("Summary of Discussion on Corporate Social Responsibility and Economics," in *Environmental Protection and the Social Responsibility of Firms: Perspectives from Law, Economics, and Business*, ed. Bruce L. Hay, Robert N. Stavins, and Richard H. K. Vietor [Washington, DC: Resources for the Future, 2005], 145, 146). Whether such activity—which is arguably contrary to the purpose of private corporations and to the interests of shareholders—exists at a meaningful level is

a matter of debate. See "Summary of Discussion," 146; Portney, "Corporate Social Responsibility," 126 (contending that almost all examples of CSR are profit-motivated).

79. Esty, "On Portney's Complaint," 141–42; Reinhardt, "Environmental Protection," 159–68. In addition, banks and insurance companies often have an interest in overseeing the risk exposure of their clients. See Esty and Winston, *Green to Gold,* 11, 94–95.

80. David J. Vogel, "Opportunities for and Limitations of Corporate Environmentalism," 199 (contending that as an empirical matter, self-regulation is inadequate to address adverse effects and that government regulation is responsible for almost all improvements in environmental quality).

81. See Knezo, *Technology Assessment,* 2–4; Chubin, "Filling the Policy Vacuum," 31; Epstein and Carter, "Dedicated Organization," 157; Sclove, *Reinventing Technology Assessment,* 18.

82. Rodemeyer, "Back to the Future," 4–5.

83. Numerous critics share this view. See, e.g., Farber, "Adaptation Planning," 10,605; Karkkainen, "Whither NEPA?," 333; Mandelker, "Thoughts," 10,640.

84. 42 U.S.C. § 4332(2)(C) (2006).

85. Ibid., §§ 4321, 4331(a) (2006).

86. On the House side, see, for example, 115 Cong. Rec. 26,569, 26,577 (1969) (statement of Rep. Farbstein) ("For too long, we have stressed technological progress, assuming that our environment could take care of itself."), 26,583 (statement of Rep. Donohue) ("[T]he advance of modern technology, however great its material benefits, has been unrestrained in its accompanying afflictions upon us through byproducts that increasingly poison our air and pollute our waters."), (statement of Rep. Cohelan) ("We are fast becoming a victim of our own technology and progress."). On the Senate side, see, for example, 115 Cong. Reg. 40,415, 40,417 (1969) (statement of Sen. Jackson) ("While the National Environmental Policy Act of 1969 is not a panacea, it is a starting point. A great deal more, however, remains to be done . . . if mankind and human dignity are not to be ground down in the years ahead by the expansive and impersonal technology modern science has created.").

87. H.R. Rep. No. 91-378, at 3 (1969).

88. S. Rep. No. 91-296, at 6 (1969); see also 8 (identifying population growth and "advancing technological developments which have enlarged man's capacity to effectuate environmental change" as primary causes for environmental concern).

89. 115 Cong. Rec. 29,069 (1969) (reprinting *A National Policy for the Environment: A Report on the Need for a National Policy for the Environment, Special Report to the Senate Committee on Interior and Insular Affairs,* 90th Cong. [1968]).

90. See Caldwell, *National Environmental Policy Act,* xvi ("[T]he procedural requirements of NEPA are intended to force attention to the policies declared in the Statement of Purpose . . . and in Title I (Section 101) of the Act.").

91. *Scientists' Inst. for Pub. Info., Inc. v. Atomic Energy Comm'n,* 481 F.2d 1079 (D.C. Cir. 1973).

92. Ibid., 1089.

93. Ibid., 1090.

94. Post-*SIPI* decisions have declined to mandate preparation of programmatic impact statements for groups of research projects that are similar but unrelated,

though courts continue to require environmental analyses of individual projects. See, e.g., *Found. on Econ. Trends v. Lyng*, 817 F.2d 882, 884–85 (D.C. Cir. 1987) (no programmatic EIS required for "diverse and discrete" animal productivity research projects); *Foundation on Econ. Trends v. Heckler*, 756 F.2d 143 (D.C. Cir. 1985) (vacating lower court injunction against GMO deliberate-release experiments in absence of a programmatic EIS).

95. 40 C.F.R. § 1508.18(a) (2010). One category of major federal actions identified in the regulation includes "[a]doption of programs, such as a group of concerted actions to implement a specific policy or plan" (§ 1508.18(b)(3)).

96. Ibid., § 1502.4.

97. 42 U.S.C. § 4331(a) (2006) (emphasis added).

98. *Strycker's Bay Neighborhood Council, Inc. v. Karlen*, 444 U.S. 223, 227–28 (1980) (stressing that NEPA establishes substantive goals but imposes only procedural duties); Caldwell, *National Environmental Policy Act*, xvi ("NEPA, as policy, is a 'template' against which decisions affecting the environment can be compared for consistency with its declared principles."); Ferester, "Revitalizing the National Environmental Policy Act," 211–17 (discussing potential substantive mandates under NEPA).

99. 42 U.S.C. § 4344 (2006). See also *National Environmental Policy: Hearing on S. 1075, S. 237, and S. 1752 before the Senate Committee on Interior and Insular Affairs*, 91st Cong. (1969) (statement of Sen. Jackson upon introducing NEPA legislation), available at 1970 NEPA Leg. Hist. 30 (LEXIS) ("The Council would . . . help the President evaluate the trends of new technologies and developments as they affect our total surroundings, and to develop broad policies, including those related to anticipatory research, to prevent future man-induced environmental changes which could have serious social and economic consequences.").

100. See, e.g., 115 Cong. Rec. 26,569, 26,575 (1969) (statement of Rep. Schadeberg) ("If such a council existed at the time of the invention of the automobile, perhaps we would have been able to realize the threat that would be presented to our atmosphere by the internal combustion of hydrocarbons before it was too late."), 26,583–84 (statement of Rep. Cohelan) ("The proposed five-man Council would provide a broad and independent overview of . . . environmental problems that have been created by advancing technology."), 26,585 (statement of Rep. Boland) (CEQ will "provide a vitally needed source for reviewing the total environmental situation—an 'early warning' system that warns us of the effect on the environment of a particular program.").

101. Exec. Order No. 11,514 § 3(k), 35 Fed. Reg. 4247, 4248 (March 7, 1970).

102. See Caldwell, *National Environmental Policy Act*, 38–42. See also 42 U.S.C. § 4344 (2006) (describing CEQ's duties and functions).

103. See 40 C.F.R. §§ 1500.1–1508.28 (2010); Forty Most Asked Questions Concerning CEQ's National Environmental Policy Act Regulations, 46 Fed. Reg. 18,026 (1981).

104. Caldwell, *National Environmental Policy Act*, 39–42; Lindstrom and Smith, *National Environmental Policy Act*, 130–31; Kraft and Vig, "Environmental Policy," 19. Indeed, several proposals to eliminate the CEQ have been put forth (Lindstrom and Smith, *National Environmental Policy Act*, 130–31).

105. Rip, Schot, and Misa, "Constructive Technology Assessment," 2. See Sare-

witz, "This Won't Hurt a Bit," 4 ("[C]onventional TA embodies a sort of hyper-rational approach to decision making whose greatest error lies not in its unrealistic expectation of accurate predictions, but in its linear view of how decisions should be made.").

106. See Klüver et al., *European Participatory Technology Assessment*, 9, 23–24; Sclove, *Reinventing Technology Assessment*, 24–25, 130, 137–39, 170.

107. See Andersen and Jæger, "Scenario Workshops," 334.

108. Crosby, "Citizens Juries," 173; Hörning, "Citizens' Panels," 358. See also Fiorino, "Citizen Participation," 234 (noting that surveys "isolate problems and issues from their social and community context"); Coote and Lenaghan, *Citizens' Juries*, 6 (describing polls as "superficial and non-interactive, designed to elicit the uninformed views of the public"); Fischhoff, "Public Values," 79 ("Polls might obtain snapshots of current beliefs; however, those beliefs should have little value for policymakers who are contemplating long-term policies or anticipating the outcome of an intensive public debate."). For a critical view of citizens' juries, see Armour, "Citizens' Jury Model," 175, 181.

109. Sclove, *Reinventing Technology Assessment*, 34.

110. See Hörning, "Citizens' Panels," 351.

111. Andersen and Jæger, "Scenario Workshops," 331–36; Hörning, "Citizens' Panels," 352; Frank Fischer, *Citizens, Experts, and the Environment*, 235–36.

112. See generally Hörning, "Citizens' Panels," 352 (identifying differences among PTA methods).

113. See Crosby, "Citizens Juries," 157–58; Joss, "Participation," 341.

114. See Dienel and Renn, "Planning Cells," 121–24; Renn, Webler, and Wiedemann, "Pursuit," 344.

115. See Sclove, *Reinventing Technology Assessment*, 4; Westrum, *Technologies and Society*, 107–27 (discussing the process of invention); Collingridge, *Social Control*, 16–19 (explaining that control of a technology is hampered at early stages by lack of knowledge about adverse consequences and at later stages by diffusion and entrenchment of the technology).

116. Schot, "Towards New Forms," 40–42; Remmen, "Pollution Prevention," 201. The terms *interactive TA* and *real-time TA* refer to similar approaches. See Guston and Sarewitz, "Real-Time Technology Assessment," 97–98 (comparing real-time technology assessment with CTA); Rip, "Nanoscience and Nanotechnologies," 145, 148.

117. See Rip, "Nanoscience and Nanotechnologies," 147.

118. Schot, "Towards New Forms," 40, 43–44; Guston and Sarewitz, "Real-Time Technology Assessment," 97–98; Remmen, "Pollution Prevention," 201; Erik Fisher, Mahajan, and Mitcham, "Midstream Modulation," 492. The concept of reflexivity is rooted in Ulrich Beck's contention in *Risk Society* that a more reflexive, self-critical approach must replace the traditional technocratic understanding of science (155–56).

119. See Guston and Sarewitz, "Real-Time Technology Assessment," 98. See also Sarewitz, "This Won't Hurt a Bit," 4–5 (describing principles of real-time TA).

120. Klüver, "Danish Board," 190–91 (describing incorporation of conference recommendations into legislation and policy). See also Joss, "Participation," 342–53 (offering practical evaluation of Danish PTA efforts).

121. See Schot, "Towards New Forms," 47–48.
122. See Nisbet and Scheufele, "What's Next," 1768.
123. See Hennen, "Impacts," 154.
124. Sunstein, *Laws of Fear*, 64–88.
125. See Frank Fischer, *Citizens, Experts, and the Environment*, 35–36. See also Delli Carpini, Cook, and Jacobs, "Public Deliberation," 315–16, 323–24, 336 (reviewing studies of public deliberation).
126. See Frank Fischer, *Citizens, Experts, and the Environment*, 42 ("[S]cience is laden with social value judgments, judgments typically hidden within the steps and phases of the research process."), 132–42 (describing concept of cultural rationality); Nisbet and Scheufele, "What's Next," 1768 (noting that values are far more important than knowledge in determining public opinion about controversial areas of science).
127. Irwin, *Citizen Science*, 172–73 (discussing citizens as a source of information and expertise); Fiorino, "Citizen Participation," 227; Frank Fischer, *Citizens, Experts, and the Environment*, 193–218 (discussing local knowledge that laypersons can contribute).
128. Hart Research Associates, *Awareness and Impressions of Synthetic Biology*, September 9, 2010, 11, http://www.synbioproject.org/library/publications/archive/6456/.
129. Gavelin and Wilson, *Democratic Technologies?*, 52–53.
130. See Dienel and Renn, "Planning Cells," 125. See also Frank Fischer, *Citizens, Experts, and the Environment*, 32 (contending that citizens "are much more capable of grappling with complex problems than generally assumed").
131. Nisbet and Scheufele, "What's Next," 1770 (reporting that participants not only learn about the technical, social, ethical, and economic aspects of the scientific topic but also become more confident and motivated to participate in science decisions); Renn, Webler, and Wiedemann, "Pursuit," 345.
132. Frank Fischer, *Citizens, Experts, and the Environment*, 148–55 (discussing citizen activists' efforts to combat AIDS, identify cancer clusters, and engage in a nuclear power plant siting decision).
133. Wynne, "Creating Public Alienation," 447, 475–76.
134. Armour, "Citizens' Jury Model," 180; Seiler, "Review of 'Planning Cells,'" 142–49.
135. See Dienel and Renn, "Planning Cells," 129.
136. See Klüver et al., *European Participatory Technology Assessment*, 171; Dienel and Renn, "Planning Cells," 129; Hörning, "Citizens' Panels," 357; Joss, "Participation," 335–36.
137. Klüver et al., *European Participatory Technology Assessment*, 134; Hörning, "Citizens' Panels," 357; Renn, Webler, and Wiedemann, "Pursuit," 353. Notwithstanding random selection, certain demographic groups, such as students and retirees, may be overrepresented on citizen panels, just as they tend to be overrepresented on ordinary juries (Dienel and Renn, "Planning Cells," 125).
138. See Genus, "Rethinking Constructive Technology Assessment," 19.
139. Klüver et al., *European Participatory Technology Assessment*, 155.
140. See Frank Fischer, *Citizens, Experts, and the Environment*, 238–39 ("Beyond merely uncovering normative assumptions and beliefs, deliberation can lead to changes in assumptions, as well as creations of new ones.").

141. See Klüver et al., *European Participatory Technology Assessment,* 154; Seiler, "Review of 'Planning Cells,'" 150; Irwin, "Constructing the Scientific Citizen," 12–13 (discussing an example in which the institutional framing of public consultation constrained the role of the public).

142. Genus, "Rethinking Constructive Technology Assessment," 19, 21–23 (suggesting the potential for TA to become a "participation trap"). See also Wynne, "Risk and Environment," 463 (suggesting that efforts to make science and technology more participatory "have perversely reinforced attention only on back-end scientific questions about consequences" and thus "exclude[d] more reflexive questions about the human purposes and visions which shape front-end innovation commitments").

143. Hörning, "Citizens' Panels," 356.

144. Klüver et al., *European Participatory Technology Assessment,* 11.

145. See Armour, "Citizens' Jury Model," 181.

146. Hörning, "Citizens' Panels," 356; Klüver et al., *European Participatory Technology Assessment,* 146. In the Netherlands, for example, PTA organizers have sought to increase the influence of lay panels by raising their profile and by facilitating communications between the panels and the general public via the Internet (van Est, "Rathenau Institute's Approach," 13, 18).

147. Schot and Rip, "Past and Future," 255–56; Sclove, *Reinventing Technology Assessment,* 6 (European industries have "come to support [PTA] as a low-stress, low-cost mechanism for gauging societal reactions to alternative research, development and innovation trajectories.").

148. See Dienel and Renn, "Planning Cells," 135–36 (describing use of planning cells in the 1980s). For a summary of U.S. PTA efforts, see Sclove, *Reinventing Technology Assessment,* 43–46.

149. According to one organizer, legislators viewed these efforts "as a bother at best and, at worst, as a real challenge to the way business is currently conducted" (Crosby, "Citizens Juries," 159; see also 167 [noting the "cultural discrepancy" between citizens' jury process and usual political procedures]).

150. White House Office of Communications, "Remarks by the President on Stem Cell Research," news release, August 9, 2001, 2001 WL 896981.

151. Executive Order No. 13,505, 74 Fed. Reg. 10,667 (March 11, 2009). For an example of criticism, see Korobkin, "Embryonic Histrionics," 1.

152. See Schroeder, "Deliberative Democracy's Attempt," 124–27.

153. Wolpe and McGee, "'Expert Bioethics,'" 185, 186. See also Gretchen Vogel, "Rumors and Trial Balloons," 186.

154. U.S. Presidential Commission, *New Directions,* vi.

155. See Barber, *Strong Democracy,* 267–98 (recommending that participatory institutions include mechanisms for improving people's competence to make reasonable political judgments); Setälä, "On the Problems of Responsibility and Accountability," 702–3 (discussing the deliberative democratic rationale for the referendum and initiative).

156. See Butler and Ranney, "Theory" (1994), 13–16 (discussing benefits from supplementing representative institutions with referendums); Setälä, "On the Problems of Responsibility and Accountability," 702.

157. Frey, "Efficiency," 219 (contending that the public discussion induced by

referendums "shapes the citizens' preferences" by confronting them with issues they have not previously considered and by encouraging them to evaluate those issues according to their basic values). See also Cronin, *Direct Democracy*, 78–79 (discussing surveys finding that citizens generally favor having a direct vote with respect to important issues and policies).

158. See Butler and Ranney, "Theory" (1994), 18; Cronin, *Direct Democracy*, 76–77.

159. See Butler and Ranney, "Theory" (1978), 23, 30 (explaining how referendums can bring policy-making and political decisions psychologically closer to the people); Cronin, *Direct Democracy*, 87–89, 178 (noting the need for ingenuity in efforts to interest and inform voters while concluding that "[v]oters who do vote on ballot measures do so more responsibly and intelligently than we have any right to expect"). See also Maija Setälä, *Referendums and Democratic Government: Normative Theory and the Analysis of Institutions* (New York: St. Martin's, 1999), 165 (suggesting that "referendums offer . . . an opportunity to participate in decision-making without the mediation of the parties").

160. See Setälä, "On the Problems of Responsibility and Accountability," 7 13–14.

161. Page and Shapiro, "Effects of Public Opinion," 188–89; Burstein, "Impact," 33–36.

162. Where candidate choices and referendum propositions are on the same ballot, there is on average a 15 percentage point drop-off in the proportion of voters casting ballots between the latter and the former. See Butler and Ranney, "Theory" (1994), 16. However, propositions involving controversial or highly visible issues may attract higher levels of voting (Cronin, *Direct Democracy*, 67–69).

163. Frodeman and Parker, "Intellectual Merit," 339. The federal government funds more than a quarter of research and development expenditures in the United States (Mark Boroush, "New NSF Estimates Indicate that U.S. R&D Spending Continued to Grow in 2008," *InfoBrief*, January 2010, http://www.nsf.gov/statistics/infbrief/nsf10312).

164. National Science Foundation, *Proposal and Award Policies and Procedures Guide*, pt. 1, ch. 2, 8; National Science Foundation, *Merit Review Broader Impacts Criterion*.

165. See Holbrook, "Assessing the Science-Society Relation," 437–48 (describing difficulties encountered in applying the BIC and suggesting that difficulties reflect basic philosophical differences regarding the relevance of broader social considerations to "pure" scientific research); Frodeman and Parker, "Intellectual Merit," 340–41 ("BIC is not simply an education and public outreach . . . criterion—but it generally gets (mis)interpreted in this way."); Frodeman and Holbrook, "Science's Social Effects," 28 (observing that such an approach to BIC "emphasize[s] a triumphalist view" of science and technology and that it "does not reflect on the larger moral, political, and policy implications of the advance of scientific knowledge and technological capabilities").

166. Frodeman and Holbrook, "Science's Social Effects," 30. See also Frodeman and Parker, "Intellectual Merit," 342 (advocating interaction between scientists and researchers on science, technology, and society at all stages of research).

167. Sarewitz, "Anticipatory Governance," 104.

168. See Owen, "Bending Nature," 609–10 (proposing liability in "situations where industry has recklessly let loose untested new technology with frightening potential consequences").

169. Albert C. Lin, "Size Matters," 398–99.

Chapter 2

1. For purposes of this book, the term *genetic engineering* refers only to conventional genetic engineering, and not synthetic biology, and the terms *genetically engineered* and *genetically modified* are used interchangeably.

2. See GAO, *Genetically Engineered Crops*, 7.

3. Jaffe, "Next Generation," 38.

4. National Research Council, *Impact of Genetically Engineered Crops*, 3–10; Cowan and Becker, *Agricultural Biotechnology*, 1–3; Miller and Conko, *Frankenfood Myth*, 32–33.

5. See National Agricultural Statistics Service, Acreage, June 29, 2012, http://usda01.library.cornell.edu/usda/current/Acre/Acre-06-29-2012.pdf. GAO, *Genetically Engineered Crops*, 1.

6. See Carey Gilliam, "U.S. Consumer Groups Demand GMO Labeling, Question Food Safety," Reuters, March 27, 2012, http://www.reuters.com/article/2012/03/27/usa-food-idUSL2E8ERK7C20120327.

7. See Cowan and Becker, *Agricultural Biotechnology*, 6; GAO, *Genetically Engineered Crops*, 1.

8. See Cowan and Becker, *Agricultural Biotechnology*, 6–7; GAO, *Genetically Engineered Crops*, 1; Mandel, "Gaps, Inexperience, Inconsistencies, and Overlaps," 2186–89.

9. See National Science Board, *Science and Engineering Indicators* (2008), 7–36 (summarizing findings of surveys conducted by Pew Initiative on Food and Biotechnology).

10. See Knight, "Perceptions, Knowledge and Ethical Concerns," 183–85.

11. See Spiroux de Vendômois et al., "Comparison," 706 (reporting signs of kidney and liver toxicity in rats who had been fed corn genetically modified to synthesize Bt toxins used as insecticides); National Research Council, *Genetically Modified Pest-Protected Plants*, 69–73.

12. See "Premarket Notice Concerning Bioengineered Foods," 66 Fed. Reg. 4706, 4728 (2001); Spök, "Suggestions," 171; Metcalfe, "What Are the Issues," 1111.

13. Van Tassel, "Genetically Modified Plants," 231 (describing the "Central Dogma" of molecular biology, a model of information transfer from DNA to RNA to protein); Krimsky, "From Asilomar to Industrial Biotechnology," 316 (distinguishing "the *Lego System* versus the *Ecosystem* models of the genome").

14. Gerstein et al., "What Is a Gene," 671–73 (discussing findings that conflict with "Central Dogma" model of gene); Van Tassel, "Genetically Modified Plants," 232 (discussing "Networked Gene" model); Krimsky, "From Asilomar to Industrial Biotechnology," 316. Conventional plant breeding can also lead to unpredictable results. See Miller and Conko, *Frankenfood Myth*, 5–6, 20–22.

15. See National Research Council, *Genetically Modified Pest-Protected Plants*, 67.

16. Engineered genes have been detected, for example, in native strains of corn in Mexico. See David Quist and Ignacio H. Chapela, "Transgenic DNA Introgressed into Traditional Maize Landraces in Oaxaca, Mexico," *Nature* 414 (2001): 541.

17. See National Research Council, *Genetically Modified Pest-Protected Plants*, 67, 70–71. For example, GM bentgrass, a GMO not approved for commercial use, escaped from field trials and has spread along irrigation canals in eastern Oregon. See Mitch Lies, "Bentgrass Eradication Plan Unveiled," *Capital Press*, June 16, 2011, http://www.capitalpress.com/newest/ml-scotts-061711.

18. National Research Council, *Impact of Genetically Engineered Crops*, 73–78; Powles, "Evolved Glyphosate-Resistant Weeds," 360–65; Scott Kilman, "Monsanto Corn Plant Losing Bug Resistance," *Wall Street Journal*, August 29, 2011, B1; William Neuman and Andrew Pollack, "Farmers Cope with Roundup-Resistant Weeds," *New York Times*, May 4, 2010, B1.

19. Torrance, "Intellectual Property," 275.

20. See Aoki, "Seeds of Dispute," 146–50.

21. Thompson, *Food Biotechnology*, 48–49, 215–16, 220; Street, "Constructing Risks," 95, 102.

22. Thompson, *Food Biotechnology*, 48–50, 105–6, 111, 265–67.

23. National Research Council, *Environmental Effects*, 15, 68.

24. See Mandel, "Gaps, Inexperience, Inconsistencies, and Overlaps," 2200–2202.

25. See National Science Board, *Science and Engineering Indicators* (2008) 7–37; Knight, "Perceptions, Knowledge and Ethical Concerns," 185–87.

26. OSTP, "Coordinated Framework for Regulation of Biotechnology," 51 Fed. Reg. 23,302 (1986).

27. See OSTP, "Proposal for a Coordinated Framework for Regulation of Biotechnology," 49 Fed. Reg. 50,856, 50,857 (1984).

28. Ibid., 50,857.

29. OSTP, "Coordinated Framework," 51 Fed. Reg. at 23,302.

30. Berg et al., *Potential Biohazards*, 303 (recommending voluntary deferral of certain types of experiments "until the potential hazards . . . have been better evaluated or until adequate methods are developed for preventing their spread").

31. Berg et al., "Asilomar Conference," 994.

32. Jasanoff, *Designs on Nature*, 46.

33. Wright, *Molecular Politics*, 148, 153.

34. Krimsky, *Genetic Alchemy*, 151–52; Roger B. Dworkin, "Science, Society, and the Expert Town Meeting," 1475. Twenty-one reporters attended the conference under some restrictions on coverage (Wright, *Molecular Politics*, 146).

35. Wright, *Molecular Politics*, 145–48.

36. Roger B. Dworkin, "Science, Society, and the Expert Town Meeting," 1481.

37. Wright, *Molecular Politics*, 148–49, 153; Roger B. Dworkin, "Science, Society, and the Expert Town Meeting," 1473; Jasanoff, *Designs on Nature*, 47; Krimsky, *Genetic Alchemy*, 103, 141.

38. "Decision of the Director, National Institutes of Health, to Release Guidelines for Research on Recombinant DNA Molecules," 41 Fed. Reg. 27,902 (1976).

39. Wright, *Molecular Politics*, 160.

40. 41 Fed. Reg. at 27,905–6, 27,921.

41. U.S. Department of Health, Education, and Welfare, NIH, "Recombinant DNA Research Guidelines Draft Environmental Impact Statement," 41 Fed. Reg. 38,426, 38,426–27 (1976) (asserting that "development of the guidelines was in large part tantamount to conducting an" EIS and that immediate issuance of guidelines was necessary to promote their acceptance by scientists).

42. Wright, "Molecular Politics," 1406.

43. *Foundation on Economic Trends v. Bowen*, 722 F. Supp. 787, 789 (D.D.C. 1989).

44. *Foundation on Economic Trends v. Heckler*, 756 F.2d 143, 153–55 (D.C. Cir. 1985) (upholding injunction against proposed experiment to release genetically altered bacteria).

45. Ibid., 158–60 (vacating injunction to block approval of other experiments involving deliberate release of GMOs).

46. Jelsma, "Learning about Learning," 147; Wright, *Molecular Politics*, 256–78; Krimsky, *Genetic Alchemy*, 198–205.

47. NIH, "Guidelines for Research Involving Recombinant DNA Molecules," 43 Fed. Reg. 60,108 (1978); NIH, "Guidelines for Research Involving Recombinant DNA Molecules," 45 Fed. Reg. 6724 (1980); NIH, "Guidelines for Research Involving Recombinant DNA Molecules," 47 Fed. Reg. 38,048 (1982); Wright, *Molecular Politics*, 281–311, 337–405.

48. Jelsma, "Learning about Learning," 148–49.

49. *Diamond v. Chakrabarty*, 447 U.S. 303, 309–10 (1980).

50. Jasanoff, *Designs on Nature*, 209; Kenney, *Biotechnology*, 190, 256.

51. *Heckler*, 756 F.2d 143.

52. Ibid., 153.

53. Jelsma, "Learning about Learning," 151. Some privately funded companies, seeking to reduce liability exposure and bolster public confidence, voluntarily submitted their testing plans for NIH or EPA review. See Jasanoff, *Designs on Nature*, 51; Krimsky, *Genetic Alchemy*, 203.

54. Sheingate, "Promotion versus Precaution," 247.

55. Ibid., 248; Jasanoff, *Designs on Nature*, 51.

56. See OSTP, "Exercise of Federal Oversight Within Scope of Statutory Authority: Planned Introductions of Biotechnology Products Into the Environment," 57 Fed. Reg. 6753, 6756 (1992).

57. 51 Fed. Reg. 23,304 ("The manufacture by the newer technologies of [genetically engineered products] will be reviewed by FDA, USDA and EPA in essentially the same manner for safety and efficacy as products obtained by other techniques.").

58. Ibid., 23,302–3.

59. Jasanoff, *Designs on Nature*, 52; Mandel, "Confidence-Building Measures," 59 (emphasizing that statutes underlying the Framework were enacted without public debate specifically relating to biotechnology); Kurt Eichenwald, "Redesigning Nature: Hard Lessons Learned; Biotechnology Food: From the Lab to a Debacle," *New York Times*, January 25, 2001, A1 (recounting how Monsanto, the leading agricultural biotech company, actively lobbied for government regulation in the hopes of reassuring consumers).

60. See 7 U.S.C. §§ 7711, 7712. The Framework cited the Plant Pest Act and the Plant Quarantine Act as the sources of USDA's authority over GMOs that might be plant pests (51 Fed. Reg. at 23,303). The 2000 Plant Protection Act (7 U.S.C. §§ 7701–72), consolidated the Plant Pest Act, Plant Quarantine Act, and other existing plant health laws into a single statute.

61. 7 C.F.R. § 340.1 (defining "regulated article" as "any organism which has been altered or produced through genetic engineering, if the donor organism, recipient organism, or vector or vector agent belongs to any genera or taxa designated in § 340.2 and meets the definition of plant pest, . . . or any other organism or product altered or produced through genetic engineering which the Administrator, determines is a plant pest or has reason to believe is a plant pest").

62. Ibid., § 340.3.

63. Ibid., § 340.3(e)(4).

64. Ibid., § 340.4.

65. National Research Council, *Environmental Effects*, 183 (estimating that APHIS conducted EAs for only 6 percent of permit applications processed between 1996 and 2002); 7 C.F.R. § 372.5(c)(3)(ii) (authorizing categorical exclusion). Through a categorical exclusion, an agency may identify "a category of actions which do not individually or cumulatively have a significant effect on the human environment . . . and for which, therefore, neither an environmental assessment nor an environmental impact statement is required" (40 C.F.R. § 1508.4).

66. 7 C.F.R. § 340.6. In theory, APHIS can bring deregulated items back under regulation, but it has never used that authority ("APHIS Biotechnology: Permitting Progress Into Tomorrow," Biotechnology Regulatory Service Factsheet, 2006, 4, http://www.aphis.usda.gov/publications/biotechnology/content/printable_version/BRS_FS_permitprogress_02-6.pdf).

67. USDA, Animal and Plant Health Inspection Service, *Introduction*, 20. APHIS is considering changes to its regulations that would expand its oversight of GM plants to those that are not plant pests but may qualify as noxious weeds (20–21).

68. 7 C.F.R. § 340.1 (defining "plant pest").

69. Kunich, "Mother Frankenstein," 840 (noting that APHIS does not mandate the performance of individual environmental evaluations for notifications). See also *Center for Food Safety v. Johanns*, 451 F. Supp. 2d 1165, 1177 n. 7 (D. Haw. 2006) (explaining that APHIS's analysis of potential spread of plant pests and noxious weeds resulting from GMO field test, as required by the Plant Protection Act, does not necessarily satisfy NEPA's broader requirements); National Research Council, *Environmental Effects*, 19 (noting concerns that "an agency with a mandate to promote U.S. agriculture may not be able to objectively assess the safety of new products of agricultural biotechnology"); Mandel, "Gaps, Inexperience, Inconsistencies, and Overlaps," 2231–33 (questioning APHIS's ability to analyze environmental effects).

70. Of the 13,000 notifications and permits for field trials approved by USDA between 1987 and 2007, more than 90 percent involved notifications. See GAO, *Genetically Engineered Crops*, 74. See also USDA, Office of Inspector General, Southwest Region, *Audit Report*, 2 (reporting that almost 97 percent of all field trials of regulated GM crops in 2004 were conducted under notifications).

71. 7 C.F.R. § 340.3(e); "USDA-APHIS Biotechnology Regulatory Services, User Guide: Notification," 21 (March 29, 2011), http://www.aphis.usda.gov/bio

technology/downloads/notification_guidance_0311.pdf; National Research Council, *Environmental Effects*, 108, 123. The agency presumes "that introductions of many regulated articles can be conducted with little or no plant pest or environmental risk" (USDA, "Genetically Engineered Organisms and Products; Notification Procedures for the Introduction of Certain Regulated Articles; and Petition for Nonregulated Status," 57 Fed. Reg. 53,036, 53,036–37 [1992]).

72. GAO, *Genetically Engineered Crops*, 18–19.

73. Ibid., 19. With respect to field tests of GM plants that are designed to produce pharmaceutical and industrial chemicals (and thus are subject to a permit rather than a notification), APHIS failed to inspect such sites as required. See USDA, Office of Inspector General, Southwest Region, Audit Report, 29.

74. USDA, Office of Inspector General, Southwest Region, Audit Report, 6–7, 13–17.

75. *Johanns*, 451 F. Supp. 2d at 1183–86. Although the court held that APHIS's lack of analysis violated NEPA, it declined to issue an injunction because the field tests had already been completed at the time of the court's decision (1195–96).

76. National Research Council, *Environmental Effects*, 193–95.

77. Ibid., 192–98.

78. *Geertson Seed Farms v. Johanns*, 2007 WL 518624 (N.D. Cal. 2007).

79. Ibid., *9. The trial court also found APHIS's analysis of the potential development of herbicide-resistant weeds to be inadequate (*9–*10). Although the trial court's injunction against any planting of GM alfalfa was subsequently overturned by the U.S. Supreme Court, its finding of a NEPA violation was not challenged on appeal and remained intact (*Monsanto Co. v. Geertson Seed Farms*, __ U.S. __, 130 S. Ct. 2743, 2752 [2010]).

80. *Center for Food Safety v. Vilsack*, 2009 WL 3047227 (N.D. Cal. 2009), at *9. *Center for Food Safety* illustrates one potential difficulty of relying on NEPA litigation after a new technology has already entered the marketplace. Although the court found a NEPA violation and a likelihood that continued cultivation of GM sugar beets would result in biological contamination, the court declined to enjoin further planting or processing of the crop (*Center for Food Safety v. Schafer*, 2010 WL 964017 [N.D. Cal. 2010]). In the five years between APHIS's decision to deregulate and the court's finding of a NEPA violation, GM sugar beets had come to account for 95 percent of the sugar beet market (*Center for Food Safety*, *3–*5). The court ultimately vacated APHIS's decision to deregulate sugar beets and remanded the matter to the agency (*Center for Food Safety v. Vilsack*, 2010 WL 3222482 [N.D. Cal. 2010]).

81. National Research Council, *Environmental Effects*, 177.

82. Ibid., 168–74.

83. 57 Fed. Reg. at 22,989–90.

84. 21 U.S.C. § 342(a)(2)(C). The term *adulterated food* also includes "any poisonous or deleterious substance which may render it injurious to health" (21 U.S.C. § 342[a][1]), and the FDA has affirmed its authority to regulate GM foods containing toxicants through this provision. See FDA, "Statement of Policy: Foods Derived from New Plant Varieties," 57 Fed. Reg. 22,984, 22,990 (1992).

85. See Bratspies, "Some Thoughts," 408 n. 70.

86. 21 U.S.C. § 342(a)(2)(C); 21 U.S.C. § 348.

87. 57 Fed. Reg. at 22,990.

88. 21 U.S.C. § 321(s) (defining *food additive* as "any substance the intended use of which results . . . in its becoming a component or otherwise affecting the characteristics of any food . . . , if such substance is not generally recognized . . . to be safe under the conditions of its intended use").

89. "Foods Derived from New Plant Varieties," 57 Fed. Reg. at 22,985.

90. See *Alliance for Bio-Integrity v. Shalala*, 116 F. Supp. 2d 166, 173–75 (D.D.C. 2000) (upholding the FDA's refusal to perform NEPA analysis on grounds that the FDA's statement of policy on GM foods "is reversible, maintains the substantive status quo, and takes no overt action"); Pelletier, "FDA's Regulation," 581; Eichenwald, "Redesigning Nature," A1.

91. 57 Fed. Reg. at 22,985. Substantial equivalence reflects the FDA's judgment that genetic engineering techniques are mere "extensions" of traditional plant breeding methods and that the product rather than the process used to make the product is what matters for purposes of regulation (22,984–85, 22,991).

92. Ibid., 22,991 (declining to require disclosure of the presence of GMOs in food labeling); McGarity, "Seeds of Distrust," 428–31, 442, 472 (the "FDA has no way of knowing whether companies are distributing or importing new GE [genetically engineered] foods."); Bratspies, "Some Thoughts," 410.

93. See 57 Fed. Reg. at 22,989; McGarity, "Seeds of Distrust," 430 ("The most disturbing aspect of the substantial equivalence doctrine to its critics is its subjectivity.").

94. "Substances Generally Recognized as Safe, Notice of Proposed Rulemaking," 62 Fed. Reg. 18,938, 18,941 (1997) ("[A] manufacturer may market a substance that the manufacturer determines is GRAS without informing the agency."); "Foods Derived from New Plant Varieties," 57 Fed. Reg. at 22,989; 21 C.F.R. § 170.35 (setting out the procedure for voluntarily obtaining FDA affirmation of GRAS status); McGarity, "Seeds of Distrust," 485. The FDA believes that most companies undergo a voluntary consultation process for GM crops being developed for food or animal feed (GAO, *Genetically Engineered Crops*, 13). See also FDA, "Guidance on Consultation Procedures: Foods Derived from New Plant Varieties," published October 1997, last modified February 22, 2010, http://www.fda.gov/food/guidancecomplianceregulatoryinformation/guidancedocuments/biotechnology/ucm096126.htm. In early 2001, the FDA proposed requiring manufacturers to notify the agency before marketing new GM foods, but this proposal was subsequently withdrawn. See "Premarket Notice Concerning Bioengineered Foods," 66 Fed. Reg. 4706 (January 18, 2001); Bratspies, "Some Thoughts," 409; GAO, *Genetically Engineered Crops*, 44.

95. 7 U.S.C. § 136a(c)(5).

96. "Regulations under the Federal Insecticide, Fungicide, and Rodenticide Act for Plant-Incorporated Protectants (Formerly Plant-Pesticides)," 66 Fed. Reg. 37,772, 37,772 (July 19, 2001). The EPA does not regulate the plant itself (40 C.F.R. § 152.20(a); 66 Fed. Reg. 37,774).

97. 40 C.F.R. Pt. 174 (setting out requirements for PIPs and reserving various subjects for future rulemaking); Angelo, "Regulating Evolution," 127–28.

98. 40 C.F.R. § 172.3. Greenhouse experiments and small-scale field tests of new pesticides generally do not require a permit (§ 172.3(b), (c)). To determine

whether an experimental use permit is required, the EPA requires notification for small-scale field testing of GM pesticidal microorganisms (§ 172.45).

99. Kunich, "Mother Frankenstein," 835.

100. 40 C.F.R. § 156.10. See also Angelo, "Regulating Evolution," 116. The EPA is currently drafting new rules under FIFRA that would establish data requirements specifically for PIPs (EPA, "Plant-Incorporated Protectants; Potential Revisions to Current Production Regulations," 72 Fed. Reg. 16,312 [2007]). This rulemaking effort, however, is not slated to overhaul the EPA's general regulatory approach to GMOs.

101. EPA, "Proposed Policy; Plant Pesticides Subject to the Federal Insecticide, Fungicide, and Rodenticide Act and the Federal Food, Drug, and Cosmetic Act," 59 Fed. Reg. 60,496, 60,507 (November 23, 1994).

102. Kunich, "Mother Frankenstein," 835–36; Angelo, "Regulating Evolution," 116–17. Likewise, limitations on where GM crops are planted may be irrelevant if the progeny of GMOs escape commercial cultivation (Kunich, "Mother Frankenstein," 835–36; Angelo, "Regulating Evolution," 116–17). Whether geographic restrictions are followed during field tests is uncertain, as neither the EPA nor the states, to which the EPA has delegated inspection authority, have made inspection of field tests a priority (GAO, *Genetically Engineered Crops*, 21).

103. Kunich, "Mother Frankenstein," 835–36; Angelo, "Regulating Evolution," 116–17.

104. 7 U.S.C. § 136(bb); *Love v. Thomas*, 858 F.2d 1347, 1357 (9th Cir. 1988); *Environmental Defense Fund, Inc. v. EPA*, 548 F.2d 998, 1005 (D.C. Cir. 1976).

105. National Research Council, *Environmental Effects*, 68, 70.

106. 21 U.S.C. § 346a(a)(1); ibid., § 342(a)(2)(B); GAO, *Genetically Engineered Crops*, 12.

107. 21 U.S.C. § 346a(d)(3) (providing for public notice of petition to establish a tolerance or for exemption); McGarity, "Seeds of Distrust," 478. See also 7 U.S.C. § 136a(c)(4) (30-day-notice period for public to comment on applications for pesticide registration). Recognizing these limitations, the EPA recently began to offer the public the opportunity to review and comment on risk assessments and proposed registration decisions in certain instances. See EPA, "Public Participation Process for Registration Actions," March 31, 2010, http://www.epa.gov/pesticides/regulating/public-participation-process.html.

108. See Chris Wozniak to author, November 30, 2009 ("To date, all PIPs registered have received an exemption from the requirement of a tolerance based upon the absence of adverse toxicological effects in an acute oral toxicity study and passing an allergenicity screen."); 21 U.S.C. § 346a(c)(2)(A); 40 C.F.R. §§ 174.500–174.530 (listing tolerance exemptions for PIP residues).

109. See Chris Wozniak to author, November 30, 2009; McGarity, "Seeds of Distrust," 418, 479–80 (noting EPA's "liberal use of the substantial equivalence doctrine to waive testing requirements that would otherwise be applicable to chemical pesticides").

110. GAO, *Genetically Engineered Crops*, 90–95 (describing six documented incidents of unauthorized release of GM crops into the food or feed supply), 16–17 (USDA records indicated 98 violations, mostly self-reported, that involved possible unauthorized releases of GM crops into the environment over a 4½-year period).

111. *In re StarLink Corn Prods. Liability Litig.*, 212 F. Supp. 2d 828, 834 (N.D. Ill. 2002).

112. Mandel, "Gaps, Inexperience, Inconsistencies, and Overlaps," 2207 (discussing commingling of corn in harvesting, storage, and shipping process and noting that the crop developer had not adequately warned farmers of crop segregation requirements); GAO, *Genetically Engineered Crops*, 90–91 (noting the likely commingling of corn approved for human consumption with corn grown adjacent to and cross-pollinated by StarLink); Segarra and Rawson, *StarLink Corn Controversy* (mentioning the possibility of StarLink pollen drifting into neighboring fields); Bratspies, "Myths," 635–36 (noting the lack of systematic monitoring by the EPA or FDA to ensure that StarLink did not enter the food supply).

113. Centers for Disease Control and Prevention, *Investigation of Human Health Effects Associated with Potential Exposure to Genetically Modified Corn*, June 11, 2001, 3, http://www.cdc.gov/nceh/ehhe/cry9creport/pdfs/cry9creport.pdf.

114. Bratspies, "Myths," 632; Dave Deegan and Martha Casey, "EPA Releases Draft Report on StarLink Corn," press release, March 7, 2001, http://yosemite.epa.gov/opa/admpress.nsf/6427a6b7538955c585257359003f0230/cd90138019732 59885256a0800710574!OpenDocument&Start=3.2&Count=5&Expand=3.2.

115. See William Neuman, "'Non-GMO' Seal Identifies Foods Mostly Biotech-Free," *New York Times*, August 29, 2009, B1.

116. See GAO, *Genetically Engineered Crops*, 91–95.

117. The Genetically Engineered Safety Act, a bill introduced by Representative Dennis Kucinich, would bar the open-air cultivation of GM pharmaceutical and industrial crops and the cultivation of plants commonly used for human food if genetically modified to produce pharmaceuticals or industrial chemicals (H.R. 5578, Title I [111th Cong. 2nd sess.]). See also Bratspies, "Myths," 647–48 (proposing independent audits).

118. See Bratspies, "Myths," 413. The EPA could regulate such GMOs under its authority to regulate chemicals under the Toxic Substances Control Act but has not indicated that it will do so.

119. GAO, *Genetically Engineered Crops*, 32 (criticizing lack of a "mechanism to monitor, evaluate, and report on the impact of the commercialization of GE crops following the completion of the agencies' evaluation procedures").

120. National Research Council, *Environmental Effects*, 193–94.

121. GAO, *Genetically Engineered Crops*, 93–94 (also noting a similar incident involving genetically modified rice that resulted in bans on rice imports from the United States).

122. 49 Fed. Reg. at 50,856–57.

123. Mark A. Pollack and Shaffer, *When Cooperation Fails*, 116.

124. Ibid., 113–16.

125. Ibid., 116; Cowan and Becker, *Agricultural Biotechnology*, 21.

126. Jasanoff, *Designs on Nature*, 79; Winickoff et al., "Adjudicating the GM Food Wars," 87.

127. European Parliament and Council Regulation 1829/2003, 2003 O.J. (L 268) 1, http://eur-lex.europa.eu/LexUriServ/LexUriServ.do?uri=OJ:L:2003:268:0001:0 023:EN:PDF; Jasanoff, *Designs on Nature*, 142–43; Winickoff et al., "Adjudicating the GM Food Wars," 88–89.

128. European Parliament and Council Regulation 1829/2003, Arts. 1, 4–6; Mark A. Pollack and Shaffer, *When Cooperation Fails*, 241.

129. Winickoff et al., "Adjudicating the GM Food Wars," 100–102; Mark A. Pollack and Shaffer, *When Cooperation Fails*, 68–77.

130. Mark A. Pollack and Shaffer, *When Cooperation Fails*, 66; Winickoff et al., "Adjudicating the GM Food Wars," 100–102; Patrycja Dabrowska, "Civil Society Involvement in the EU Regulations on GMOs: From the Design of a Participatory Garden to Growing Trees of European Public Debate," *Journal of Civil Society* 3 (2007): 290–92.

131. See WTO, *Panel Report;* Winickoff et al., "Adjudicating the GM Food Wars," 82.

132. See Winickoff et al., "Adjudicating the GM Food Wars," 90–91, 106–7. Further analysis of these trade disputes can be found in Mark A. Pollack and Shaffer, *When Cooperation Fails;* McMahon, "*EC-Biotech* Decision," 337–54; Boisson de Chazournes and Mbengue, "Trade, Environment, and Biotechnology," 205, 237–44; Kysar, "Preference for Processes," 564–69.

133. See WTO, *Panel Report*, ¶¶ 4.130–4.135.

134. Agreement on the Application of Sanitary and Phytosanitary Measures, April 15, 1994, Arts. 3.1, 5.1, 1867 U.N.T.S. 493 (1994).

135. WTO, *Panel Report*, ¶¶ 4.374–4.375; Agreement on the Application of Sanitary and Phytosanitary Measures, Art. 5.7.

136. WTO, *Panel Report*, ¶¶ 4.358, 4.524; Cartagena Protocol on Biosafety to the Convention on Biological Diversity, January 29, 2000, Arts. 10.6, 11.8, 15, 39 I.L.M. 1027; Mark A. Pollack and Shaffer, *When Cooperation Fails*, 152–58. The Cartagena Protocol is a subsidiary agreement to the Convention on Biological Diversity (CBD), a treaty regime concerned with the conservation and sustainable use of biological diversity for present and future generations. See Convention on Biological Diversity, June 5, 1992, 1760 U.N.T.S. 143, http://www.cbd.int/convention/convention.shtml. The United States is not a party to the CBD or the Cartagena Protocol ("List of Parties," Convention on Biological Diversity, http://www.cbd.int/convention/parties/list/; "Parties to the Protocol and Signatories to the Supplementary Protocol," Convention on Biological Diversity, http://bch.cbd.int/protocol/parties/).

137. David Hunter, Salzman, and Zaelke, *International Environmental Law and Policy*, 1046.

138. WTO, *Panel Report*, ¶¶ 7.1384–7.1395; Mark A. Pollack and Shaffer, *When Cooperation Fails*, 177–234.

139. Winickoff et al., "Adjudicating the GM Food Wars," 84–85.

140. Scott H. Segal, "Environmental Regulation of Nanotechnology: Avoiding Big Mistakes for Small Machines," *Nanotechnology Law and Business* 1 (2004): 302.

141. Segarra and Rawson, *StarLink Corn Controversy*, 9 n. 20; Torrance, "Intellectual Property," 269–72.

142. Domingo, "Toxicity Studies," 731; Bratspies, "Myths," 639 ("[T]he agbiotech industry has no body of solid scientific evidence to back up safety claims for GM crops."); Pelletier, "Science, Law, and Politics," 211 (noting a "dearth of empirical research" to support claims regarding safety of GM foods). See also Committee on Identifying and Assessing Unintended Effects, *Safety*, 180, 183 (noting lack of

documented adverse health effects while recommending postmarket surveillance); National Research Council, *Environmental Effects*, 168 ("APHIS' regulatory process has never led to the release of a transgenic plant that clearly caused environmental damage. However, without systematic monitoring, the lack of evidence of damage is not necessarily lack of damage.").

143. Van Tassel, "Genetically Modified Plants," 242–43; Committee on Identifying and Assessing Unintended Effects, *Safety*, 145–46, 148. Conventional plant breeding can also produce unintended modifications. See Committee on Identifying and Assessing Unintended Effects, *Safety*, 145.

144. Séralini et al., "Long Term Toxicity," 4221–29; Séralini et al., "Answers to Critics"; Séralini et al., "How Subchronic and Chronic Health Effects Can Be Neglected," 442; Spiroux de Vendômois et al., "Comparison"; Dona and Arvanitoyannis, "Health Risks," 169–70 (reviewing studies).

145. Committee on Identifying and Assessing Unintended Effects, *Safety*, 147; Lemaux, "Genetically Engineered Plants and Foods," 777.

146. Séralini et al., "Long Term Toxicity," 4221.

147. GAO, *Genetically Modified Foods*, 30–31.

148. Committee on Identifying and Assessing Unintended Effects, *Safety*, 12–14; GAO, *Genetically Modified Foods*, 6 (urging federal agencies to conduct postmarket risk-based monitoring).

149. Segarra and Rawson, *StarLink Corn Controversy*, 27. Legislation has been proposed to require labeling. See, e.g., Genetically Engineered Food Right to Know Act, H.R. 5577, 111th Cong. 2nd Sess.

150. See Degnan, "Biotechnology and the Food Label," 25–27.

151. FDA, *Guidance for Industry: Voluntary Labeling Indicating Whether Foods Have or Have Not Been Developed Using Bioengineering*, Draft Guidance, January 2001, http://www.fda.gov/food/guidancecomplianceregulatoryinformation/guidance documents/foodlabelingnutrition/ucm059098.htm. See also Peters and Lambert, "Regulatory Barriers," 163–69 (discussing impediments to voluntary labeling). Organic and natural food producers nonetheless have launched a campaign to test and label their products as "non-GMO." See Neuman, "'Non-GMO' Seal," B1.

152. McGarity, "Frankenfood Free," 129, 131–32.

153. See Kysar, "Preference for Processes," 531; Caswell, "Labeling Policies," 53, 56.

154. Pelletier, "FDA's Regulation," 582.

155. See Gilliam, "U.S. Consumer Groups Demand GMO Labeling" (reporting poll finding that 91 percent of Americans support labeling of GMO foods); National Science Board, *Science and Engineering Indicators* (2010), 7–40 (reporting that 87 percent of respondents in 2008 poll supported labeling); Huffman and Rousu, "Consumer Attitudes," 201, 209.

156. McGarity, "Frankenfood Free," 132–34; Crespi and Marette, "'Does Contain' vs. 'Does Not Contain,'" 330; Rechtschaffen, "Warning Game," 313.

157. See Heslop, "If We Label It, Will They Care?," 224; Amy Harmon and Andrew Pollack, "Battle Brewing over Labeling of Genetically Modified Food," *New York Times*, May 25, 2012, A1. See also Pollan, *Omnivore's Dilemma*.

158. Kysar, "Preference for Processes," 590.

159. See Batte et al., "Putting Their Money Where Their Mouths Are," 158

(concluding that national labeling program has had a significant impact on organic market).

160. See Grossman, "European Community Legislation," 25–27; Center for Food Safety, Genetically Engineered Food Labeling Laws, http://www.centerfor foodsafety.org/ge-mapl.

161. See Teisl and Caswell, *Information Policy*, 11–16 (noting that cost estimates range from "very modest to significant increases in costs"); Gruère and Rao, "Review," 56–57 (discussing cost estimates from several studies ranging from less than $1 to $48 per person per year). For purposes of comparison, organic farms and processors pay certification fees ranging from several hundred to a few thousand dollars, depending on size. See California Certified Organic Farmers, CCOF Certification Services Fee Schedule (effective October 1, 2011), http://www.ccof.org/ certification/fees.

162. Carter and Gruère, "Mandatory Labeling" (describing a shift in ingredients away from GMOs among food processors selling to the EU and Japan). See also Rechtschaffen, "Warning Game," 341 (discussing product reformulation driven by the warning requirement of California's Proposition 65).

163. McGarity, "Seeds of Distrust," 494–96. See also Krueger, "Public Debate."

164. Busch and Lloyd, "What Can Nanotechnology Learn," 269.

165. See Bijker, *Of Bicycles, Bakelites, and Bulbs*, 85.

166. Andrew Pollack, "Genetically Altered Salmon Get Closer to the Table," *New York Times*, June 26, 2010, A1.

167. Lulu Liu, "Genetically Engineered Salmon Could Be on Plates Soon," *Sacramento Bee*, July 21, 2010, D1.

168. FDA, *Guidance for Industry*, 5. The Food Safety and Inspection Service of the USDA, through its authority to prohibit adulterated meat and meat food products (21 U.S.C. §§ 451–72, 601–25), enforces tolerances (maximum allowable amounts) set by the FDA on new animal drug residues in food (10).

169. Ibid., 11. See also 21 U.S.C. § 360b(d).

170. Even the FDA admits that "application of some of the statutory and regulatory requirements for new animal drug applications to [genetically engineered] animals may not be obvious" (FDA, *Guidance for Industry*, 12).

171. Tony Reichhardt, "Will Souped Up Salmon Sink or Swim?," 10.

172. FDA, *Guidance for Industry*, 11.

173. 21 C.F.R. § 312.23 (rule governing content and format of investigational new drug application), § 314.50 (rule governing content and format of new drug application), § 25.40(b). The FDA does retain responsibility for the scope and content of the EA.

174. See Smythe and Isber, "NEPA in the Agencies," 290.

175. 21 C.F.R. § 25.40(a); FDA, *Guidance for Industry*, 24.

176. Mandel, "Gaps, Inexperience, Inconsistencies, and Overlaps," 2233 (noting the FDA's lack of expertise in analyzing environmental impacts).

177. 21 C.F.R. § 514.11.

178. See "FDA's Response to Public Comments," updated June 25, 2009, http:// www.fda.gov/AnimalVeterinary/DevelopmentApprovalProcess/GeneticEngineer ing/GeneticallyEngineeredAnimals/ucm113612.htm.

179. FDA, *Guidance for Industry*, 12, 23–24.

180. National Science Board, *Science and Engineering Indicators* (2010), 7–40; Einsiedel, "Public Perceptions," (reporting 2004 survey finding that only 21 percent of American consumers approved of genetically modifying animals); Kuzma, Najmaie, and Larson, "Evaluating Oversight Systems," 565 (discussing studies finding particular discomfort with GM animals in food).

181. See Pew Initiative on Food and Biotechnology, *Public Sentiment About Genetically Modified Food,* 2005, 6–7, http://www.pewtrusts.org/news_room_detail.aspx?id=32804 (reporting strong public support for incorporating moral and ethical considerations into regulation of cloned and genetically modified animals); Pew Initiative on Food and Biotechnology, *A Future for Animal Biotechnology?*, 2005, 14–15, http://www.pewtrusts.org/news_room_detail.aspx?id=36770.

182. See, e.g., AquaBounty Technologies, *Environmental Assessment.*

183. Andrew Pollack, "Genetically Altered Salmon," A1; *AquaBounty Technologies, Frequently Asked Questions,* http://www.aquabounty.com/technology/faq-297.aspx.

184. FDA, "Food Labeling; Labeling of Food Made from AquAdvantage Salmon," 75 Fed. Reg. 52,602 (August 26, 2010); FDA, Veterinary Medicine Advisory Committee, "Notice of Meeting," 75 Fed. Reg. 52,605 (August 26, 2010).

185. Andrew Zajac, "No Agreement Imminent on Salmon Labeling," *Los Angeles Times,* September 22, 2010, 20. The U.S. House of Representatives voted in June 2011 to block the FDA from approving AquaBounty's application, and Senate opposition is mounting. See Andrew Seidman, "FDA Faces Opposition over Genetically Engineered Salmon," *Los Angeles Times,* July 31, 2011, http://articles.latimes.com/2011/jul/31/nation/la-na-congress-salmon-20110731.

186. Mark A. Pollack and Shaffer, *When Cooperation Fails,* 70, 79.

187. See, e.g., Mandel, "Gaps, Inexperience, Inconsistencies, and Overlaps"; Angelo, "Regulating Evolution"; Kunich, "Mother Frankenstein."

188. Mandel, "Gaps, Inexperience, Inconsistencies, and Overlaps," 2244.

189. Bernstein, "When New Technologies Are Still New," 938–39.

190. See Hathaway, "Path Dependence," 609; Mark A. Pollack and Shaffer, *When Cooperation Fails,* 77; Lyndon, "Tort Law," 152–53.

191. See Collingridge, *Social Control,* 19 (describing "dilemma of [technology] control").

192. Hathaway, "Path Dependence," 610; Pierson, "Increasing Returns," 254; Arthur, *Increasing Returns,* 112; Lyndon, "Tort Law," 152.

193. Bauer and Gaskell, "Biotechnology Movement," 388; Kenney, *Biotechnology,* 28–89.

194. Krimsky, "From Asilomar to Industrial Biotechnology," 309.

195. Kenney, *Biotechnology,* 91.

196. See ibid., 94–106 (describing professors' active participation in the commercialization of biotechnology).

197. 35 U.S.C. §§ 200–212; Jasanoff, *Designs on Nature,* 235–37.

198. Wright, *Molecular Politics,* 398; Kenney, *Biotechnology,* 33.

199. Burkhardt, "Ethics of Agri-Food Biotechnology." See also Kenney, *Biotechnology,* 190–216, 222–27, 256.

200. Eichenwald, "Redesigning Nature," A1; Lambrecht, *Dinner,* 49–50.

201. Sunstein, *Laws of Fear,* 1–2, 44, 54; Frank Fischer, "Technological Deliberation," 300. Commercial researchers' desire to preserve a competitive advantage

over their rivals provides a further incentive to secrecy (Boucher, *Nanotechnology*, 70).

202. Geerlings and David, "Engagement and Translation," 200. See also Bernstein, "When New Technologies Are Still New," 927 (discussing justification for resistance to early intervention).

203. Jasanoff, *Science at the Bar*, 157.

204. See Busch and Lloyd, "What Can Nanotechnology Learn," 264.

205. Kysar, "Preference for Processes," 531.

206. For example, the documentary film *Who Killed the Electric Car?* alleges that the automobile and oil industries undermined the commercialization of the electric car.

207. Annika K. Martin, "Why McDonalds Pulled Frankenfries from Menu," *Time.com*, April 28, 2000, http://archives.cnn.com/2000/US/04/28/fries4_28.a.tm/index.html. See also Thompson, "Nano and Bio," 148; Busch and Lloyd, "What Can Nanotechnology Learn," 270.

208. Wilkins, "Introduction," 12–13.

209. Eichenwald, "Redesigning Nature," A1.

210. See generally Olson, *Logic of Collective Action*, 53–65.

211. Sheingate, "Promotion versus Precaution," 252.

212. Mandel, "Gaps, Inexperience, Inconsistencies, and Overlaps," 2243.

213. GAO, *Genetically Engineered Crops*, 4–6.

214. See Manning, "Competing Presumptions," 2042–43.

215. Feenberg, "Democratic Rationalization," 656. See also Feenberg, *Transforming Technology*, 3 ("The design of technology is thus an ontological decision fraught with political consequences. The exclusion of the vast majority from participation in this decision is profoundly undemocratic.").

216. See Bauer, "Distinguishing *Red* and *Green* Biotechnology," 84.

217. Ibid., 65.

218. Miller and Conko, *Frankenfood Myth*, 2, 8.

219. See Krimsky and Wrubel, *Agricultural Biotechnology*, 217, 231.

220. Eichenwald, "Redesigning Nature," A1.

221. Priest, "Biotechnology," 228–29; Eichenwald, "Redesigning Nature," A1.

222. Torgersen et al., "Promises, Problems, and Proxies," 21, 23.

223. Mellon, "View," 81, 87; Bratspies, "Biotechnology, Sustainability, and Trust," 282 ("[T]he bitter controversy will continue until and unless the public trusts in the social, as well as environmental sustainability of agricultural biotechnology.").

Chapter 3

1. Miles, "Nanotechnology Captured," 94; GAO, *Nanotechnology*, 1; Royal Society and Royal Academy of Engineering, *Nanoscience and Nanotechnologies*, 5.

2. See, e.g., Roco, "Introduction," 2; Karn and Bergeson, "Green Nanotechnology," 9 (Nanotechnology's "seemingly limitless potential will continue to inspire innovations in a dizzying array of beneficial applications and briskly transform society.").

3. GAO, *Nanotechnology*, 5; Scown, van Aerle, and Tyler, "Review," 654.

4. National Research Council, *Research Strategy*, 3; Rejeski, "Molecular Economy," 38.

5. See Davies, *Oversight*, 9–11; Tour, "Nanotechnology," 361–62; EPA, *Nanotechnology White Paper*, 12–13; Phoenix and Drexler, "Safe Exponential Manufacturing," 870; Jacobstein, "Foresight Guidelines," 2–4.

6. See Tour, "Nanotechnology," 361–62. See also European Commission, Scientific Committee, *Modified Opinion*, 11–12; EPA, *Nanotechnology White Paper*, 12–13.

7. See Davies, *Oversight*, 17–19; European Commission, Scientific Committee, *Modified Opinion*, 11–12; Phoenix and Drexler, "Safe Exponential Manufacturing," 870; Posner, *Catastrophe*, 35–36. The danger that self-replicating nanomachines might proliferate uncontrollably and ultimately consume the Earth, popularized in the Michael Crichton novel *Prey*, is increasingly dismissed as improbable. See Rip and Van Amerom, "Emerging De Facto Agendas," 136–43.

8. See Davies, *Oversight*, 9, 11; Jacobstein, "Foresight Guidelines," 2.

9. See Royal Society and Royal Academy of Engineering, *Nanoscience and Nanotechnologies*, 5; Rick Weiss, "Nanotech Poses Big Unknown to Science," *Washington Post*, February 1, 2004, A1.

10. EPA, *State of the Science Literature Review*, 10, 27–29.

11. Ibid., 22–23; Klaine et al., "Nanomaterials," 1827; GAO, *Nanotechnology*, 15.

12. See Oberdörster, Oberdörster, and Oberdörster, "Nanotoxicology," 835.

13. Klaine et al., "Nanomaterials," 1826–27.

14. Ibid., 1826; GAO, *Nanotechnology*, 9–10. Carbon nanotubes are basically tubes consisting of rolled-up sheets of graphite (9 n.10).

15. Klaine et al., "Nanomaterials," 1826; GAO, *Nanotechnology*, 18–19, 22.

16. Klaine et al., "Nanomaterials," 1827.

17. GAO, *Nanotechnology*, 10–23.

18. Ibid., 10–23; Fleischer and Grunwald, "Making Nanotechnology Developments Sustainable," 890–92.

19. See Gee and Greenberg, "Asbestos," 49.

20. Jim Morris, "Asbestos' U.S. Legacy May Be Half Million Deaths," *McClatchy Newspapers*, July 21, 2010, http://www.mcclatchydc.com/2010/07/21/97624/asbestos-us-legacy-may-be-half.html; Gee and Greenberg, "Asbestos," 49–63.

21. See U.S. Department of Health and Human Services, National Institute for Occupational Safety and Health, *Approaches;* Jasmine Li et al., "Nanoparticle-Induced Pulmonary Toxicity," 1026; Oberdörster, Oberdörster, and Oberdörster, "Nanotoxicology," 824; Nel et al., "Toxic Potential," 622; Royal Society and Royal Academy of Engineering, *Nanoscience and Nanotechnologies*, 41–42.

22. See Williams, "Scientific Basis," 117; Royal Society and Royal Academy of Engineering, *Nanoscience and Nanotechnologies*, 36, 79–80.

23. See Jasmine Li et al., "Nanoparticle-Induced Pulmonary Toxicity," 1026.

24. See Ning Li, Xia, and Nel, "Role of Oxidative Stress," 1695; Nel et al., "Understanding Biophysicochemical Interactions," 551.

25. See Nel et al., "Understanding Biophysicochemical Interactions," 554; Scown, van Aerle, and Tyler, "Review," 661. But see Mutlu et al., "Biocompatible Nanoscale Dispersion," 1669 (concluding that "aggregation of SWCNTs rather than their large aspect ratio" accounts for their toxicity).

26. See Chaudhry, Bouwmeester, and Hertel, "Current Risk Assessment Paradigm," 128–29.

27. See Baroli, "Penetration of Nanoparticles," 29–30, 41; Jasmine Li et al., "Nanoparticle-Induced Pulmonary Toxicity," 1025.

28. See Oberdörster, Oberdörster, and Oberdörster, "Nanotoxicology," 90–91, table 1.

29. Trickler et al., "Silver Nanoparticle Induced Blood-Brain Barrier Inflammation," 160. See also Jasmine Li et al., "Nanoparticle-Induced Pulmonary Toxicity," 1026–27.

30. See Royal Society and Royal Academy of Engineering, *Nanoscience and Nanotechnologies*, 35; European Commission, Scientific Committee, *Modified Opinion*, 34.

31. See Royal Society and Royal Academy of Engineering, *Nanoscience and Nanotechnologies*, 38 (summarizing human defenses against small particles in the lungs, skin, and digestive tract).

32. See Scown, van Aerle, and Tyler, "Review," 656 (mentioning rodent studies reporting lung damage from inhaled nanosized particles); Maynard and Kuempel, "Airborne Nanostructured Particles," 592–93 (discussing epidemiology and pathology studies involving nanosized particles); Nel et al., "Toxic Potential," 622, 625; Oberdörster, Oberdörster, and Oberdörster, "Nanotoxicology," 825.

33. Poland et al., "Carbon Nanotubes," 423. See also Xia, Li, and Nel, "Potential Health Impact," 139.

34. See Oberdörster, "Safety Assessment," 89; Xia, Li, and Nel, "Potential Health Impact," 139.

35. See Swiss Re, *Nanotechnology*, 13.

36. See Jasmine Li et al., "Nanoparticle-Induced Pulmonary Toxicity," 1696; Xia, Li, and Nel, "Potential Health Impact," 139 (observing that ultrafine particles and engineered nanoparticles "differ in many aspects such as sources, composition, homo- or heterogeneity, size distribution, oxidant potential, and potential routes of exposure").

37. Jasmine Li et al., "Nanoparticle-Induced Pulmonary Toxicity," 1690 (explaining that oxidative stress is a result of nanoparticle-induced overproduction of reactive oxygen species and/or a weakening of antioxidant defense mechanisms). See also Park et al., "Pro-Inflammatory and Potential Allergic Responses," 113.

38. For areas for further research, see, e.g., EPA, *Nanomaterials Research Strategy;* Klaine et al., "Nanomaterials," 1845. About 5 percent of federal nanotechnology funding falls under the rubric of research on health and environmental effects—a category that includes a wide range of projects such as the use of nanotechnology to remediate environmental pollution—and only about one-third of that modest amount supports research aimed at addressing nanotechnology's hazards. See Davies, *Nanotechnology Oversight*, 7.

39. Bennett and Sarewitz, "Too Little, Too Late?," 322. For an inventory of nanotechnology products, see *The Project on Emerging Nanotechnologies, Consumer Products*, http://www.nanotechproject.org/inventories/consumer/.

40. Helland et al., "Reviewing the Environmental and Human Health Knowledge Base," 1129–30; GAO, *Nanotechnology*, 29; Scown, van Aerle, and Tyler, "Review," 661.

41. Scown, van Aerle, and Tyler, "Review," 661.

42. GAO, *Nanotechnology*, 29.

43. Ibid., 30; EPA, *Nanomaterials Research Strategy*, 15, 20–23.

44. See Scown, van Aerle, and Tyler, "Review," 666 (noting the impracticality of "testing every single nanoparticle type"); Kulinowski and Colvin, "Environmental Impact," 21, 22.

45. Klaine et al., "Nanomaterials," 1832–34. See also Du et al., "TiO_2 and ZnO Nanoparticles," 822; Scown, van Aerle, and Tyler, "Review," 658–60 (suggesting that exposure to nanoparticles in aquatic environments may harm microorganisms, invertebrates, and vertebrates).

46. Klaine et al., "Nanomaterials," 1826 (estimating annual worldwide production of single-walled carbon nanotubes at more than 1,000 tons).

47. Aitken, "Regulation," 205–37; GAO, *Nanotechnology*, 24; Helland et al., "Reviewing the Environmental and Human Health Knowledge Base," 1125 (concluding that the size, shape, and other properties of carbon nanotubes determine their effects on environmental health); Royal Society and Royal Academy of Engineering, *Nanoscience and Nanotechnologies*, 43.

48. Helland et al., "Reviewing the Environmental and Human Health Knowledge Base," 1130.

49. Trouiller et al., "Titanium Dioxide Nanoparticles," 8784, 8788.

50. See National Research Council, *A Research Strategy*, 8–10 (discussing critical research gaps); Oberdörster, Oberdörster, and Oberdörster, "Nanotoxicology," 835 (concluding that the "lack of toxicology data on engineered NPs [nanoparticles] does not allow for adequate risk assessment"); Corley, Scheufele, and Hu, "Of Risks and Regulations," 1574.

51. EPA, *Nanomaterials Research Strategy*, 3.

52. See OECD, *List of Manufactured Nanomaterials and List of Endpoints for Phase One of the OECD Testing Programme*, 2008, http://www.oecd.org/officialdocuments/displaydocumentpdf/?cote=env/jm/mono(2008)13/rev&doclanguage=en; Kearns, "Nanomaterials."

53. OECD, "Six Years of OECD Work on the Safety of Manufactured Nanomaterials: Achievements and Future Opportunities," (2012), http://www.oecd.org/env/ehs/nanosafety/Nano%20Brochure%20Sept%202012%20for%20Web site%20%20(2).pdf.

54. *Strategy for Nanotechnology-Related Environmental, Health, and Safety Research National Nanotechnology Initiative*, 2008, www.nano.gov/NNI_EHS_Research_Strategy.pdf.

55. National Research Council, *Review*, 6–7.

56. Office of the U.S. President, National Science and Technology Council, Committee on Technology, *National Nanotechnology Initiative Environmental, Health, and Safety Research Strategy* (2011), www.nano.gov/sites/default/files/pub_resource/nni_2011_ehs_research_strategy.pdf.

57. EPA, *Nanomaterials Research Strategy*.

58. Beveridge and Diamond, "Update on Developments in EPA Regulation of Nanotechnology," April 5, 2010, http://www.bdlaw.com/news-843.html.

59. EPA, *Nanoscale Materials Stewardship Program*, 2010, http://www.epa.gov/oppt/nano/stewardship.htm; GAO, *Nanotechnology*, 33. A voluntary program in the

United Kingdom yielded similarly paltry results. See Marchant, Sylvester, and Abbott, "New Soft Law Approach," 135.

60. GAO, *Nanotechnology*, 31–33; EPA, *Nanoscale Materials Stewardship Program: Interim Report*, 27.

61. Bowman and Hodge, "'Governing' Nanotechnology," 479–83.

62. See 15 U.S.C. §§ 7501–9.

63. See GAO, *Nanotechnology*, 47–48 (discussing information requests by the state of California and the city of Berkeley's hazardous materials ordinance). See also, e.g., letter from chief scientist Jeffrey Wong, California Department of Toxic Substances Control, December 21, 2010), http://www.dtsc.ca.gov/TechnologyDevelopment/Nanotechnology/upload/Round_Two_Call-in_Letter.pdf (requesting chemical information and analytical test methods for six specified nanomaterials from 40 manufacturers and importers).

64. 15 U.S.C. §§ 2601–92.

65. See Percival et al., *Environmental Regulation*, 93 ("In theory, [the TSCA] is the broadest source of EPA's regulatory authority.").

66. See, e.g., Denison, "Ten Essential Elements," 10,020; GAO, *Chemical Regulation*, 3–6; Haemer, "Reform," 99; Applegate, "Worst Things First," 277; Applegate, "Perils," 261.

67. 15 U.S.C. § 2603.

68. Ibid., 2603(a).

69. GAO, *Nanotechnology*, 37; EPA, *Fall 2011 Regulatory Agenda* (January 20, 2012), 234, http://www.regulations.gov/#!documentDetail;D=EPA-HQ-OA-2012-0077-0001. EPA also intends to issue a rule that would require manufacturers of nanomaterials to report on production volume, methods of manufacture and processing, and exposure and release, but the rule would not require the generation of new health and safety information. See EPA, *Fall 2012 Regulatory Agenda* (December 24, 2012), 187, http://www.regulations.gov/#!documentDetail;D=EPA-HQ-OA-2012-0987-0001.

70. GAO, *Nanotechnology*, 38. See also Applegate, "Perils," 318–19 (suggesting that infrequent promulgation of test rules under the TSCA "is best explained by the elaborate procedural barriers that confine the test rules").

71. 15 U.S.C. § 2604(a).

72. Ibid., § 2602(9).

73. Ibid., § 2604(a), (b). The EPA takes no action on the vast majority of PMNs. See Applegate, Laitos, and Campbell-Mohn, *Regulation*, 611.

74. 15 U.S.C. § 2604(f). If there is insufficient information for the EPA to make a reasoned evaluation of health and environmental effects and if the chemical substance may present an unreasonable risk or will be produced or enter the environment in substantial quantities, the EPA may issue an administrative order to prohibit or limit manufacture or use (§ 2604(e)).

75. Ibid., § 2604(a); Elizabeth C. Brown et al., *TSCA Deskbook*, 36–37.

76. EPA, *TSCA Inventory Status*, 5. The EPA has declared, for example, that it generally considers carbon nanotubes to be distinct from graphite and other allotropes of carbon listed on the TSCA Inventory (EPA, "Toxic Substances Control Act Inventory Status of Carbon Nanotubes," 73 Fed. Reg. 64,946 [2008]).

77. See *Petition Requesting FDA Amend Its Regulations for Products Composed of*

Engineered Nanoparticles Generally and Sunscreen Drug Products Composed of Engineered Nanoparticles Specifically" (2006), 64–68, http://www.icta.org/doc/Nano%20FDA%20petition%20final.pdf.

78. EPA, *TSCA Inventory Status*, 6. See also Bergeson and Plamondon, "TSCA and Engineered Nanoscale Substances," 65–70. This position is consistent with the EPA's historical practice in defining "chemical substance." See American Bar Association, *Regulation of Nanoscale Materials*, 8–10, 12.

79. EPA, *Fall 2012 Regulatory Agenda*, 187. The TSCA authorizes the EPA to issue a significant new use rule for categories of chemical substances, and not just individual substances (15 U.S.C. § 2625(c)).

80. GAO, *Nanotechnology*, 36.

81. 15 U.S.C. § 2604(d).

82. See GAO, *Chemical Regulation*, 11. In response to a PMN or an SNUN, EPA sometimes enters into a consent order with a manufacturer limiting production until the manufacturer conducts testing (Monica and Monica, "Examples," 391–92).

83. 40 C.F.R. § 723.50(c)(1) (Low Volume Exemption), § 723.50(c)(2) (Low Release and Exposure Exemption).

84. Ibid., § 723.50(d), (e). Thus far, EPA has authorized only the use of the latter exemption for nanomaterials. See Beveridge and Diamond, "Update on Developments."

85. 15 U.S.C. § 2605(a).

86. *Corrosion Proof Fittings v. EPA*, 947 F.2d 1201, 1222 (5th Cir. 1991) (striking down the bulk of an EPA rule prohibiting almost all uses of asbestos despite the agency's reliance on more than 100 studies demonstrating health risks). See also Applegate, "Perils," 263; Flournoy, "Legislating Inaction," 340.

87. See Percival et al., *Environmental Regulation*, 407–8; *Corrosion Proof Fittings*, at 1222 ("EPA must balance the costs of its regulations against their benefits").

88. *Corrosion Proof Fittings*, at 1215–17, 1220.

89. 15 U.S.C. § 2607(e).

90. See EPA, "TSCA Section 8(e); Notification of Substantial Risk; Policy Clarification and Reporting Guidance," 68 Fed. Reg. 33,129, 33,131 (2003). A 1996 EPA offer of reduced penalties for noncompliance with Section 8(e) led to thousands of notifications, suggesting that the reporting requirement had been widely disregarded. Wagner, "Commons Ignorance," 1648. Nonetheless, the EPA has also imposed substantial penalties for noncompliance (Bergeson and Plamondon, "TSCA and Engineered Nanoscale Substances," 64).

91. Applegate, "Worst Things First," 311.

92. See ibid. (attributing the paucity of rules produced under the TSCA to the difficulty of establishing unreasonable risk). See also Albert C. Lin, "Beyond Tort," 1445–52 (describing the difficulties faced by toxic tort plaintiffs in proving causation).

93. See Davies, *Managing the Effects*, 11; Kuzma, *Nanotechnology-Biology Interface*, 23 (summarizing comments of workshop participants that the federal oversight process increasingly will have trouble keeping pace with nanotechnology product development and market entry).

94. See Haemer, "Reform," 115–16. See also Albert C. Lin, "Beyond Tort," 1441 n. 2 (noting that relatively few chemicals have been subject to toxicity testing).

95. See Davies, *Managing the Effects*, 11.

96. Safe Chemicals Act of 2010, S. 3209, 111th Cong. §§ 5, 7; Toxic Chemicals Safety Act of 2010, H.R. 5820, 111th Cong. §§ 4, 6.

97. Safe Chemicals Act of 2010 § 4(1)(B), 4(9); Toxic Chemicals Safety Act § 3(a)(2)(H), 3(a)(3). The legislation would go beyond the significant new use rule being drafted by the EPA in that it potentially would subject all ongoing uses of nanomaterials to notice and safety data requirements.

98. See, e.g., 33 U.S.C. § 1342 (establishing National Pollutant Discharge Elimination System to issue permits under the Clean Water Act); 42 U.S.C. § 7475 (requiring preconstruction permits for "major emitting facilities" in certain areas under the Clean Air Act).

99. See Davies, *Managing the Effects*, 14. Even where permit limits are feasible, application of the Clean Air Act or Clean Water Act to nanomaterials will face other hurdles. Both statutes generally require significant amounts of risk data as a prerequisite for regulatory action. See, e.g., 42 U.S.C. § 7412(b)(2) (authorizing regulation of hazardous air pollutants upon a finding that they "present or may present . . . a threat of adverse human health effects . . . or adverse environmental effects"). See also Barker et al., *Nanotechnology Briefing Paper*, 3 (concluding that to exercise Clean Water Act authority to regulate nanoparticles, the EPA would have to demonstrate adverse effects on human health or the environment). Both statutes also rely heavily on monitoring, which will likely require sophisticated equipment that is not yet available. See GAO, *Nanotechnology*, 40–41; Davies, *EPA and Nanotechnology*, 26.

100. 40 C.F.R. § 261.4(b)(1).

101. GAO, *Nanotechnology*, 41.

102. See, e.g., Davies, *Managing the Effects*, 18; Cormack, *CERCLA Nanotechnology Issues*, 3.

103. 15 U.S.C. § 2051(b)(1); 15 U.S.C. § 2052(a)(5); 15 U.S.C. §§ 2051–84.

104. Felcher, *Consumer Product Safety Commission*, 1; 15 U.S.C. § 2502(a)(1) (defining "consumer product" under CPSA); 15 U.S.C. § 1261(f)(2) (excluding pesticides, foods, drugs, cosmetics, and other substances from definition of "hazardous substances" under Federal Hazardous Substances Act). See also 15 U.S.C. § 2080 (listing additional limitations on jurisdiction of the Commission).

105. 15 U.S.C. §§ 2053–56.

106. See ibid., § 2056(b).

107. Ibid., § 2056. The commission may ultimately ban products that create an unreasonable risk that cannot be addressed by a "feasible consumer product safety standard" (15 U.S.C. § 2057).

108. The commission also has authority under the Federal Hazardous Substances Act to ban or regulate toxic substances that "may cause substantial personal injury or substantial illness" as a result of consumer use, but this authority has proven to be of limited use as well. See 15 U.S.C. § 1261(f), (q); Thomas et al., "Research Strategies," 18.

109. Felcher, *Consumer Product Safety Commission*, 9. The commission has approximately 500 employees and is responsible for the safety of more than 15,000 kinds of consumer products. See U.S. Consumer Product Safety Commission, *Frequently Asked Questions*, http://www.cpsc.gov/about/faq.html#wha.

110. 15 U.S.C. § 2055 (barring disclosure of trade secrets).

111. Felcher, *Consumer Product Safety Commission*, 8–10.

112. 21 U.S.C. § 355. Medical devices for which there is insufficient information to show that general or special controls (such as notification, labeling, or registration requirements) will provide a reasonable assurance of safety and effectiveness also require premarket approval (21 U.S.C. § 360c).

113. 7 U.S.C. § 136a.

114. See 40 C.F.R. Part 158. See also American Bar Association, *Adequacy of FIFRA;* Sumner, Luempert, and Stevens, "Agricultural Chemicals," 137–41 (summarizing testing requirements).

115. EPA, "Petition for Rulemaking"; International Center for Technology Assessment et al., "Citizen Petition for Rulemaking to the United States Environmental Protection Agency," Docket No. EPA-HQ-OPP-2008-650.

116. *Natural Resources Defense Council v. U.S. Environmental Protection Agency*, No. 12-70268 (9th Cir. filed Jan. 26, 2012).

117. 21 U.S.C. § 355(a).

118. 21 C.F.R. §§ 312.20, 312.23.

119. Ibid., § 312.21. See also 21 C.F.R. §§ 314.1–314.650 (regulations governing approval of new drugs).

120. See Kuzma, *Nanotechnology-Biology Interface*, 21 (suggesting that the "FDA may be unaware that nanotechnology is being used in a particular product" because it "can only regulate products based on the claims of the sponsor"). Because different FDA divisions (centers) are responsible for regulating new drugs, biologics, and medical devices, some commentators have expressed concern that the classification of a product will influence the regulatory procedures and requirements that are applied. See Mandel, "Nanotechnology Governance," 1359–61; Fender, "FDA and Nano," 1076–80; Kuzma, *Nanotechnology-Biology Interface*, 21 (noting that medical devices that are of low risk are regulated less stringently than are drugs).

121. See FDA, "Sunscreen Drug Products for over-the-Counter Human Use," 28,195.

122. 21 U.S.C. § 321(p) (defining "new drug"); 21 U.S.C. § 355(a) (requiring approval of new drugs).

123. FDA, "Sunscreen Drug Products for over-the-Counter Human Use: Final Monograph," 27,671.

124. Complaint, *International Center for Technology Assessment et al. v. Hamburg*, Case No. CV-11-6592, (N.D. Cal.), December 21, 2011, http://www.centerforfoodsafety.org/wp-content/uploads/2011/12/1-Pls-Complaint.pdf.

125. Letter from Leslie Kux, Assistant Commissioner for Policy, Food and Drug Administration, to Andrew Kimball, April 20, 2012, http://www.icta.org/doc/Andrew%20Kimbrell-FDA-2006-P-0213-Citizen%20Petition.pdf.

126. Fender, "FDA and Nano," 1074.

127. FDA, "FDA Authority over Cosmetics," March 3, 2005, http://www.fda.gov/Cosmetics/GuidanceComplianceRegulatoryInformation/ucm074162.htm; Taylor, *Regulating the Products*, 28. If the safety of a product has not been substantiated, the manufacturer must include a warning to that effect on the cosmetic product's label (21 C.F.R. § 740.10). Color additives are the only cosmetic ingredient for which premarket approval is required (21 U.S.C. § 379e).

128. 21 C.F.R. Pts. 710, 720; 62 Fed. Reg. 43,071, 43,073–74 (1997) (revoking regulation providing for collection of voluntarily filed information on adverse reactions but maintaining availability of adverse reaction reporting forms). Per the Fair Packaging and Labeling Act, the FDA does require cosmetic manufacturers to provide a list of ingredients. See 21 C.F.R. § 701.3; 15 U.S.C. § 1454(c)(3).

129. See Taylor, *Regulating the Products*, 28; van Calster and Bowman, "Good Foundation?," 268, 272.

130. GAO, *FDA Should Strengthen Its Oversight*, 29.

131. FDA, Draft Guidance for Industry: Assessing the Effects of Significant Manufacturing Process Changes, Including Emerging Technologies on the Safety and Regulatory Status of Food Ingredients and Food Contact Substances, Including Food Ingredients that are Color Additives, April 2012, http://www.fda.gov/Food/GuidanceComplianceRegulatoryInformation/GuidanceDocuments/FoodIngredientsandPackaging/ucm300661.htm#manufacturing.

132. Ibid.

133. 29 U.S.C. § 654(a)(1); see also §§ 651–78.

134. Ibid., § 652(8); see also § 655(b).

135. Ibid., § 655(b)(5), (7).

136. See Sidney A. Shapiro and McGarity, "Reorienting OSHA," 2–12.

137. See, e.g., ibid., 2; Mendeloff, *Dilemma*, 100–102; Kniesner and Leeth, "Abolishing OSHA," 46.

138. See Bartis and Landree, *Nanomaterials in the Workplace*, 8 (reporting concerns expressed at workshop that "the formal process of establishing [occupational exposure limits] would overwhelm NIOSH and OSHA capabilities").

139. Sarahan, "Nanotechnology Safety," 195. See also Lyndon, "Information Economics," 1826, n. 119 (discussing OSHA regulations).

140. *Industrial Union Dept., AFL-CIO v. American Petroleum Inst.*, 448 U.S. 607, 639, 652–53 (1980) (plurality statement that OSHA can promulgate a new standard only if it demonstrates that regulation is "reasonably necessary and appropriate to remedy a significant risk of material health impairment"); *American Textile Mfrs. Inst. v. Donovan*, 452 U.S. 490, 505 n. 25 (1981).

141. See Bartis and Landree, *Nanomaterials in the Workplace*, 8; Sarahan, "Nanotechnology Safety," 198–201.

142. See Shapo, *Experimenting with the Consumer*, 210 (discussing nanotechnology as example of uncontrolled experimentation).

143. See Dan B. Dobbs, *The Law of Torts* (St. Paul, MN: West, 2000), 969–77.

144. See Albert C. Lin, "Beyond Tort," 1446–52.

145. See Dana, "When Less Liability May Mean More Precaution," 170–71 (discussing hypothetical case of injury caused by nanoparticles in skin creams).

146. Davies, *EPA and Nanotechnology*, 41; Albert C. Lin, "Beyond Tort," 1446. See also Gifford, "Peculiar Challenges," 613–98.

147. Albert C. Lin, "Deciphering the Chemical Soup," 963; Davies, *EPA and Nanotechnology*, 18.

148. See Tolan, "Natural Resource Damages," 425–26 (discussing difficulties faced by the government in recovering natural resources damages); Doremus et al., *Environmental Policy Law*, 61 (considering obstacles to recovery by private parties for environmental harms to their property).

149. See Wood, Geldart, and Jones, "Crystallizing," 17–18. See also Kenneth

David and Paul B. Thompson, eds., *What Can Nanotechnology Learn from Biotechnology?: Social and Ethical Lessons for Nanoscience from the Debate over Agrifood Biotechnology and GMOs* (Burlington, MA: Academic, 2008).

150. See, e.g., Gavelin and Wilson, *Democratic Technologies?*, 13–22, 113–35 (listing public engagement processes involving nanotechnology). For a description of the relationship between the National Nanotechnology Program and the National Nanotechnology Initiative, see John F. Sargent Jr., *The National Nanotechnology Initiative: Overview, Reauthorization, and Appropriations Issues* (Washington, DC: Congressional Research Service, 2012).

151. Chris Toumey, "Democratizing Nanotech, Then and Now," *Nature Nanotechnology* 6 (2011): 605.

152. 15 U.S.C. § 7501(b)(10).

153. Ibid., § 7501(b)(10)(D).

154. See National Science Foundation, "NSF Renews Centers for Nanotechology in Society," news release, October 12, 2010, http://www.nsf.gov/news/news_summ.jsp?cntn_id=117862.

155. See Guston, "Center for Nanotechnology," 377, 380–84. See also *Center for Nanotechnology in Society at Arizona State University*, http://cns.asu.edu/index.htm; Barben et al., "Anticipatory Governance," 979–1000. For a description of research activities at CNS-UCSB, see *Research at CNS-UCSB*, http://www.cns.ucsb.edu/index.php?option=com_content&task=view&id=20&Itemid=54. See also, e.g., Pidgeon et al., "Deliberating the Risks," 95 (discussing results of public workshops on energy and health nanotechnologies held at Santa Barbara and in United Kingdom).

156. Hamlett, Cobb, and Guston, *National Citizens' Technology Forum;* Laurent, *Replicating Participatory Devices*, 5. See also Wilsdon, "Paddling Upstream," 1 (describing use of citizens' jury in United Kingdom to consider nanotechnology and resulting recommendations).

157. See Powell and Kleinman, "Building Citizen Capacities," 341, 344 (reporting that several participants in 2005 Wisconsin nanotechnology consensus conference expressed the view that neither scientists nor government would be responsive to their concerns and that the conference would not influence policy); Perez, "Precautionary Governance," 61 (noting "general frustration in the literature from the continuing failure of participatory processes to influence the regulatory output"); Gavelin and Wilson, *Democratic Technologies?*, 62, 81–82 (finding little evidence that public engagement on nanotechnology has affected decision making in the United Kingdom).

158. See Laurent, *Replicating Participatory Devices*, 12. See also Bennett and Sarewitz, "Too Little, Too Late?," 319 (noting that the Nanotechnology Act does not specify "the processes by which research results are to enhance decision making").

159. See Laurent, *Replicating Participatory Devices*, 10–11.

160. See ibid., 10–11, 25.

161. See Guston, "Center for Nanotechnology," 379, 389; *Nano-Scale Informal Science Education Network*," http://www.nisenet.org/.

162. See *Hart Research Associates*, *Nanotechnology*, *Synthetic Biology*, *and Public Opinion*, September 22, 2009, http://www.nanotechproject.org/process/assets/files/8286/nano_synbio.pdf (reporting that nearly 7 in 10 Americans have heard little or nothing about nanotechnology); Guston, "Center for Nanotechnology,"

389 (noting that polls reflect modest public knowledge of and engagement in nanotechnology); see also Corley and Scheufele, "Outreach Gone Wrong?," 22 (finding widening gaps in nanotech knowledge between the least educated and the most educated citizens).

163. Guston, "Center for Nanotechnology," 389 (noting that "a $1 billion/per year NNI in the United States overwhelms the $3-million/year nanotechnology-in-society network"). See also Kurath and Gisler, "Informing, Involving, or Engaging?," 568 (contending that public outreach efforts in Europe with respect to nanotechnology have occurred after major investment decisions were already made and "tend to limit public engagement to matters of values and social and ethical aspects, rather than to expose expertise to scrutiny").

164. See Guston, "Center for Nanotechnology," 384–85; CNS-ASU, Annual Report to the National Science Foundation for the Period September 15, 2011 to September 14, 2012, NSF 0937591 (2012), 50–51, http://cns.asu.edu/cns-library/year/?action=getfile&file=552§ion=lib; Jameson Wetmore (assistant professor, Arizona State University), interview by author, March 24, 2010.

165. See Erik Fisher and Mahajan, "Contradictory Intent?," 5 (comparing 15 U.S.C. § 7501(b)(1)-(9) with 15 U.S.C. § 7501(b)(10)).

166. See Ebbesen, "Role of the Humanities and Social Sciences," 2–3. See also Fisher and Mahajan, "Contradictory Intent?," 13 (noting potential for Nanotechnology Act to "emerge as a shrewd piece of legislative rhetoric, reducing societal research and related activities to a sideshow in order to push rapid nanotechnology development past a potentially wary public"); Rogers-Hayden and Pidgeon, "Moving Engagement 'Upstream'?," 345 (discussing the danger that public engagement on nanotechnology in the United Kingdom will "serve only token purposes").

167. See, e.g., Davies, *Nanotechnology Oversight*, 7–8; Environmental Law Institute, *Securing the Promise of Nanotechnology*, 19; Mandel, "Nanotechnology Governance," 1371–72; Maynard and Rejeski, "Too Small to Overlook," 174.

168. Holdren et al., *Policy Principles.*

169. Bergeson, *Nanotechnology*, 2. See also Holdren et al., *Policy Principles*, 3 ("existing regulatory statutes provide a firm foundation for the regulation and oversight of nanomaterials").

170. Mandel, "Nanotechnology Governance," 1364.

171. Holdren et al., *Policy Principles*, 5.

172. Rhitu Chatterjee, "Insurers Scrutinize Nanotechnology," *Environmental Science and Technology* 43 (2009): 1240. See also Berube, *Nano-Hype*, 498.

173. Chartis, "Lexington Insurance Company Introduces LexNanoShield," news release, March 30, 2010, http://www.chartisinsurance.com/ncglobalweb/internet/US/en/files/PR_LexNanoShield_03_30_10_tcm295-253662.pdf.

174. Berube, *Nano-Hype*, 498; Rakhlin, "Regulating Nanotechnology," ¶ 28. In addition, insurance can play a role in facilitating technology development by encouraging companies to undertake activities in spite of uncertain hazards. See David A. Fischer and Jerry, "Teaching Torts," 865. With nanotechnology research and development proceeding at full speed, however, the role of insurance as technology facilitator is of minimal concern.

175. Dana, "When Less Liability May Mean More Precaution,"198; Rakhlin, "Regulating Nanotechnology," ¶¶ 37, 43.

176. Hansen and Tickner, "Challenges," 343; Marchant, Sylvester, and Abbott, "Risk Management Principles," 53; Mandel, "Nanotechnology Governance," 1363.

177. See Davies, *EPA and Nanotechnology*, 35–39; Marchant, Sylvester, and Abbott, "New Soft Law Approach," 132–33.

178. Marchant, Sylvester, and Abbott, "New Soft Law Approach," 132; Mandel, "Nanotechnology Governance," 1363 ("Not only would designing a new statutory scheme to regulate nanotechnology appear impossible due to the substantial current unknowns, but any such regulation would not confront the problems of insufficient science and limited detection capabilities."). See also Bowman and Hodge, "'Governing' Nanotechnology?," 477–78 (stating that soft law mechanisms provide "scope for innovation, creativity and flexibility").

179. See EPA, *Nanoscale Materials Stewardship Program; Interim Report;* Environmental Defense–DuPont Nano Partnership, *Nano Risk Framework*, 11–14. Another voluntary effort resulted in the creation of the Responsible NanoCode, a set of general principles for nanotechnology companies to follow. See *Responsible NanoCode*, http://www.responsiblenanocode.org/index.html.

180. Marchant, Sylvester, and Abbott, "New Soft Law Approach," 147–52; Mandel, "Nanotechnology Governance," 1376.

181. See, e.g., Darnall and Sides, "Assessing the Performance," 110 (finding "little evidence that *overall* VEP participation is associated with improved environmental performance"); Kurt A. Strasser, "Do Voluntary Corporate Efforts Improve Environmental Performance?," 548–50.

182. Hansen and Tickner, "Challenges," 354.

183. Maynard and Rejeski, "Too Small to Overlook," 174.

184. Daniel J. Fiorino, *Voluntary Initiatives, Regulation, and Nanotechnology Oversight: Charting a Path* (2010), 25–26, www.nanotechproject.org/publications/archive/voluntary; Mandel, "Nanotechnology Governance," 1377.

185. Davies, *EPA and Nanotechnology*, 36 (noting that major trade associations with interests in nanotechnology include few small firms, that many different trade associations represent users or potential users of nanotechnology, and that even manufacturers of nanomaterials are not represented by a single trade association); Patrick Lin, "Nanotechnology Bound," 111–12 (contending that cooperation in self-regulation across all nanotechnology companies is unlikely because nanotechnology is not a single "industry").

186. Darnall and Sides, "Assessing the Performance," 111; Strasser, "Do Voluntary Corporate Efforts Improve Environmental Performance?," 551.

187. Environmental Defense–DuPont Nano Partnership, *Nano Risk Framework*. The Framework itself states that it is not a substitute for compliance with applicable government regulations (14).

188. See, e.g., Coglianese, "Assessing Consensus," 1278–86 (finding that negotiated rulemaking efforts by EPA took nearly as long as traditional rulemaking).

189. EPA, *Nanoscale Materials Stewardship Program: Interim Report*.

190. See Marchant, Sylvester, and Abbott, "New Soft Law Approach," 135 (noting that only four companies have pledged to participate in depth in the NMSP and concluding that "limited industry participation has precluded any comprehensive risk assessment"); Fiorino, *Voluntary Initiatives*, 36.

191. Meili and Widmer, "Voluntary Measures," 449.

192. Kimbrell, "Governance," 708.

193. Ibid.; Bowman and Hodge, "'Governing' Nanotechnology?," 478; Meili and Widmer, "Voluntary Measures," 453, 457; Marchant, Sylvester, and Abbott, "New Soft Law Approach," 148 ("[F]ull public disclosure is required to promote public trust."); Kuzma, *Nanotechnology-Biology Interface*, 39.

194. Kimbrell, "Governance," 709.

195. Marchant, Sylvester, and Abbott, "New Soft Law Approach," 136–51.

196. Ibid., 151–52. See also Hansen and Tickner, "Challenges," 355 (noting that start-up and small companies in particular may lack resources to generate and disclose information voluntarily).

197. *Consumer Products Drop Nanotechnology Claims for Fear of Consumer Recoil*, June 15, 2009, http://www.nanowerk.com/news/newsid=11181.php.

198. Dana, "When Less Liability May Mean More Precaution," 154–55, 157–58, 195.

199. Ibid., 183–87 (acknowledging that conditioning liability relief on premarket and postmarket testing raises the cost of obtaining such relief).

200. Ibid., 157, 167.

201. *GoodNanoGuide*, http://goodnanoguide.org/HomePage.

202. Karn and Bergeson, "Green Nanotechnology," 9, 11.

203. Davies, *Nanotechnology Oversight*, 18.

204. Mandel, "Nanotechnology Governance," 1363. See also Miles, "Nanotechnology Captured," 90–105 (discussing efforts to establish standards for measuring and characterizing nanomaterials).

205. Patrick Lin, "Nanotechnology Bound," 118.

206. For a discussion of such practices, see U.S. Department of Health and Human Services, National Institute for Occupational Safety and Health, *Approaches*, 35–51.

207. Wagner, "Triumph," 94–103; Sidney A. Shapiro and McGarity, "Not So Paradoxical," 745–51. See also Marchant, Sylvester, and Abbott, "Risk Management Principles," 45 (noting that feasibility standards may overregulate or underregulate risks).

208. Berube, *Nano-Hype*, 493; Matsuura, *Nanotechnology Regulation and Policy*, 126–40.

209. Patrick Lin, "Nanotechnology Bound," 116.

210. Office of the U.S. President, National Science and Technology Council, *National Nanotechnology Initiative Strategic Plan*, 23.

211. Roco, "Introduction," 2. See also Drexler, "Nanotechnology," 24–25.

212. Nordmann and Schwarz, "Lure of the 'Yes,'" 255, 257.

213. See Ramey, "Promise," 11.

214. See, e.g., "Down on the Farm: Genetic Engineering Meets Ecology," *Technology Review*, January 1987, 24; Cornish, "Products of Genetic Engineering," 11.

215. See Berube, *Nano-Hype*, 353–54 ("[T]he whole [nanotechnology] enterprise promises to eradicate poverty and overcome the limitation of resources by providing material goods—pollution free—to all the world's people."); Nordmann and Schwarz, "Lure of the 'Yes,'" 272.

216. See Rechtschaffen, "Warning Game," 313–14; Lyndon, "Information Economics," 1830–31. See also Mendeloff, *Dilemma*, 209–10 (discussing economic efficiency and rights-based rationales for information disclosure).

217. See Bowman, van Calster, and Friedrichs, "Nanomaterials and Regulation of Cosmetics," 92; Plater et al., *Environmental Law and Policy*, 539 (noting the effect of California's Proposition 65 warning requirement on product reformulation); Barsa, "Note," 1239–43.

218. European Union, *Regulation of the European Parliament*, Arts. 16, 19(1)(g).

219. Cf. Noah, "Imperative to Warn," 364–65 (emphasizing the importance of ensuring that warnings do not overstate or understate risks); Golan et al., "Economics of Food Labeling," 139 ("Consumers are more likely to read and understand labels that are clear and concise.").

220. See, e.g., Blackwelder, "Comment," 10,520; ETC Group, "Nanotech Product Recall Underscores Need for Nanotech Moratorium: Is the Magic Gone?," news release, 2006, http://www.etcgroup.org/en/node/14.

221. See Satterfield et al., "Anticipating the Perceived Risk," 754 (meta-analysis of surveys finding that the public generally perceives the benefits of nanotechnology as outweighing risks).

222. See Drexler and Wejnert, "Nanotechnology and Policy," 15–16 (contending that "opting out of nanotechnology would relegate a nation to no longer being a player in technological civilization").

223. Davies, *Managing the Effects*, 18–19.

224. Berube, "Regulating Nanoscience," 493.

225. See Fleischer and Grunwald, "Making Nanotechnology Developments Sustainable," 893.

226. Albert C. Lin, "Size Matters," 397–404. The financial assurance could take the form of collateral bonds such as cash deposits, legally binding promissory obligations, or surety bonds by third-party guarantors. See Kysar, "Ecologic," 243 (discussing types of environmental assurance bonds).

227. See Costanza and Perrings, "Flexible Assurance Bonding System," 71; Cornwell and Costanza, "Environmental Bonds," 220–21.

228. Kysar, "Ecologic," 208.

229. Lyndon, "Information Economics," 1813–17 (discussing disincentives for chemical producers to identify and publicize toxic effects).

230. Perrings, "Environmental Bonds," 106–7.

231. See Wagner, "Commons Ignorance," 1738–39; Glicksman, Markell, and Mandelker, *Environmental Protection Law*, 764; Applegate, "Worst Things First," 308.

232. See Shogren, Herriges, and Govindasamy, "Limits to Environmental Bonds," 113; Bohm and Russell, "Comparative Analysis," 432.

233. Kysar, "Ecologic," 242.

234. Ibid., 208.

235. See ibid., 245–46.

236. See Wätzold, "Efficiency and Applicability," 307 (noting the difficulty of measuring the value of environmental damage in monetary terms and the potential for actual damage to be greater than amount of the bond). One of the primary difficulties encountered in implementing bonding requirements under the Surface Mining Control and Reclamation Act and similar statutes has been the inadequacy of bonds to cover damages actually incurred. See, e.g., Giffin, "West Virginia's Seemingly Eternal Struggle," 133–34.

237. See Wätzold, "Efficiency and Applicability," 307; Cornwell and Costanza,

"Environmental Bonds," 235. In light of these concerns, the imposition of bonding requirements should not serve as a justification for capping tort liability or for preempting liability under other statutes. See David Gerard, "The Law and Economics of Reclamation Bonds," *Resources Policy* 26 (2000): 189.

238. See Shogren, Herriges, and Govindasamy, "Limits to Environmental Bonds," 115–18 (discussing liquidity constraints).

239. See Kysar, "Ecologic," 248; Shogren, Herriges, and Govindasamy, "Limits to Environmental Bonds," 116–17. But see also James Boyd, "Financial Responsibility," 438–39 (contending that opponents of financial assurance requirements often express unwarranted fears of mass disruption and insurance unavailability).

240. See Cornwell and Costanza, "Environmental Bonds," 231.

241. Sclove, *Reinventing Technology Assessment*, 38–39.

242. Kaufmann et al., "Why Enrol Citizens," 201, 210–11.

243. Perez, "Precautionary Governance," 63–64.

244. Abbott, Marchant, and Sylvester, "Framework Convention," 10,507; Marchant and Sylvester, "Transnational Models," 717; Abbott, Marchant, and Sylvester, "Transnational Regulation," 528.

245. Abbott, Marchant, and Sylvester, "Framework Convention," 10,513.

246. Bergeson, "Regulation, Governance, and Nanotechnology," 10,516.

247. Ibid.

248. Abbott, Marchant, and Sylvester, "Transnational Regulation," 530–31.

249. Rejeski, "Comment," 10,518.

250. Abbott, Marchant, and Sylvester, "Framework Convention," 10,513–14.

Chapter 4

1. Intergovernmental Panel, *Climate Change 2007: Impacts, Adaptation, and Vulnerability*, 11–18.

2. See ibid., 869, 878.

3. See Kyoto Protocol, Annex B.

4. See John M. Broder, "Climate Talks Yield Commitment to Ambitious, but Unclear, Actions," *New York Times*, Dec. 9, 2012, A13; "The Cancun Climate-Change Conference: A Sort of Progress," *The Economist*, December 18, 2010, 41.

5. See Keith, "Geoengineering," *Nature*, 420. Geoengineering proposals can be distinguished from carbon capture and sequestration techniques in that the latter seek to control CO_2 emission into the atmosphere, whereas geoengineering seeks to control atmospheric CO_2 postemission. See Keith, "Geoengineering the Climate," 248–49.

6. GAO, *Technology Assessment*, v–vi (rating all geoengineering techniques as unready for full-fledged production or use); GAO, *Climate Change*, 13–15.

7. GAO, *Technology Assessment*, 43.

8. T. J. Blasing, *Carbon Dioxide Information Analysis Center, Recent Greenhouse Gas Concentrations*, http://cdiac.ornl.gov/pns/current_ghg.html.

9. United Nations Framework Convention, Copenhagen Accord, ¶ 2, FCCC/CP/2009/11/Add.1 ("We agree that deep cuts in global emissions are required.").

10. Royal Society, *Geoengineering the Climate,* 10; Keith, "Geoengineering," *Nature,* 420 (suggesting that "[g]eoengineering has become a label for technologically overreaching proposals that are omitted from serious consideration in climate assessments").

11. For a more detailed discussion of ocean fertilization, see Royal Society, *Geoengineering the Climate,* 16–19.

12. Chisholm, Falkowski, and Cullen, "Dis-Crediting Ocean Fertilization," 309.

13. See Buesseler and Boyd, "Will Ocean Fertilization Work?," 67 (discussing ocean fertilization experiments that generated algal blooms over several weeks but found limited evidence of carbon transport to deep ocean); Smetacek and Naqvi, "Next Generation," 3958 (noting that the turnover time of atmospheric CO_2 by the biosphere is approximately four years).

14. See Royal Society, *Geoengineering the Climate,* 17; Peterson, "Can Algae Save Civilization?," 69–70, 76 (reporting estimated costs for iron fertilization that range from $0.5 billion to $3 billion per billion tons of atmospheric carbon transferred to the deep ocean, less than the cost of reducing equivalent emissions or removing equivalent emissions from power plant smokestacks).

15. See GAO, *Technology Assessment,* 29; Royal Society, *Geoengineering the Climate,* 17; Zahariev, Christian, and Denman, "Preindustrial, Historical, and Fertilization Simulations," 79 (reporting modeling results predicting that even if the entire Southern Ocean were fertilized with iron, such efforts at best could stimulate ocean uptake of only 11 percent of 2004 anthropogenic CO_2 emissions).

16. See Aumont and Bopp, "Globalizing Results," 2017 (concluding that factors other than iron also influence effectiveness of sequestration and that fertilization outside the Southern Ocean is relatively ineffective); Philip W. Boyd et al., "Mesoscale Iron Enrichment Experiments," 612 (summarizing results of small-scale iron fertilization experiments); Buesseler and Boyd, "Will Ocean Fertilization Work?," 68.

17. Royal Society, *Geoengineering the Climate,* 18; Blain et al., "Effect of Natural Iron Fertilization," 1073 (noting the "complex interplay between the iron and carbon cycles" and cautioning against the assumption that iron fertilization will work based on observed natural phytoplankton bloom); Richard Black, "Setback for Climate Technical Fix," *BBC News,* March 23, 2009, http://news.bbc.co.uk/2/hi/7959570.stm (reporting that much phytoplankton growth in one experiment entered the food chain rather than sinking to the bottom of the oceans).

18. Royal Society, *Geoengineering the Climate,* 19.

19. Silver et al., "Toxic Diatoms and Domoic Acid."

20. See Strong et al., "Ocean Fertilization," 347 (contending that properly field testing ocean fertilization would entail fertilizing "an enormous swath of ocean" and assessing the effects "for between decades and a century or so" to demonstrate sequestration and to document adverse effects on ecosystems); Royal Society, *Geoengineering the Climate,* 17–18; Peterson, "Can Algae Save Civilization?," 77–78.

21. See Royal Society, *Geoengineering the Climate,* 15–16; Workman, "Assessment of Options," 2878; Intergovernmental Panel, *Carbon Dioxide Capture and Storage.* For a summary of ongoing air capture research efforts, see American Physical Society, *Direct Air Capture;* Keith, "Why Capture CO_2," 1655.

22. Royal Society, *Geoengineering the Climate*, 15.

23. GAO, *Technology Assessment*, 21; Keith, Heidel, and Cherry, "Capturing CO_2 from the Atmosphere," 107.

24. Global CCS Institute, *The Global Status of CCS: 2012* (2012), 3, 30, 32, at http://cdn.globalccsinstitute.com/sites/default/files/publications/47936/global-status-ccs-2012.pdf; Royal Society, *Geoengineering the Climate*, 15.

25. Ibid., 16; "Net Benefits," *The Economist*, March 17, 2012, 89 (noting that the estimated cost of capturing and storing carbon using direct capture, between $600 and $800 per ton, is approximately 80 times the current price of carbon credits). See also Workman et al., "Assessment of Options," 2879–80. Proponents nonetheless contend that with further technical advances, direct capture could ultimately prove competitive with the cost of controlling GHG emissions from mobile or relatively dispersed sources. See Keith, Heidel, and Cherry, "Capturing CO_2 from the Atmosphere," 108; Lackner, "Capture of Carbon Dioxide," 104–5.

26. GAO, *Technology Assessment*, 21.

27. American Physical Society, *Direct Air Capture*, iii.

28. Schuiling and Krijgsman, "Enhanced Weathering," 352–54 (suggesting the land application of olivine to capture CO_2); Harvey, "Mitigating the Atmospheric CO_2 Increase" (discussing the addition of calcium carbonate powder to the ocean to absorb CO_2 and to counter ocean acidification).

29. Royal Society, *Geoengineering the Climate*, 14.

30. Ibid., 13–15.

31. Ibid., 21.

32. See Intergovernmental Panel, *Climate Change 2007: The Physical Science Basis*, 97, 100.

33. Royal Society, *Geoengineering the Climate*, 23.

34. See, e.g., ibid., v; U.K. House of Commons Science and Technology Committee, *Regulation of Geoengineering*, 38.

35. For a more detailed discussion of these scenarios, see Kintisch, *Hack the Planet*, 39–52.

36. See, e.g., Wigley, "Combined Mitigation/Geoengineering Approach," 452 (proposing the use of geoengineering in the near term, in combination with emission reductions, to reduce amount of mean temperature increase).

37. See Crutzen, "Albedo Enhancement," 211–12.

38. See Robock, "Benefits, Risks, and Costs," 3–7.

39. Royal Society, *Geoengineering the Climate*, 29. The release of engineered nanoparticles has also been proposed. See David W. Keith, "Phosphoretic Levitation of Engineered Aerosols for Geoengineering," *Proceedings of the National Academy of Sciences* 107 (2010): 16,428–31.

40. See Crutzen, "Albedo Enhancement," 212.

41. See GAO, *Technology Assessment*, 33; Royal Society, *Geoengineering the Climate*, 29; Crutzen, "Albedo Enhancement," 212.

42. Royal Society, *Geoengineering the Climate*, 29–31; Rasch et al., "Overview," 4030; Caldeira and Wood, "Global and Arctic Climate Engineering," 4052–53 (discussing "highly idealized numerical simulations of extremely simple climate engineering").

43. Royal Society, *Geoengineering the Climate*, 31.

44. See McClellan, Keith, and Apt, "Cost Analysis," 034019. Crutzen, "Albedo Enhancement," 212–13. See also Robock et al., "Benefits, Risks, and Costs," 3–7 (estimating costs of various methods); Royal Society, *Geoengineering the Climate*, 32; Turco, "Geoengineering the Stratospheric Sulfate Layer"; "10 Questions: Richard Turco and the Nuttiness of Climate Engineering," *UCLA Today*, http://www.today.ucla.edu/portal/ut/PRN-10-questions-richard-turco-and-161052.aspx.

45. Royal Society, *Geoengineering the Climate*, 31.

46. Ibid. (calling for further development of models to support more reliable assessments); Bala, "Problems with Geoengineering Schemes," 46 (urging caution in interpreting climate modeling results, particularly because feedbacks in the natural climate system are not fully represented); Turco and Yu, "Geoengineering the Stratospheric Sulfate Aerosol Layer" (explaining that properties of injected aerosol would be far from ideal for blocking radiation).

47. Turco, "Geoengineering the Stratospheric Sulfate Layer," 6.

48. See Oliver Morton, "Is This What It Takes to Save the World?," *Nature* 447 (2007): 134 (reporting modeling results finding shifts in temperature and precipitation); Schneider, "Geoengineering," 297–98.

49. Robock, Oman, and Stenchikov, "Regional Climate Responses," D16101.

50. See Tilmes, Müller, and Salawitch, "Sensitivity of Polar Ozone Depletion," 1203–4; Crutzen, "Albedo Enhancement," 215–16. Other adverse effects could result from the interaction of sulfur aerosols with poorly understood processes. See Royal Society, *Geoengineering the Climate*, 31 (noting the "range of so far unexplored feedback processes"); Morton, "Is This What It Takes," 135 (remarking that the stratosphere "is tied to the troposphere below in complex ways that greenhouse warming is already changing").

51. Turco, "Geoengineering the Stratospheric Sulfate Layer," 1, 3–4.

52. Royal Society, *Geoengineering the Climate*, 27; Latham et al., "Global Temperature Stabilization," 3970. Under certain conditions, however, cloud seeding might decrease albedo. See Rasch, Latham, and Chen, "Geoengineering by Cloud Seeding," 2.

53. See Philip Rasch, "Geoengineering IIe: The Scientific Basis and Engineering Challenges," written testimony before the House Committee on Science and Technology, February 4, 2010, 11–12, http://science.house.gov/sites/republicans.science.house.gov/files/documents/hearings/020410_Rasch.pdf.

54. Latham et al., "Global Temperature Stabilization," 3970–71; Royal Society, *Geoengineering the Climate*, 27.

55. Royal Society, *Geoengineering the Climate*, 28.

56. Rasch, Latham, and Chen, "Geoengineering by Cloud Seeding," 6.

57. Latham et al., "Global Temperature Stabilization," 3985; Ben Webster, "Bill Gates Pays for 'Artificial' Clouds to Beat Greenhouse Gases," *Times* (London), May 8, 2010, http://technology.timesonline.co.uk/tol/news/tech_and_web/article7120011.ece.

58. Royal Society, *Geoengineering the Climate*, 28.

59. Ibid.; Latham et al., "Global Temperature Stabilization," 3982–83.

60. See Royal Society, *Geoengineering the Climate*, 32; Angel, "Feasibility of Cooling the Earth," 17,184.

61. Royal Society, *Geoengineering the Climate*, 33.

62. See Angel, "Feasibility of Cooling the Earth," 17,188–89.

63. Royal Society, *Geoengineering the Climate*, 32. With respect to Roger Angel's proposal to deploy a fleet of discs, for example, approximately 16 trillion discs would need to be manufactured and placed in orbit, and the cost of the proposal has been estimated at $5 trillion. See Morton, "Is This What It Takes," 136.

64. Royal Society, *Geoengineering the Climate*, 33.

65. See Barrett, "Incredible Economics," 47 ("Geoengineering is a stopgap measure, a 'quick fix,' a 'Band-Aid.'").

66. See Bengtsson, "Geo-Engineering to Confine Climate Change," 231 (explaining that the dissolution of CO_2 into oceans sequesters 70–80 percent of CO_2 over several hundred years and that cessation of a sulfur release project would quickly lead to renewed warming).

67. See Matthews and Caldeira, "Transient Climate-Carbon Simulations," 9951–52 (describing how temperatures, previously suppressed by aerosols, would quickly rebound to the levels they would have reached had no geoengineering been implemented).

68. See Burns, "Climate Geoengineering," 49.

69. For further discussion regarding the problem of ocean acidification, see Caldeira and Wickett, "Ocean Model Predictions"; Kolbert, "Darkening Sea," 66, 69–74. Higher GHG concentrations in the atmosphere may also affect terrestrial ecosystems by changing the competitive balances between different plant species. See Bengtsson, "Geo-Engineering to Confine Climate Change," 231.

70. See Bala, "Problems with Geoengineering Schemes," 45–46.

71. See Rasch, "Geoengineering IIe," 8, 14–15 (describing research on stratospheric aerosols and cloud whitening); see also 4 (estimating total U.S. research budget for geoengineering at $1 million per year or less); U.K. House of Commons Science and Technology Committee, *Regulation of Geoengineering*, 23 (noting that Russian scientists have carried out subscale field testing).

72. Roger B. Dworkin, "Science, Society, and the Expert Town Meeting," 1481.

73. GAO, *Technology Assessment*, 49 (noting that a majority of experts consulted in one study "advocated starting significant climate engineering research now or in the very near future"); Leinen, "Asilomar International Conference," 2–3. See also Bipartisan Policy Center Task Force, *Geoengineering*, 3; Gordon, *Engineering the Climate*, 38 ("[B]road consideration of comprehensive and multi-disciplinary climate engineering research at the federal level [should] begin as soon as possible in order to ensure scientific preparedness for future climate events[.]"); GAO, *Climate Change*, 39 (recommending development of a "clear, defined, and coordinated approach to geoengineering research in the context of a federal strategy to address climate change"); U.K. House of Commons Science and Technology Committee, *Regulation of Geoengineering*, 35–38; Kintisch, *Hack the Planet*, 132–48; Hester, *Remaking the World*, 11–12.

74. Henry Fountain, "A Rogue Climate Experiment Has Ocean Experts Outraged," *New York Times*, October 19, 2012, A1.

75. See Jamieson, "Can Space Reflectors Save Us?" ("[A] dedicated geoengineering research program risks creating a self-amplifying cycle of interest groups and lobbies, building momentum toward eventual deployment as a way of justifying the research.").

76. See Gardiner, "Is 'Arming the Future' with Geoengineering Really the Lesser Evil?" (discussing ethical issues raised by proposed research into stratospheric aerosols as an emergency measure); Fleming, "Weather as a Weapon" (urging study of historical, ethical, legal, political, and societal aspects of geoengineering as a priority over technically oriented research).

77. See Solar Radiation Management Governance Initiative, *Solar Radiation Management*, 24.

78. See GAO, *Technology Assessment*, 43; Claire L. Parkinson, *Coming Climate Crisis?: Consider the Past, Beware the Big Fix* (Lanham, MD: Rowman and Littlefield, 2010), 235–58.

79. See, e.g., Keith, Parson, and Morgan, "Research on Global Sun Block," 426 (urging field testing of SRM techniques at "scales big enough to produce barely detectable climate effects and reveal unexpected problems, yet small enough . . . to limit risks").

80. Bunzl, "Researching Geoengineering," 2. See also Solar Radiation Management Governance Initiative, *Solar Radiation Management*, 21.

81. Robock et al., "Test for Geoengineering?," 530–31. Local effects would be especially hard to predict.

82. Ibid., 531.

83. See Parkinson, *Coming Climate Crisis?*, 194–200; Kintisch, *Hack the Planet*, 85–88.

84. Royal Society, *Geoengineering the Climate*, 41 ("In some cases . . . it is not clear that field trials can usefully be conducted on a limited scale."); Alan Robock, testimony before the House Committee on Science and Technology, November 5, 2009, 8 (contending that meaningful test of geoengineering via stratospheric aerosols "is not possible without full-scale deployment," given variability in weather and difficulty of determining cause and effect), http://climate.envsci.rutgers.edu/pdf/HouseTestimonyRobock.pdf); but see also Jane Long, testimony before the House Committee on Science and Technology, March 18, 2010, 16 (acknowledging that determining effects of geoengineering tests presents "a big challenge" but is "not hopeless"), http://science.house.gov/sites/republicans.science.house.gov/files/documents/031810_Long.pdf.

85. See Barrett, *Why Cooperate?*, 6, 101. See also Olson, *Logic of Collective Action*, 53–65.

86. See Barrett, *Why Cooperate?*, 93, 101; Keith, "Geoengineering," in *Encyclopedia of Global Change*, 495, 500; Schelling, "Economic Diplomacy of Geoengineering," 306 (contending that, compared to reducing emissions, geoengineering "is certainly way ahead in administrative simplicity").

87. See Victor et al., "Geoengineering Option," 71–72.

88. See U.K. House of Commons Science and Technology Committee, *Regulation of Geoengineering*, 23.

89. Victor et al., "Geoengineering Option," 73; Robock, "20 Reasons," 14, 17.

90. Schelling, "Economic Diplomacy of Geoengineering," 305. See also Barrett, "Incredible Economics," 49.

91. Oliver Burkeman, "'Asking People to Reduce Their Carbon Emissions Is a Noble Invitation, but as Incentives Go, It Isn't a Strong One,'" *Guardian*, October 12, 2009, 6. See also Steven D. Levitt and Stephen J. Dubner, *Superfreakonomics:*

Global Cooling, Patriotic Prostitutes, and Why Suicide Bombers Should Buy Life Insurance (New York: William Morrow, 2009), 190–203.

92. See Keith, "Geoengineering the Climate," 276. Geoengineering proposals may seem "painless" only because adverse effects have yet to be identified.

93. See, e.g., Bunzl, "Researching Geoengineering," 3 (characterizing moral hazard concern as "far-fetched since, at least among policy makers, nobody believes that geoengineering offers anything but a relatively short stopgap to buy time for other action"); U.K. House of Commons Science and Technology Committee, *Regulation of Geoengineering*, 23 (arguing that it is "equally plausible" that geoengineering research would persuade people that global warming presents a serious threat necessitating mitigation).

94. See, e.g., Barrett, "Incredible Economics," 47; Victor et al., "Geoengineering Option," 66 (characterizing geoengineering as an "emergency shield"); Royal Society, *Geoengineering the Climate*, v.

95. See Gardiner, "Is 'Arming the Future' with Geoengineering Really the Lesser Evil?," 292 (raising questions regarding emergency rationale).

96. Williamson, "Climate Geoengineering," 18, 21; MacCracken, "On the Possible Use," 11 (suggesting geoengineering should be viewed as "a complement to adaptation and the building of resilience"); Wigley, "Combined Mitigation/Geoengineering Approach," 452.

97. See Burns, "Climate Geoengineering," 46–47.

98. See Dale Griffin and Amos Tversky, "The Weighing of Evidence and the Determinants of Confidence," in *Heuristics and Biases: The Psychology of Intuitive Judgment*, ed. Thomas Gilovich, Dale Griffin, and Daniel Kahneman (Cambridge: Cambridge University Press, 2002), 230–32; Keith, Heidel, and Cherry, "Capturing CO_2 from the Atmosphere," 107.

99. Weber, "Experience-Based and Description-Based Perceptions," 109.

100. For further discussion of the moral hazard issue, see Albert C. Lin, "Does Geoengineering Present a Moral Hazard?," __ *Ecol. L. Q.*__ (forthcoming 2013), http://papers.ssrn.com/sol3/papers.cfm?abstract_id=2152131.

101. See Todd LaPorte to author, July 11, 2011.

102. "Lift-Off: Research into the Possibility of Engineering a Better Climate Is Progressing at an Impressive Rate—and Meeting Strong Opposition," *Economist*, November 4, 2010, 102.

103. See Gardiner, "Is 'Arming the Future' with Geoengineering Really the Lesser Evil?," 291–304. See also Jamieson, "Ethics and Intentional Climate Change," 332 (suggesting that even if geoengineering were successful, "it would still have the bad effect of reinforcing human arrogance and the view that the proper human relationship to nature is one of domination").

104. Gordon, *Engineering the Climate;* U.K. House of Commons Science and Technology Committee, *Regulation of Geoengineering.*

105. Wilfried Rickels, et al., *Large-Scale Intentional Interventions into the Climate System? Assessing the Climate Engineering Debate* (2011), iii.

106. For analyses of options for regulating geoengineering under U.S. law, see Hester, *Remaking the World;* GAO, *Climate Change*, 26–30.

107. See, e.g., Robock, "Whither Geoengineering?," 1166–67; Victor, "On the Regulation of Geoengineering," 330.

108. United Nations Framework Convention.

109. See Albert C. Lin, "Geoengineering Governance," 15, 19.

110. Alister Doyle, "Futuristic Climate Schemes to Get U.N. Hearing," *Reuters*, October 27, 2010, http://www.reuters.com/article/2010/10/27/us-climate-geoengineering-interview-idUSTRE69Q4BJ20101027.

111. CBD, Preamble. A total of 193 countries are parties to the CBD. See CBD, List of Parties, http://www.cbd.int/convention/parties/list/. The United States signed the treaty but has not ratified it.

112. Ninth Meeting.

113. United Nations Intergovernmental Oceanographic Commission, *Report*. See also "Law of the Sea."

114. Report of the Tenth Meeting.

115. Solar Radiation Management Governance Initiative, *Solar Radiation Management*, 24.

116. Convention on the Prevention of Marine Pollution; 1996 Protocol to the Convention on the Prevention of Marine Pollution.

117. 1996 Protocol to the Convention on the Prevention of Marine Pollution, Art. 1.4.1; Convention on the Prevention of Marine Pollution, Art. III.1(a).

118. 1996 Protocol to the Convention on the Prevention of Marine Pollution, Art. 1.4.2.2; Convention on the Prevention of Marine Pollution, Art. III.1(b)(ii).

119. International Maritime Organization, *Report*, ¶ 4.23. The meeting also endorsed a "Statement of Concern" prepared by scientific working groups declaring that "knowledge about the effectiveness and potential environmental impacts of ocean iron fertilization currently was insufficient to justify large-scale operations and that this could have negative impacts on the marine environment and human health" (¶¶ 4.14, 4.23 [quoting LC/SG 30/14, ¶¶ 2.23 to 2.25]).

120. International Maritime Organization, Resolution LC-LP.1.

121. See International Maritime Organization, Resolution LC-LP.2.

122. United Nations Convention on the Law of the Sea. The Law of the Sea Convention recognizes a general duty of member states "to protect and preserve the marine environment" (Art. 192). The FCCC focuses specifically on greenhouse gases in the atmosphere. Royal Society, *Geoengineering the Climate*, 40.

123. Royal Society, *Geoengineering the Climate*, 40.

124. Montreal Protocol.

125. Montreal Protocol (as amended), Arts. 2A–2I, Annexes A–C.

126. Convention on Long-Range Transboundary Air Pollution, November 13, 1979, 18 I.L.M. 1442.

127. Michael MacCracken to author, November 13, 2009.

128. See Leinen, "Asilomar International Conference," 3.

129. *Asilomar International Conference on Climate Intervention Technologies* (brochure), http://climateresponsefund.org/index.php?option=com_content&view=article&id=137&Itemid=90.

130. See *Statement by the Board of Directors of the Climate Response Fund*, March 19, 2010, http://www.climateresponsefund.org/index.php?option=com_content&view=article&id=147&Itemid=87; Kintisch, "March Geoengineering Confab."

131. ETC Group et al., *Open Letter to the Climate Response Fund and the Scientific Organizing Committee of the Asilomar Conference on Climate Intervention*, March 4, 2010, http://www.etcgroup.org/en/node/5080.

132. Ibid.

133. Krimsky, *Genetic Alchemy*, 103, 109–12.

134. See *Asilomar International Conference* (brochure) ("Because of the effectiveness of the ultimate guidelines and procedures, there have been no dangerous releases of organisms modified with recombinant DNA technologies."); Jim Rendon, "Who Eats Geoengineering Risk?," *Mother Jones*, March 29, 2010, http://motherjones.com/blue-marble/2010/03/geoengineering-risk-asilomar-climate-intervention (noting that almost all participants came from the United States and the United Kingdom).

135. See Goodell, "Hard Look."

136. Asilomar International Conference, *Statement from the Conference's Scientific Organizing Committee.*

137. Asilomar Scientific Organizing Committee, *Asilomar Conference Recommendations*, 17–24.

138. Tollefson, "Geoengineers Get the Fear," 656.

139. See Asilomar Scientific Organizing Committee, *Asilomar Conference Recommendations*, 23. One group of commentators has suggested looking to the Belmont Report as a source of ethical guidance for geoengineering research. Issued by the U.S. Department of Health, Education, and Welfare in 1979, the Belmont Report sets out three principles for research involving human subjects: respect for persons, beneficence, and justice. In the context of a geoengineering experiment, the principle of respect for persons could be the basis of a requirement of informed consent, most likely from representatives of persons affected. See Morrow, Kopp, and Oppenheimer, "Toward Ethical Norms," 4–6.

140. See Leiserowitz, *International Public Opinion*, 3.

141. Asilomar Scientific Organizing Committee, *Asilomar Conference Recommendations*, 23.

142. See Sunstein, *Laws of Fear;* Ellis and FitzGerald, "Precautionary Principle," 781; Dana, "Behavioral Economic Defense," 1315–45.

143. See United Nations Framework Convention, Art. 3(1).

144. The FCCC Conference of the Parties meets regularly to review implementation of the FCCC. See United Nations Framework Convention, Art. 7. Technical bodies include the Intergovernmental Panel on Climate Change, which reviews and assesses scientific, technical, and socioeconomic information relevant to the understanding of climate change. See Intergovernmental Panel on Climate Change, http://www.ipcc.ch/organization/organization.shtml.

145. See Esty, "Case for a Global Environmental Organization," 287, 292 (advocating comprehensive approaches to solving environmental problems rather than ad hoc issue-by-issue management).

146. See Bipartisan Policy Center Task Force, *Geoengineering*, 30.

147. See Kyoto Protocol, Annex B. The European Union and several nations collectively responsible for approximately 15 percent of global GHG emissions have agreed to an extension of emission limits beyond 2012. See Juliet Eilperin, "U.N. Opts to Extend Kyoto Protocol," *Washington Post*, December 9, 2012, A4.

148. United Nations Framework Convention, Copenhagen Accord; Eric J. Lyman, "After Marathon Talks, Countries Set Goal for New Climate Deal in Effect around 2020," *Environment Reporter* 42 (2011): 2859. See also Benedick, "Considerations on Governance," 7–8 (discussing difficulties encountered in climate negotiations).

149. See Barrett, *Why Cooperate?*, 6, 101; Victor et al., "Geoengineering Option," 71.

150. See Jamieson, "Ethics and Intentional Climate Change," 329 (noting that decision to undertake geoengineering would likely be made by the same wealthy countries that have caused climate change).

151. Scott Barrett, testimony before House Committee on Science and Technology, March 18, 2010, 8 (imagining scenario in which India might opt to implement geoengineering to counter declining agricultural production), http://science.house.gov/sites/republicans.science.house.gov/files/documents/031210_Barrett.pdf.

152. See ibid., 11–12.

153. Victor et al., "Geoengineering Option," 74–75; Royal Society, *Geoengineering the Climate*, 51–52 (recommending development of a code and framework). See also Rappert, "Pacing Science and Technology," 118.

154. Rayner et al., *Memorandum.*

155. U.K. House of Commons Science and Technology Committee, *Regulation of Geoengineering*, 29.

156. Ibid., 31.

157. Rappert, "Pacing Science and Technology," 115–17; Kourany, "Integrating the Ethical," 376–77.

158. See Blackstock and Long, "Politics of Geoengineering," 527.

159. See M. Granger Morgan, testimony before House Committee on Science and Technology, March 15, 2010, 4, http://gop.science.house.gov/Media/hearings/full10/mar18/Morgan.pdf.

160. See Frank Rusco, testimony before House Committee on Science and Technology, March 18, 2010, 7 (noting expert views that most CDR approaches, with the exception of ocean fertilization, could be addressed domestically), http://science.house.gov/sites/republicans.science.house.gov/files/documents/031810_Rusco.pdf.

161. See Barrett, testimony, 2 (defining geoengineering to refer only to SRM techniques); Blackstock and Long, "Politics of Geoengineering," 527 (calling for development of norms and formal frameworks for addressing SRM).

162. Solar Radiation Management Governance Initiative, *Solar Radiation Management*, 26. The initiative is sponsored by the Royal Society, the Academy of Sciences for the Developing World, and the Environmental Defense Fund.

163. Blackstock and Long, "Politics of Geoengineering," 527; Robock et al., "Test for Geoengineering?," 531 ("[E]ven a major disruption to agricultural output would be difficult to attribute to geoengineering."); Schneider, "Geoengineering," 294 (predicting "very high" potential for disputes over causation of any weather disasters that occur during geoengineering experiments).

164. For one such proposal, see Banerjee, "Limitations of Geoengineering Governance," 33–34.

165. See Dobbs, *Law of Remedies*, 86–91.

166. Various reports and commentators agree on this point. See, e.g., Asilomar Scientific Organizing Committee, *Asilomar Conference Recommendations*, 23; Rayner et al., *Memorandum* ("Wherever possible, those conducting geoengineering research should be required to notify, consult, and ideally obtain the prior informed

consent of, those affected by the research activities."); U.K. House of Commons Science and Technology Committee, *Regulation of Geoengineering*, 31; Royal Society, *Geoengineering the Climate*, 42.

167. See, e.g., U.K. House of Commons Science and Technology Committee, *Regulation of Geoengineering*, 31 (identifying the need "to spell out what consultation means and whether, and how, those affected can veto or alter proposed geoengineering tests"); Asilomar Scientific Organizing Committee, *Asilomar Conference Recommendations*, 24 (noting that "conference discussions were not intended to specify the organization or authority of the needed governance systems"). One paper recommends "best practices" for promoting deliberative processes on geoengineering, based on a consideration of case studies of local participation in environmental decision making. See Shobita Parthasarathy, Lindsay Rayburn, Mike Anderson, Jessie Mannisto, Molly Maguire, and Dalal Najib, *Geoengineering in the Arctic: Defining the Governance Dilemma* (Ann Arbor: University of Michigan Science, Technology, and Public Policy Program, 2010).

168. Buchanan and Keohane, "Legitimacy of Global Governance Institutions," 416.

169. Banerjee, "Limitations of Geoengineering Governance," 22 ("[T]he current conversation about geoengineering governance lacks input from those who may be most affected.").

170. Albert C. Lin, "Geoengineering Governance," 22.

171. See, e.g., Nye, "Globalization's Democratic Deficit," 2–6 (discussing how to reconcile global institutions with democratic accountability).

172. See Held, "Democratic Accountability," 374 ("[T]hose whose life expectancy and life chances are significantly affected by social forces and processes ought to have a stake in the determination of the conditions and regulation of these, either directly or indirectly through political representatives.").

173. Nanz and Steffek, "Global Governance," 326–28.

174. Nanz and Steffek, "Global Governance," 328; Buchanan and Keohane, "Legitimacy of Global Governance Institutions," 427–28.

175. See Van Rooy, *Global Legitimacy Game*, 128–29; Buchanan and Keohane, "Legitimacy of Global Governance Institutions," 416–17 ("[T]he social and political conditions for democracy are not met at the global level and there is no reason to think that they will be in the foreseeable future."); Dahl, "Can International Organizations Be Democratic?," 19.

176. Van Rooy, *Global Legitimacy Game*, 137 ("*[V]oice* is more valuable to democracy than *vote*."); Buchanan and Keohane, "Legitimacy of Global Governance Institutions," 417 ("Democracy worth aspiring to is more than elections; it includes a complex web of institutions, including a free press and media, an active civil society, and institutions to check abuses of power by administrative agencies and elected officials.").

177. Betsill and Corell, "Introduction," 1; Alkoby, "Global Networks," 388–99 (discussing involvement of nongovernmental organizations in climate change negotiations).

178. Betsill, "Reflections," 178; Raustiala, "States, NGOs, and International Environmental Institutions," 726–30; Ebbesson, "Public Participation," 689.

179. Nanz and Steffek, "Global Governance," 323. See also Scholte, "Civil So-

ciety," 213–15 ("Civil society groups bring citizens together non-coercively in deliberate attempts to mould the formal laws and informal norms that regulate social interaction."); Held, *Democracy and the Global Order*, 181–82; Habermas, *Between Facts and Norms*, 367.

180. Nanz and Steffek, "Global Governance," 323.

181. See ibid., 330–31 (proposing a similar mechanism within the WTO). NGO representation on such a board also would facilitate the expression and consideration of alternative viewpoints, but the legitimacy of such an arrangement would be open to question. See Van Rooy, *Global Legitimacy Game*, 138.

182. Buchanan and Keohane, "Legitimacy of Global Governance Institutions," 428.

183. See Alkoby, "Global Networks," 382, 400; Van Rooy, *Global Legitimacy Game*, 137–40. NGOs may nevertheless be held accountable internally by their members and externally by other institutional actors (Spiro, "Non-Governmental Organizations and Civil Society," 788–89).

184. Spiro, "Non-Governmental Organizations and Civil Society," 773.

185. Nanz and Steffek, "Global Governance," 333; Alkoby, "Global Networks," 382, 401.

186. Thorsten Benner, Reinicke, and Witte, "Multisectoral Networks," 201; Scholte, "Civil Society," 230–32; Nanz and Steffek, "Global Governance," 323.

187. See Van Rooy, *Global Legitimacy Game*, 157–58 (advocating proportional accountability—that is, the idea "that those most affected by a decision should have the greatest claims on the decision-maker and the greatest means of redress").

188. Cf. Nanz and Steffek, "Global Governance," 331 (discussing inequalities faced by stakeholders from developing countries seeking to participate in the WTO).

189. Ibid., 332.

Chapter 5

1. ETC Group, *Extreme Genetic Engineering*, 1.

2. See "What's in a Name?," *Nature Biotechnology* 27 (2009): 1071; Steven A. Benner and Sismour, "Synthetic Biology," 533; Ball, "Starting from Scratch," 624, 625; Tucker and Zilinskas, "Promise and Perils," 26. See also Haseloff and Ajioka, "Synthetic Biology," 1.

3. Check, "Designs on Life," 417. See also Royal Academy, *Synthetic Biology*, 18–19; Endy, "Foundations," 449.

4. Robert F. Service, "Algae's Second Try," *Science* 333 (2011): 1238; Warren C. Ruder, Ting Lu, and James J. Collins, "Synthetic Biology Moving into the Clinic," *Science* 333 (2011): 1248.

5. See Schmidt, *Synthetic Biology;* Royal Academy, *Synthetic Biology*, 38–41; Khalil and Collins, "Synthetic Biology," 367; Gibbs, "Synthetic Life," 75, 81; Ball, "Starting from Scratch," 625; Yearley, "Ethical Landscape," 2; Genya V. Dana, Todd Kuiken, David Rejeski, and Allison A. Snow, "Four Steps to Avoid a Synthetic-Biology Disaster," *Nature* 483 (2012): 29.

6. U.S. Presidential Commission, *New Directions*, 70.

7. Tucker and Zilinskas, "Promise and Perils," 25; Rodemeyer, *New Life, Old Bottles*, 18.

8. See Ball, "Starting from Scratch," 626; Gibbs, "Synthetic Life," 81; Rodemeyer, *New Life, Old Bottles*, 17–18.

9. Pennisi, "Synthetic Genome," 958. See also Nicholas Wade, "Genetic Code of E. Coli Is Hijacked by Biologists," *New York Times*, July 15, 2011, A14 (describing another recent breakthrough involving large-scale alteration of a bacterial genome).

10. Pennisi, "Synthetic Genome," 958–59. Experts note that this work did not truly synthesize an organism, as the synthesized genome was a mere copy of an existing genome and the genome was inserted into an existing cell (959). See also Jim Collins, "Got Parts, Need Manual," *Nature* 465 (2010): 424.

11. Pennisi, "Synthetic Genome," 958. One critical difficulty involves the tendency of genetic circuits to mutate and become nonfunctional (Tucker and Zilinskas, "Promise and Perils," 31). See also Check, "Designs on Life," 418 (noting the complexity and unpredictability of biological systems); Kwok, "Five Hard Truths," 288 (summarizing key challenges).

12. See Yearley, "Ethical Landscape," 2 (comparing potential for informal experiments in synthetic biology to information technology breakthroughs initiated in garages and other informal settings); Gibbs, "Synthetic Life," 81 (noting ease of access to information and synthetic DNA); Mooallem, "Do-It-Yourself Genetic Engineering," 40 (noting synthetic biologists' goal of enabling "the most sophisticated custom-built life forms [to] be assembled from a catalog of standardized parts").

13. Mooallem, "Do-It-Yourself Genetic Engineering," 40; Yearley, "Ethical Landscape," 2; Tucker and Zilinskas, "Promise and Perils," 28, 43; ETC Group, *Extreme Genetic Engineering*, 7–10 (discussing the low cost of DNA synthesis and of DNA synthesizing machines).

14. Tucker and Zilinskas, "Promise and Perils," 33.

15. See Bette Hileman, "Chemicals Can Turn Genes On and Off; New Tests Needed, Scientists Say," *Environmental Health News*, August 3, 2009, http://www.environmentalhealthnews.org/ehs/news/epigenetics-workshop.

16. International Risk Governance Council, *Guidelines*, 26.

17. Rodemeyer, *New Life, Old Bottles*, 26; Dana et al., "Four steps," 29.

18. Tucker and Zilinskas, "Promise and Perils," 35.

19. U.S. Presidential Commission, *New Directions*, 71–74; Tucker and Zilinskas, "Promise and Perils," 37; ETC Group, *Extreme Genetic Engineering*, 23–24.

20. See Yearley, "Ethical Landscape," 3; Tucker and Zilinskas, "Promise and Perils," 35–42.

21. See Thomas Douglas and Savulescu, "Synthetic Biology," 690.

22. Zhang, Marris, and Rose, *Transnational Governance*, 27.

23. See Rodemeyer, *New Life, Old Bottles*, 29–45 (reviewing potential applicability of existing laws and guidelines to synthetic biology), 9 (characterizing new legislation specific to synthetic biology as "an unlikely option"); U.S. Presidential Commission, *New Directions*, 80–102 (summarizing existing laws and regulations that apply to synthetic biology).

24. For a concise summary of the reviews associated with different types of experiments, see Pei et al., *Regulatory Frameworks*, 160.

25. U.S. Department of Health and Human Services, National Institutes of Health, "Final Action under the NIH Guidelines for Research Involving Recombinant DNA Molecules (NIH Guidelines)," 77 Fed. Reg. 54,584 (2012).

26. Rodemeyer, *New Life, Old Bottles*, 30, 34. These committees are established by each research institution that receives funding from the NIH (30).

27. See U.S. Presidential Commission, *New Directions*, 80–81, 101–2; Rodemeyer, *New Life, Old Bottles*, 38–45 (analyzing potentially applicable laws).

28. 40 C.F.R. pt. 725.

29. Pei et al., *Regulatory Frameworks*, 166.

30. 7 U.S.C. § 8401; 42 U.S.C. § 262a; 7 C.F.R. pt. 331; 9 C.F.R. pt. 121; 42 C.F.R. pt. 73.

31. 42 C.F.R. § 73.3(c).

32. APHIS/CDC, *Applicability of the Select Agent Regulations*, 4.

33. See U.S. Presidential Commission, *New Directions*, 83–85; National Research Council, Committee on Advances in Technology and the Prevention of Their Application to Next Generation Biowarfare Threats, *Globalization, Biosecurity, and the Future of the Life Sciences* (Washington, DC: National Academies Press, 2006), 232.

34. 15 C.F.R. Pts. 730–74. See also U.S. Presidential Commission, *New Directions*, 86.

35. Bar-Yam et al., *Regulation of Synthetic Biology*, 10.

36. Department of Health and Human Services, "Screening Framework Guidance." See also U.S. Presidential Commission, *New Directions*, 88.

37. Sarah Kellogg, "The Rise of DIY Scientists: Is It Time for Regulation?," *Washington Lawyer*, May 2012, 21, 25–27 (discussing FBI outreach efforts as well as a partnership between the Woodrow Wilson International Center for Scholars and DIYbio.org to develop a code of ethics for amateur synthetic biologists); National Science Advisory Board for Biosecurity, *Strategies to Educate Amateur Biologists and Scientists in Non–Life Science Disciplines about Dual Use Research in the Life Sciences* (2011), 8–9, http://oba.od.nih.gov/biosecurity/pdf/finalnsabbreport-amateurbiologist-nonlifescientists_june-2011.pdf.

38. Convention on Biological Diversity, Art. 8(g).

39. Cartagena Protocol, Art. 7.

40. Ibid., Art. 3(g). The protocol defines "[m]odern biotechnology" as the application of "[i]n vitro nucleic acid techniques" or "[f]usion of cells beyond the taxonomic family that overcome natural physiological reproductive or recombination barriers and that are not techniques used in traditional breeding and selection" (Art. 3(i)).

41. The Nagoya–Kuala Lumpur Supplementary Protocol, signed in 2011 but not yet in effect, would require member states to provide for measures to respond to significant damage to biodiversity resulting from living modified organisms that find their origin in a transboundary movement. See Nagoya–Kuala Lumpur Supplementary Protocol on Liability and Redress to the Cartagena Protocol on Biosafety, http://bch.cbd.int/protocol/NKL_text.shtml.

42. Tenth Meeting of the Conference of the Parties to Convention on Biological Diversity, Decision X/37: Biofuels and Biodiversity, § 16, UNEP/CBD/COP/DEC/X/37 (October 29, 2010), http://www.cbd.int/decision/cop/?id=12303.

43. Agreement on the Application of Sanitary and Phytosanitary Measures, Arts. 3.1, 5.1.

44. See Christopher J. Borgen, "Resolving Treaty Conflicts," George *Washington International Law Review* 37 (2005): 573; Thomas Gehring, "The Institutional Complex of Trade and Environment: Toward an Interlocking Governance Structure and a Division of Labor," in *Managing Institutional Complexity: Regime Interplay and Global Environmental Change*, ed. Sebastian Oberthür and Olav Schram Stokke (Cambridge, MA: MIT Press, 2011), 227, 240.

45. Protocol for the Prohibition of the Use of Asphyxiating, Poisonous, or Other Gases, and of Bacteriological Methods of Warfare, June 17, 1925, 26 U.S.T. 571, 94 L.N.T.S. 65.

46. Convention on the Prohibition of the Development, Production and Stockpiling of Bacteriological (Biological) and Toxin Weapons and on Their Destruction, Art. I, April 10, 1972, 26 U.S.T. 583, 1015 U.N.T.S. 163. As of February 2013, 167 states were parties to the BWC. See United Nations Office at Geneva, http://www.unog.ch/80256EE600585943/(httpPages)/7BE6CBBEA0477B52C1257186 0035FD5C?OpenDocument.

47. European Group on Ethics, *Ethics of Synthetic Biology*, 35.

48. Atlas and Dando, "Dual-Use Dilemma," 276, 277.

49. Convention on the Prohibition of the Development, Production and Stockpiling of Bacteriological (Biological) and Toxin Weapons, Art. III; Australia Group, *Background Paper*, http://www.australiagroup.net/en/background.html.

50. Australia Group, *List of Biological Agents for Export Control*, http://www.australiagroup.net/en/biological_agents.html.

51. Pei et al., *Regulatory Frameworks*, 183.

52. See e.g., Rio Declaration on Environment and Development, June 14, 1992, U.N. Doc. A/CONF. 151/5, 31 I.L.M. 8744 (1992).

53. See generally Parens, Johnston, and Moses, *Ethical Issues*, 19–22.

54. Friends of the Earth, International Center, and ETC Group, *Principles*.

55. Erickson, Singh, and Winters, "Synthetic Biology," 1254–56. See also Parens, Johnston, and Moses, *Ethical Issues*, 20 (discussing failed efforts at a 2006 meeting of synthetic biology researchers, Synbio 2.0, to agree on a system of self-regulation).

56. Garfinkel et al., *Synthetic Genomics*, 21–37; Tucker and Zilinskas, "Promise and Perils," 43–44; Kelle, "Security Issues," 105, 111–12; International Risk Governance Council, *Guidelines*, 32.

57. Garfinkel et al., *Synthetic Genomics*, 21–37.

58. Kellogg, "Rise of DIY Scientists," 27.

59. Synthetic Biology Engineering Research Center, Research Program, http://www.synberc.org/research; Rabinow and Bennett, *Designing Human Practices*, 16–20, 40.

60. Rabinow and Bennett, *Designing Human Practices*, 8, 28–29. See also Jennifer Gollan, "Lab Fight Raises U.S. Security Issues," *New York Times*, October 23, 2011, A25A.

61. Tucker and Zilinskas, "Promise and Perils," 43–44; Schmidt, "Do I Understand," 96.

62. U.S. Presidential Commission, *New Directions*, 129–30; Ball, "Starting from Scratch," 626; Schmidt, "Do I Understand," 93.

63. See, e.g., Abbott, "International Framework Agreement," 127–56 (proposing a framework agreement on scientific and technological innovation and regulation).

64. Zhang, Marris, and Rose, *Transnational Governance*, 27.

65. Ibid., 26.

66. A 2010 survey found that Americans are increasingly aware of synthetic biology although most still know little or nothing about it. See Hart Research Associates, *Awareness and Impressions of Synthetic Biology*, September 9, 2010, 3–4, http://www.synbioproject.org/library/publications/archive/6456/. When respondents were given information about synthetic biology, large majorities supported the continuation of research if subject to public disclosure and government regulation (10–12). Highly religious and conservative persons, who tend to be relatively unconcerned about environmental risks, are among those most worried by the risks of synthetic biology (Kahan, Braman, and Mandel, *Risk and Culture*, 4–5; Hart Research Associates, *Awareness and Impressions*, 1 [finding that evangelicals are more likely than the average respondent to support a ban on synthetic biology]).

67. Anna Deplazes, Agomoni Ganguli-Mitra, and Nikola Biller-Andorno, "The Ethics of Synthetic Biology: Outlining the Agenda," in *Synthetic Biology: The Technoscience and Its Societal Consequences*, ed. Markus Schmidt, Alexander Kelle, Agomoni Ganguli-Mitra, and Huib de Vriend (Dordrecht: Springer, 2009), 65, 67.

68. Thomas Douglas and Savulescu, "Ethics of Knowledge," 688.

69. See Boldt and Müller, "Newtons," 388 (suggesting that the shift from manipulation of existing genetic material to the creation of new genetic material "is decisive because it involves a fundamental change in our way of approaching nature"); Preston, "Synthetic Biology," 23, 34 (objecting to synthetic biology because it "departs from the fundamental principle of Darwinian evolution").

70. U.S. Presidential Commission, *New Directions*, 135; Cho et al., "Ethical Considerations," 2087–90.

71. Thomas Douglas and Savulescu, "Ethics of Knowledge," 688–89; Deplazes, Ganguli-Mitra, and Biller-Andorno, "Ethics of Synthetic Biology," 67–68.

72. Boldt and Müller, "Newtons," 388; Cho et al., "Ethical Considerations," 2087–90; Arthur L. Caplan, testimony before the Presidential Commission for the Study of Bioethical Issues, September 13, 2010, 7–8, http://www.bioethics.gov/cms/sites/default/files/Testimony-of-Arthur-L-Caplan.pdf.

73. Bryan G. Norton, "Synthetic Biology: Some Concerns of a Biodiversity Advocate," remarks to the Presidential Commission on Bioethics, September 13, 2010, 6, http://www.bioethics.gov/documents/ synthetic-biology/Synthetic-Biology.pdf.

74. Boldt and Müller, "Newtons," 388.

75. See, e.g., Buchanan and Powell, *Ethics of Synthetic Biology*, 6.

76. Preston, "Synthetic Biology," 35.

77. Kaebnick, "Should Moral Objections," 1106, 1107; Thomas Douglas and Savulescu, "Ethics of Knowledge," 689 ("moral status is conferred not by life, but by characteristics that some living things possess").

78. See John Bohannon, "The Life Hacker," *Science* 333 (2011): 1236–37; George Church, "Synthetic Biology," *Nature* 463 (2010): 28; Nicholas Wade, "Regenerating a Mammoth for $10 Million," *New York Times*, November 19, 2008, A1.

79. National Research Council, *Research Frontiers in Bioinspired Energy: Molecular-Level Learning from Natural Systems* (2012), 13.

80. See, e.g., van den Belt, "Playing God," 257–68.

81. Thomas Douglas and Savulescu, "Ethics of Knowledge," 688; Kaebnick, "Should Moral Objections," 1107.

82. U.S. Presidential Commission, *New Directions*, 138.

83. See Hart Research Associates, *Awareness and Impressions*, 14–15 (reporting that when asked to select which of four concerns raised about synthetic biology concerned them the most, 45 percent of evangelical respondents to a survey identified the concern that it is morally wrong to create artificial life).

84. U.S. Presidential Commission, *New Directions*, 151–60.

Chapter 6

1. Parens, Johnston, and Moses, *Ethical Issues*, 9–11; ETC Group, *Extreme Genetic Engineering*, 5.

2. See National Research Council, *Human Performance Modification*, 5; Nordmann, *Converging Technologies*, 27–29; Roco and Bainbridge, *Converging Technologies*, 179–82.

3. Roco and Bainbridge, *Converging Technologies*, ix, 13.

4. Ibid., 5.

5. Ibid., 6.

6. Kurzweil, *Singularity Is Near*, 9.

7. The Transhumanist Declaration, http://www.transhumanism.org/resources/TenQuestions.pdf. See, e.g., Bainbridge, "Converging Technologies," 197, 198 (contending that technological convergence can lay "an entirely fresh foundation for our culture and institutions" and enable "transcendence of the traditional human condition"), 210–12 (suggesting use of biotechnology or nanotechnology to make the environment of other planets or moons more hospitable, followed by colonization by humans "in transformed bodies suitable for life on other planets and moons"). See also Hughes, "Contradictions," 622.

8. Roco and Bainbridge, *Converging Technologies*, 104–12.

9. E.g., Hughes, *Citizen Cyborg*, xii.

10. Warwick, "Cybernetic Enhancements," 123, 124–25; European Technology Assessment Group, *Technology Assessment*, 37.

11. Sandberg and Bostrom, "Cognitive Enhancement."

12. European Technology Assessment Group, *Technology Assessment*, iv; Garreau, *Radical Evolution*, 210–12 (discussing the limits of futuristic technology predictions). For one example of such projections, see Hughes, *Citizen Cyborg*, 23–32.

13. Coenen, "Deliberating Visions," 73, 77–79.

14. See, e.g., Warwick, "Cybernetic Enhancements," 127; Kevin Warwick, "Future Issues with Robots and Cyborgs," *Studies in Ethics, Law, and Technology* 4 (2011), DOI: 10.2202/1941-6008.1127; National Research Council, *Human Performance Modification*, 31.

15. Garreau, *Radical Evolution*, 28–38.

16. See Keim, "Gene Therapy"; Allhoff et al., *Ethics of Human Enhancement*, 19; U.S. President's Council on Bioethics, *Beyond Therapy*, 280.

17. See Hartens and Kuipers, "Effects of Androgenic-Anabolic Steroids."

18. See Allhoff et al., *Ethics of Human Enhancement*, 20; Fukuyama, *Our Posthuman Future*, 62–71.

19. See Allhoff et al., *Ethics of Human Enhancement*, 24–25.

20. See Roco, "Possibilities," 11, 12.

21. 45 C.F.R. §§ 46.101–409. See also Rice, "Historical, Ethical, and Legal Background," 1325.

22. U.S. Department of Health, Education, and Welfare, National Commission for the Protection of Human Subjects of Biomedical and Behavioral Research, *Belmont Report;* U.S. Department of the Army, Office of the Judge Advocate General, *Trials of War Criminals before the Nuremberg Military Tribunals* (Washington, DC: U.S. Government Printing Office, 1949), 2:181–82.

23. The Federal Food, Drug, and Cosmetic Act defines *drugs* as "articles (other than food) intended to affect the structure or any function of the body of man or other animals" and *medical device* as "an instrument . . . or other similar or related article . . . intended to affect the structure or any function of the body of man or other animals" (21 U.S.C. § 321(g)(1), (h)). See Greely, "Remarks," 1155.

24. See 21 U.S.C. §§ 355, 360c.

25. Dupuy, "Some Pitfalls," 238–39 (distinguishing between ethical analysis and cost-benefit analysis and contending that the ethical questions raised by human enhancement technologies are not susceptible to cost-benefit analysis).

26. See, e.g., Khushf, "Ethic for Enhancing Human Performance," 263–65.

27. See, e.g., Agar, *Liberal Eugenics;* Ronald M. Dworkin, *Sovereign Virtue*, 452. See also Allhoff et al., *Ethics of Human Enhancement*, 18.

28. See Chan and Harris, "In Support of Human Enhancement," 1. One attempt to distinguish the two suggests that enhancement "boost[s] our capabilities beyond the species-typical level or statistically-normal range," whereas therapy is aimed at treating pathological conditions (Allhoff et al., *Ethics of Human Enhancement*, 8). For a nuanced discussion of the concept of enhancement, see Bess, "Enhanced Humans," 641.

29. Chan and Harris, "In Support of Human Enhancement," 3.

30. U.S. President's Council on Bioethics, *Beyond Therapy*, 284.

31. Fukuyama, *Our Posthuman Future*, 4–7. See also Sarewitz, "Governing Our Future Selves," 217–18 (noting the lack of formal sanction in most societies for using human enhancement technology to bring about happiness).

32. Allhoff et al., *Ethics of Human Enhancement*, 27; U.S. President's Council on Bioethics, *Beyond Therapy*, 286–301.

33. U.S. President's Council on Bioethics, *Beyond Therapy*, 140. For human activities that are valued "more for their outcomes than for what they reveal about their participants," one might argue that human enhancement is less problematic. See Thomas Douglas, "Enhancement in Sport," 11.

34. Allhoff et al., *Ethics of Human Enhancement*, 21–22; U.S. President's Council on Bioethics, *Beyond Therapy*, 135–36.

35. Allhoff et al., *Ethics of Human Enhancement*, 21–22; European Technology Assessment Group, *Technology Assessment*, 72. Access to medical resources ultimately

may be limited for individuals who will never be able to afford enhancements, as advances in basic care are sacrificed for enhancement technology research (Selgelid, "Argument against Arguments," 3).

36. European Technology Assessment Group, *Technology Assessment*, 72; Koch, "Enhancing Who? Enhancing What?," 696. Francis Fukuyama even warns that human enhancement threatens the fundamental tenet of liberal democracy that all persons are created equal and entitled to equal rights. See Fukuyama, "Transhumanism," 42; Fukuyama, *Our Posthuman Future*, 9–10.

37. Sandel, "Case against Perfection," 54. Evoking themes of human limitations and hubris, Jean-Pierre Dupuy similarly describes NBIC convergence as "a Promethean project if ever there was one" ("Some Pitfalls," 250).

38. Fukuyama, *Our Posthuman Future*, 101, 125–28.

39. Koch, "Enhancing Who? Enhancing What?," 69. Roco and Bainbridge's ideally enhanced human, for example, is a militaristic creature who possesses superhuman strength, sensory abilities, and information-processing capacities (Schummer, "From Nano-Convergence to NBIC-Convergence," 65–66).

40. Norman Daniels suggests, for example, that a person endowed with "great intuition about the feelings of others" and a keen sense of foresight could succeed equally well as a social worker or a con artist ("Can Anyone Really Be Talking," 25, 40–41).

41. Koch, "Enhancing Who? Enhancing What?," 694; McCarthy, "Liberal Freedoms," 11.

42. Fuller, "Research Trajectories," 26. See also Koch, "Enhancing Who? Enhancing What?," 691–95.

43. Caplan, "Good, Better, or Best?," 195, 197–99.

44. Ibid., 201–3.

45. Hughes, "Beyond Human Nature."

46. Parens, Johnston, and Moses, *Ethical Issues*, 12.

47. Kahan, Braman, and Mandel, *Risk and Culture*, 7–8.

48. Kahan et al., "Cultural Cognition."

49. Nordmann, *Converging Technologies*, 4; Schummer, "From Nano-Convergence to NBIC-Convergence," 69 ("science policy makers have used the concept of convergence to mask their own goals, to articulate the alleged goals of the scientists, and the alleged goals of society").

50. The America COMPETES Act of 2007 specifies that proposals for NSF grants must include "a plan to provide appropriate training and oversight in the responsible and ethical conduct of research to undergraduate students, graduate students, and postdoctoral researchers participating in the proposed research project" (H.R. 2272, 110th Cong. 1st Sess. (2007), § 7009).

51. Center for Engineering, Ethics, and Society, *Ethics Education*, 30, 34–35. See also Rejeski, "Public Policy," 55 (discussing efforts to increase interaction between social scientists and natural scientists during research).

52. Weir and Selgelid, "Professionalization," 91–97.

53. Ibid., 92. A different approach has been taken by SynBERC, a multi-institutional synthetic biology research effort. Through a "Human Practices" initiative, SynBERC has sought to foster ethical reflection by encouraging collaboration between synthetic biologists and social scientists. See Rabinow, "Prosperity, Amelioration, Flourishing," 301, 303–5; Edmond and Mercer, "Norms and Irony,"

456. Observers have questioned, however, whether the initiative will have any practical impact, and the director of the initiative himself has acknowledged that the initiative "is in a dominated position" within SynBERC. See Rabinow, "Prosperity, Amelioration, Flourishing," 315; Edmond and Mercer, "Norms and Irony," 459–63. See also Caudill, "Synthetic Science," 441 (questioning Rabinow's idealized account of science).

54. Zhang, Marris, and Rose, *Transnational Governance*, 27.

55. Carrier, "Knowledge, Politics, and Commerce," 18.

56. Jasanoff, "Introduction," 1, 22.

57. Khushf, "Ethic for Enhancing Human Performance," 261–62. Khushf justifies such an approach in part by declaring "that we are going in the direction of enhancement whether we like it or not" and that an ethical discussion can shape the course of technological development (271).

58. Judy Dempsey and Jack Ewing, "In Reversal, Germany Announces Plans to Close All Nuclear Plants by 2022," *New York Times*, May 30, 2011, A4.

59. Convention on the Prohibition of the Development, Production, and Stockpiling of Bacteriological (Biological) and Toxin Weapons; Convention on the Prohibition of the Development, Production, Stockpiling and Use of Chemical Weapons; Tait, "Governing Synthetic Biology," 151.

60. See Fukuyama, *Our Posthuman Future*, 188 (contending that "pessimism about the inevitability of technological advance is wrong, and it could become a self-fulfilling prophecy"); McKibben, *Enough*, 162–73 (discussing examples of societal decisions to reject technologies).

61. Gaudet and Marchant, "Administrative Law Tools," 178–79; Sunstein, "Irreversible and Catastrophic," 841; Finn, "Sunset Clauses and Democratic Deliberation," 446–50.

62. Mooney, "Short History," 68–69.

63. Dupuy, "Some Pitfalls," 242 ("men dream science before doing it and . . . these dreams, which can take the form of science fiction, have a causal effect on the world").

64. Cf. Yearley, "Ethical Landscape," S564; Bratspies, "Regulatory Trust," 606 (discussing attributes of regulatory trust).

65. National Science Board, *Science and Engineering Indicators* (2010), O-5; Wise, "Thoughts," 286–89; Krimsky, *Science in the Public Interest*, 27–52.

66. Krimsky, *Science in the Public Interest*, 3–7.

67. Wise, "Thoughts," 283–86; Michaels, *Doubt Is Their Product;* McGarity and Wagner, *Bending Science*.

68. Patrik Jonsson, "Climate Scientists Exonerated in 'Climategate' but Public Trust Damaged," *Christian Science Monitor*, July 7, 2010, http://www.csmonitor.com/Environment/2010/0707/Climate-scientists-exonerated-in-climategate-but-public-trust-damaged.

69. National Science Board, *Science and Engineering Indicators* (2010), chap. 7; Woods Institute for the Environment, "Majority of Americans Continue to Believe That Global Warming Is Real," March 13, 2010, http://woods.stanford.edu/docs/surveys/Krosnick-20090312.pdf.

70. Resnik, *Playing Politics*, 203 (proposing the establishment of the Office of Science and Technology Advice).

71. David Ferris, "Science Cafes Tap Nation's Fascination with Research and

Discoveries," *Wired*, December 15, 2007, http://www.wired.com/science/discover ies/news/2007/12/science_cafe.

72. Nicole Farkas, "Dutch Science Shops: Matching Community Needs with University R&D," *Science Studies* 12 (1999): 33–47.

73. Sclove, *Reinventing Technology Assessment*, 38–41.

74. *About Us*, FactCheck.org, http://factcheck.org/about/.

75. *About Snopes.com*, Snopes.com, http://snopes.com/info/aboutus.asp.

76. Morrison, "Environmental Scanning," 86–99; Day and Schoemaker, *Peripheral Vision.*

77. See Health and Safety Executive, *Current Issues*, http://www.hse.gov.uk/ho rizons/library.htm.

78. Swart, Raskin, and Robinson, "Problem of the Future," 139–40; Slaughter, *Foresight Principle*, 38; Cornish, *Futuring*, 93–107; Deb Bennett-Woods, "Integrating Ethical Considerations into Funding Decisions for Emerging Nanotechnologies," *Nanotechnology Law and Business* 4 (2007): 81.

79. Rejeski, "Molecular Economy," 40–41. The EPA's Office of Research and Development identified the analysis of future environmental stressors as an organizational goal in 2001 and issued a handbook on futures analysis in 2005, but the agency's current activity in this area appears minimal. EPA, Office of Science Policy, Office of Research and Development, *Shaping Our Environmental Future.*

80. Duinker and Greig, "Scenario Analysis."

81. 40 C.F.R. § 1502.22 (1985) (amended 1986). For policy problems involving great uncertainty, Daniel A. Farber proposes application of an "α-precautionary principle" that considers both worst-case and best-case scenarios ("Uncertainty," 929–32).

82. Flatt, "'Worst Case' May Be the Best," 181, 189–98.

83. Gaudet and Marchant, "Administrative Law Tools," 176–78.

84. Mandel, "Regulating Emerging Technologies," 9.

85. Davies, *Oversight*, 25–26.

86. See Light, "Homeland Security Hash." Defending the Department of Homeland Security, Secretary Michael Chertoff remarked in 2007 that the Department of Defense took 40 years "to get configured properly" (Daniela Deane, "Chertoff Defends Bush's 2008 Budget Proposal," *Washington Post*, February 8, 2007, http://www.washingtonpost.com/wp-dyn/content/article/2007/02/08/AR2 007020801037.html).

87. Dara Kay Cohen, Cuéllar, and Weingast, "Crisis Bureaucracy," 718.

88. Davies, *Oversight*, 27.

89. Lempert, "Low Probability/High Consequence Events."

90. See, e.g., Robert T. Stafford Disaster Relief and Emergency Assistance Act, 42 U.S.C. §§ 5121–5207 (generally authorizing federal disaster response activities); 40 C.F.R. § 1506.11 (providing for potential waivers of the EIS requirement under emergency circumstances).

91. Richard Posner has suggested more narrowly that "scientific research projects in specified areas" be forbidden if they "would create an undue risk to human survival" (*Catastrophe*, 221).

92. Sabel and Simon, "Contextualizing Regimes," 1278–85.

93. Ibid., 1266–67.

94. Negotiated Rulemaking Act of 1990, 5 U.S.C. §§ 561–70 (2012); Lobel, "Renew Deal," 342, 377; Bingham, "Collaborative Governance," 269.

Conclusion

1. Schnaiberg, *Environment*, 279–81.

2. See Robert N. Proctor, "Agnotology: A Missing Term to Describe the Cultural Production of Ignorance (and Its Study)," in *Agnotology: The Making and Unmaking of Ignorance*, ed. Robert N. Proctor and Londa Schiebinger (Stanford, CA: Stanford University Press, 2008), 11–18; Naomi Oreskes and Erik M. Conway, "Challenging Knowledge: How Climate Science Became a Victim of the Cold War," in *Agnotology*, ed. Proctor and Schiebinger, 55–89.

3. U.S. Senate Permanent Subcommittee on Investigations, *Wall Street and the Financial Crisis: Anatomy of a Financial Collapse* (2011), 1, http://hsgac.senate.gov/public/_files/Financial_Crisis/FinancialCrisisReport.pdf. See also Financial Crisis Inquiry Commission, *The Financial Crisis Inquiry Report* (2011), http://fcic-static.law.stanford.edu/cdn_media/fcic-reports/fcic_final_report_full.pdf.

4. U.S. Senate Permanent Subcommittee on Investigations, *Wall Street*, 12; Financial Crisis Inquiry Commission, *Financial Crisis Inquiry Report*, xvii–xviii.

5. Declaration, Summit on Financial Markets and the World Economy, November 15, 2008, http://www.g20.org/Documents/g20_summit_declaration.pdf. The G-20, a forum for international economic governance, is comprised of the finance ministers and central bank governors of nineteen countries and the European Union (*What Is the G-20*, http://www.g20.org/about_what_is_g20.aspx).

6. Declaration, Summit on Financial Markets and the World Economy, ¶ 8.

7. Nassim Nicholas Taleb, *The Black Swan*, 2nd ed. (New York: Random House, 2010), xx (discussing three defining characteristics of a "black swan" event: "rarity, extreme impact, and retrospective (though not prospective) predictability").

8. Financial Crisis Inquiry Commission, *Financial Crisis Inquiry Report*, xvii, xxi; Robert J. Shiller, "Challenging the Crowd in Whispers, Not Shouts," *New York Times*, November 2, 2008, BU5 (describing how regulators ignored concerns about housing bubble).

9. David Scharfstein and Adi Sunderam, "The Economics of Housing Finance Reform: Privatizing, Regulating, and Backstopping Mortgage Markets," in *The Future of Housing Finance*, ed. Martin Neil Bailey (Washington, DC: Brookings Institution Press, 2011), 181–83.

10. David Colander, Hans Föllmer, Armin Haas, Michael Goldberg, Katarina Juselius, Alan Kirman, Thomas Lux, and Brigitte Sloth, *The Financial Crisis and the Systemic Failure of Academic Economics* (2009), http://www.ifw-members.ifw-kiel.de/publications/the-financial-crisis-and-the-systemic-failure-of-academic-economics/KWP_1489_ColanderetalFinancial%20Crisis.pdf (attributing economists' failure to anticipate the financial crisis "to the profession's insistence on constructing models that, by design, disregard the key elements driving outcomes in real-world markets").

11. Betsey Stevenson and Justin Wolfers, "Trust in Public Institutions over the Business Cycle," March 8, 2011, http://www.brookings.edu/~/media/Files/rc/papers/2011/0308_institutions_trust_stevenson_wolfers/0308_institutions_trust_stevenson_wolfers.pdf; "*New York Times*/CBS News Poll," October 2011, http://s3.documentcloud.org/documents/259646/the-new-york-times-cbs-news-poll-oct-2011.pdf (reporting that only 10 percent of respondents trust the federal government to do what is right always or most of the time).

12. For example, a petition protest by a certain percentage of nearby property owners may trigger a requirement that rezoning be approved by a supermajority. See Nicole Stelle Garnett, "Relocating Disorder," *Virginia Law Review* 91 (2005): 1132; Julian Conrad Juergensmeyer and Thomas E. Roberts, *Land Use Planning and Development Regulation Law*, 2nd ed. (St. Paul, MN: Thomson/West, 2007), § 5.7.

13. Marc B. Mihaly, "Citizen Participation in the Making of Environmental Decisions: Evolving Obstacles and Potential Solutions through Partnership with Experts and Agents," *Pace Environmental Law Review* 27 (2009): 151; Daniel R. Mandelker, *Land Use Law*, 5th ed. (Newark, NJ: LexisNexis, 2003), § 6.75. Various commentators note, however, that growing project complexity and the increased use of development agreements have hindered effective public participation. See Mihaly, "Citizen Participation," 179–86; Daniel P. Selmi, "The Contract Transformation in Land Use Regulation," *Stanford Law Review* 63 (2011): 640–41.

Selected Bibliography

Selected Books, Articles, Reports, and Government Publications

Abbott, Kenneth W. "An International Framework Agreement on Scientific and Technological Innovation and Regulation." In *The Growing Gap between Emerging Technologies and Legal-Ethical Oversight*, ed. Gary E. Marchant, Braden R. Allenby, and Joseph R. Herkert, 127–56. New York: Springer Science and Business Media, 2011.

Abbott, Kenneth W., Gary E. Marchant, and Douglas J. Sylvester. "A Framework Convention for Nanotechnology?" *Environmental Law Reporter* 38 (2008): 10,507–17.

Abbott, Kenneth W., Gary E. Marchant, and Douglas J. Sylvester. "Transnational Regulation of Nanotechnology: Reality or Romanticism?" In *International Handbook on Regulating Nanotechnologies*, ed. Graeme A. Hodge, Diana M. Bowman, and Andrew D. Maynard, 525–44. Cheltenham: Elgar, 2010.

Adler, Robert. "Are We on Our Way Back to the Dark Ages?" *New Scientist*, July 2, 2005, 26.

Agar, Nicholas. *Liberal Eugenics: In Defence of Human Enhancement.* Oxford: Blackwell, 2005.

Ahearne, John, and Peter Blair. "Expanded Use of the National Academies." In *Science and Technology Advice for Congress*, ed. M. Granger Morgan and Jon M. Peha, 118–33. Washington, DC: Resources for the Future, 2003.

Aitken, Robert J. "Regulation of Carbon Nanotubes and Other High Aspect Ratio Nanoparticles: Approaching This Challenge from the Perspective of Asbestos." In *International Handbook on Regulating Nanotechnologies*, ed. Graeme A. Hodge, Diana M. Bowman, and Andrew D. Maynard, 205–37. Cheltenham: Elgar, 2010.

Alkoby, Asher. "Global Networks and International Environmental Lawmaking: A Discourse Approach." *Chicago Journal of International Law* 8 (2008): 377–407.

Allenby, Braden. *Reconstructing Earth: Technology and Environment in the Age of Humans.* Washington, DC: Island, 2005.

Allhoff, Fritz, Patrick Lin, James Moor, and John Weckert. *Ethics of Human Enhancement: 25 Questions and Answers.* Version 1.0.1, 2009. http://www.humanenhance.com/NSF_report.pdf.

American Bar Association. *The Adequacy of FIFRA to Regulate Nanotechnology-Based*

Pesticides. May 2006. http://www.americanbar.org/content/dam/aba/migrated/environ/nanotech/pdf/FIFRA.authcheckdam.pdf.

American Bar Association. *Regulation of Nanoscale Materials under the Toxic Substances Control Act*. June 2006. http://www.americanbar.org/content/dam/aba/migrated/environ/nanotech/pdf/TSCA.authcheckdam.pdf.

American Physical Society. *Direct Air Capture of CO2 with Chemicals*. 2011. http://www.aps.org/policy/reports/assessments/upload/dac2011.pdf.

Andersen, Ida-Elisabeth, and Birgit Jæger. "Scenario Workshops and Consensus Conferences: Towards More Democratic Decision-Making." *Science and Public Policy* 26 (1999): 331–40.

Andrews, Clinton S. *Humble Analysis: The Practice of Joint Fact-Finding*. Westport, CT: Praeger, 2002.

Angel, Roger. "Feasibility of Cooling the Earth with a Cloud of Small Spacecraft near the Inner Lagrange Point." *Proceedings of the National Academy of Sciences* 103 (2006): 17,184–89.

Angelo, Mary Jane. "Regulating Evolution for Sale: An Evolutionary Biology Model for Regulating the Unnatural Selection of Genetically Modified Organisms." *Wake Forest Law Review* 42 (2007): 93–165.

Aoki, Keith. "Seeds of Dispute: Intellectual Property Rights and Agricultural Biodiversity." *Golden Gate University Environmental Law Journal* 3 (2009): 79–160.

Applegate, John S. "The Perils of Unreasonable Risk: Information, Regulatory Policy, and Toxic Substances Control." *Columbia Law Review* 91 (1991): 261–333.

Applegate, John S. "Worst Things First: Risk, Information, and Regulatory Structure in Toxic Substances Control." *Yale Journal on Regulation* 9 (1992): 277–354.

Applegate, John S., Jan G. Laitos, and Celia Campbell-Mohn. *The Regulation of Toxic Substances and Hazardous Wastes*. New York: Foundation, 2000.

Aqua Bounty Technologies. *Environmental Assessment for AquAdvantage Salmon*. 2010. http://www.fda.gov/downloads/AdvisoryCommittees/CommitteesMeetingMaterials/VeterinaryMedicineAdvisoryCommittee/UCM224760.pdf.

Armour, Audrey. "The Citizens' Jury Model of Public Participation: A Critical Evaluation." In *Fairness and Competence in Citizen Participation: Evaluating Models for Environmental Discourse*, ed. Ortwin Renn, Thomas Webler, and Peter Wiedemann, 175–87. Boston: Kluwer Academic, 1995.

Arthur, W. Brian. *Increasing Returns and Path Dependence in the Economy*. Ann Arbor: University of Michigan Press, 1994.

Asilomar International Conference on Climate Intervention Technologies. Statement from the Conference's Scientific Organizing Committee. 2010. http://climateresponsefund.org/index.php?option=com_content &view=article&id=152&Itemid=89.

Asilomar Scientific Organizing Committee. *The Asilomar Conference Recommendations on Principles for Research into Climate Engineering Techniques*. 2010. http://www.climate.org/PDF/AsilomarConferenceReport.pdf.

Atlas, Ronald M., and Malcolm Dando. "The Dual-Use Dilemma for the Life Sciences: Perspectives, Conundrums, and Global Solutions." *Biosecurity and Bioterrorism* 4 (2006): 276–86.

Aumont, O., and L. Bopp. "Globalizing Results from Ocean in Situ Iron Fertilization Studies." *Global Biogeochemical Cycles* 20 (2006): GB2017, DOI: 10.1029/2005GB002591.

Ayres, Robert U. "Turning Point: The End of Exponential Growth." *Technological Forecasting and Social Change* 73 (2006): 1188–1203.

Bainbridge, William Sims. "Converging Technologies and Human Destiny." *Journal of Medicine and Philosophy* 32 (2007): 197–216.

Bala, Govindswamy. "Problems with Geoengineering Schemes to Combat Climate Change." *Current Science* 96 (2009): 41–48.

Ball, Philip. "Starting from Scratch." *Nature* 431 (2004): 624–26.

Banerjee, Bidisha. "The Limitations of Geoengineering Governance in a World of Uncertainty." *Stanford Journal of Law, Science and Policy* 4 (2011): 15–36.

Barben, Daniel, Erik Fisher, Cynthia Selin, and David H. Guston. "Anticipatory Governance of Nanotechnology: Foresight, Engagement, and Integration." In *The New Handbook of Science and Technology Studies*, 3rd ed., ed. Edward J. Hackett, Olga Amsterdamska, Michael E. Lynch, and Judy Wajcman, 979–1000. Cambridge: MIT Press, 2008.

Barber, Benjamin R. *Strong Democracy: Participatory Politics for a New Age*. Berkeley: University of California Press, 2003.

Barker, Pamela E., Timothy Butler, Joseph M. Dawley, Paul Herran, Brian King, Kirsten L. Nathanson, Kavita Patel, Jim Wedeking, Harry Weiss, Jack Wubinger, and Steven Ziesmann. *Nanotechnology Briefing Paper: Clean Water Act*. 2006. http://www.abanet.org/environ/nanotech/pdf/CWA.pdf.

Barnosky, Anthony D., Nicholas Matzke, Susumu Tomiya, Guinevere O. U. Wogan, Brian Swartz, Tiago B. Quental, Charles Marshall, Jenny L. McGuire, Emily L. Lindsey, Kaitlin C. Maguire, Ben Mersey, and Elizabeth A. Ferrer. "Has the Earth's Sixth Mass Extinction Already Arrived?" *Nature* 471 (2011): 51–57.

Baroli, Biancamaria. "Penetration of Nanoparticles and Nanomaterials in the Skin: Fiction or Reality?" *Journal of Pharmaceutical Sciences* 99 (2010): 21–50.

Barrett, Scott. "The Incredible Economics of Geoengineering." *Environmental and Research Economics* 39 (2008): 45–54.

Barrett, Scott. *Why Cooperate?: The Incentive to Supply Global Public Goods*. Oxford: Oxford University Press, 2007.

Barsa, Michael. "Note: California's Proposition 65 and the Limits of Information Economics." *Stanford Law Review* 49 (1997): 1223–47.

Bartis, James T., and Eric Landree. *Nanomaterials in the Workplace: Policy and Planning Workshop on Occupational Safety and Health*. Santa Monica, CA: Rand, 2006.

Bar-Yam, Shlomiya, Jennifer Byers-Corbin, Rocco Casagrande, Florentine Eichler, Allen Lin, Martin Oesterreicher, Pernilla Regardh, R. Donald Turlington, and Kenneth A. Oye. *The Regulation of Synthetic Biology: A Guide to United States and European Union Regulations, Rules and Guidelines, SynBERC and iGEM*. Version 9.1, 2012). http://synberc.org/sites/default/files/Concise%20Guide%20to%20Synbio%20Regulation%20OYE%20Jan%202012_0.pdf.

Batte, Marvin T., Neal H. Hooker, Timothy C. Haab, and Jeremy Beaverson. "Putting Their Money Where Their Mouths Are: Consumer Willingness to Pay for Multi-Ingredient, Processed Organic Food Products." *Food Policy* 32 (2007): 145–59.

Bauer, Martin W. "Distinguishing *Red* and *Green* Biotechnology: Cultivation Effects of the Elite Press." *International Journal of Public Opinion Research* 17 (2005): 63–89.

Bauer, Martin W., and George Gaskell. "The Biotechnology Movement." In *Biotechnology: The Making of a Global Controversy*, ed. Martin W. Bauer and George Gaskell, 379–404. Cambridge: Cambridge University Press, 2002.

Baumol, William J. *The Free-Market Innovation Machine: Analyzing the Growth Miracle of Capitalism*. Princeton: Princeton University Press, 2002.

Beck, Ulrich. *The Risk Society: Towards a New Modernity*. Trans. Mark Ritter. Newbury Park, CA: Sage, 1992.

Benedick, Richard Elliot. "Considerations on Governance for Climate Remediation Technologies: Lessons from the 'Ozone Hole.'" *Stanford Journal of Law, Science, and Policy* 4 (2011): 6–9.

Bengtsson, Lennart. "Geo-Engineering to Confine Climate Change: Is It at All Feasible?" *Climatic Change* 77 (2006): 229–34.

Benner, Steven A., and A. Michael Sismour. "Synthetic Biology." *Nature Reviews* 6 (2005): 533–43.

Benner, Thorsten, Wolfgang H. Reinicke, and Jan Martin Witte. "Multisectoral Networks in Global Governance: Towards a Pluralistic System of Accountability." *Government and Opposition* 39 (2004): 191–210.

Bennett, Ira, and Daniel Sarewitz. "Too Little, Too Late?: Research Policies on the Societal Implications of Nanotechnology in the United States." *Science as Culture* 15 (2006): 309–25.

Berg, Paul, David Baltimore, Herbert W. Boyer, Stanley N. Cohen, Ronald W. Davis, David S. Hogness, Daniel Nathans, Richard Roblin, James D. Watson, Sherman Weissman, and Norton D. Zinder. "Potential Biohazards of Recombinant DNA Molecules." *Science* 185 (1974): 303.

Berg, Paul, David Baltimore, Sydney Brenner, Richard O. Roblin III, and Maxine F. Singer. "Asilomar Conference on Recombinant DNA Molecules." *Science* 188 (1975): 991–94.

Bergeson, Lynn L., ed. *Nanotechnology: Environmental Law, Policy, and Business Considerations*. Chicago: American Bar Association, 2010.

Bergeson, Lynn L. "Regulation, Governance, and Nanotechnology: Is a Framework Convention for Nanotechnology the Way to Go?" *Environmental Law Reporter* 38 (2008): 10,515–17.

Bergeson, Lynn L., and Joseph E. Plamondon. "TSCA and Engineered Nanoscale Substances." *Nanotechnology Law and Business* 4 (2007): 51–74.

Bernstein, Gaia. "When New Technologies Are Still New: Windows of Opportunity for Privacy Protection." *Villanova Law Review* 51 (2006): 921–50.

Berube, David M. *Nano-Hype: The Truth behind the Nanotechnology Buzz*. Amherst, NY: Prometheus, 2006.

Berube, David M. "Regulating Nanoscience: A Proposal and a Response to J. Clarence Davies." *Nanotechnology Law and Business* 3 (2006): 485–506.

Bess, Michael. "Enhanced Humans versus 'Normal People': Elusive Definitions." *Journal of Medicine and Philosophy* 35 (2010): 641–55.

Betsill, Michele M. "Reflections on the Analytic Framework and NGO Diplomacy." In *NGO Diplomacy: The Influence of Nongovernmental Organizations in International Environmental Negotiations*, ed. Michele M. Betsill and Elisabeth Corell, 177–206. Cambridge: MIT Press, 2008.

Betsill, Michele M., and Elisabeth Corell. "Introduction to NGO Diplomacy." In *NGO Diplomacy: The Influence of Nongovernmental Organizations in International Environmental Negotiations*, ed. Michele M. Betsill and Elisabeth Corell, 1–17. Cambridge: MIT Press, 2008.

Bijker, Wiebe E. *Of Bicycles, Bakelites, and Bulbs: Toward a Theory of Sociotechnical Change*. Cambridge: MIT Press, 1995.

Binder, Denis. "NEPA, NIMBYs, and New Technology." *Land and Water Law Review* 25 (1990): 11–42.

Bingham, Lisa Blomgren. "Collaborative Governance: Emerging Practices and the Incomplete Legal Framework for Public and Stakeholder Voice." *Journal of Dispute Resolution* 2009 (2009): 269–325.

Bipartisan Policy Center Task Force on Climate Remediation Research. *Geoengineering: A National Strategic Plan for Research on the Potential Effectiveness, Feasibility, and Consequences of Climate Remediation Technologies*. 2011. http://bipartisanpolicy.org/sites/default/files/BPC%20Climate%20Remediation%20Final%20Report.pdf.

Blackstock, Jason J., and Jane C. S. Long. "The Politics of Geoengineering." *Science* 327 (2010): 527.

Blackwelder, Brent. "Comment on *A Framework Convention for Nanotechnology?*" *Environmental Law Reporter* 38 (2008): 10,520.

Blain, Stephane, et al. "Effect of Natural Iron Fertilization on Carbon Sequestration in the Southern Ocean." *Nature* 446 (2007): 1070–75.

Bodansky, Daniel. "May We Engineer the Climate?" *Climatic Change* 33 (1996): 309–21.

Bohm, Peter, and Clifford S. Russell. "Comparative Analysis of Alternative Policy Instruments." In *Handbook of Natural Resource and Energy Economics*, ed. Allen V. Kneese and James L. Sweeney, 1:395–460. Amsterdam: North-Holland, 1985–93.

Boisson de Chazournes, Laurence, and Makane Moise Mbengue. "Trade, Environment, and Biotechnology: On Coexistence and Coherence." In *Genetic Engineering and the World Trade System*, ed. Daniel Wüger and Thomas Cottier, 205–46. Cambridge: Cambridge University Press, 2008.

Boldt, Joachim, and Oliver Müller. "Newtons of the Leaves of Grass." *Nature Biotechnology* 26 (2008): 387–89.

Boucher, Patrick. *Nanotechnology: Legal Aspects*. Boca Raton, FL: CRC, 2008.

Bowman, Diana M., and Graeme A. Hodge. "'Governing' Nanotechnology without Government?" *Science and Public Policy* 35 (2008): 475–87.

Bowman, Diana M., G. van Calster, and S. Friedrichs. "Nanomaterials and Regulation of Cosmetics." *Nature Nanotechnology* 5 (2010): 92.

Boyd, James. "Financial Responsibility for Environmental Obligations: Are Bonding and Assurance Rules Fulfilling Their Promise?" *Research in Law and Economics* 20 (2002): 417–85.

Boyd, Philip W., et al. "Mesoscale Iron Enrichment Experiments, 1993–2005: Synthesis and Future Directions." *Science* 315 (2007): 612–17.

Bratspies, Rebecca M. "Biotechnology, Sustainability, and Trust." *Kansas Journal of Law and Public Policy* 18 (2009): 273–91.

Bratspies, Rebecca M. "Myths of Voluntary Compliance: Lessons from the StarLink Corn Fiasco." *William and Mary Environmental Law and Policy Review* 27 (2003): 593–649.

Bratspies, Rebecca M. "Regulatory Trust." *Arizona Law Review* 51 (2009): 575–631.

Bratspies, Rebecca M. "Some Thoughts on the American Approach to Regulating Genetically Modified Organisms." *Kansas Journal of Law and Public Policy* 16 (2007): 393–423.

Brown, Elizabeth C., Carolyne R. Hathaway, Julia A. Hatcher, William K. Rawson, and Robert M. Sussman. *TSCA Deskbook*. Washington, DC: Environmental Law Institute, 1999.

Brown, Lester. "Eradicating Hunger: A Growing Challenge." In *State of the World 2001*, ed. Lester R. Brown et al., 43–62. New York: W. W. Norton, 2001.

Brown, Mark B. *Science in Democracy: Expertise, Institutions, and Representation*. Cambridge: MIT Press, 2009.

Bryner, Gary C. "Economic Policy, International Competitiveness, and the Role of Technology Policy." In *Science, Technology, and Politics: Policy Analysis in Congress*, ed. Gary C. Bryner, 179–206. Boulder, CO: Westview, 1992.

Bryner, Gary C. "Science, Technology, and Policy Analysis in Congress: An Introduction." In *Science, Technology, and Politics: Policy Analysis in Congress*, ed. Gary C. Bryner, 3–12. Boulder, CO: Westview, 1992.

Buchanan, Allen, and Robert O. Keohane. "The Legitimacy of Global Governance Institutions." *Ethics and International Affairs* 20 (2006): 405–37.

Buchanan, Allen, and Russell Powell. *The Ethics of Synthetic Biology: Suggestions for a Comprehensive Approach*. http://www.bioethics.gov/cms/sites/default/files/The-Ethics-of-Synthetic-Biology-Suggestions-for-a-Comprehensive-Approach.pdf.

Buesseler, Ken O., and Philip W. Boyd. "Will Ocean Fertilization Work?" *Science* 300 (2003): 67–68.

Bunzl, Martin. "Researching Geoengineering: Should Not or Could Not?" *Environmental Research Letters* 4 (2009): 045104. DOI: 10.1088/1748-9326/4/4/045104.

Burkhardt, Jeffrey. "The Ethics of Agri-Food Biotechnology: How Can an Agricultural Technology be so Important?" In *What Can Nanotechnology Learn from Biotechnology?: Social and Ethical Lessons for Nanoscience from the Debate over Agrifood Biotechnology and GMOs*, ed. Kenneth David and Paul B. Thompson, 55–80. Amsterdam and Boston: Elsevier/Academic, 2008.

Burns, William C. G. "Climate Geoengineering: Solar Radiation Management and Its Implications for Intergenerational Equity." *Stanford Journal of Law, Science, and Policy* 4 (2011): 37–55.

Burstein, Paul. "The Impact of Public Opinion on Public Policy: A Review and an Agenda." *Political Research Quarterly* 56 (2003): 33–36.

Busch, Lawrence, and John R. Lloyd. "What Can Nanotechnology Learn from Biotechnology?" In *What Can Nanotechnology Learn from Biotechnology?: Social and Ethical Lessons for Nanoscience from the Debate over Agrifood Biotechnology and GMOs*, ed. Kenneth David and Paul B. Thompson, 261–76. Amsterdam and Boston: Elsevier/Academic, 2008.

Butler, David, and Austin Ranney. "Theory." In *Referendums: A Comparative Study of Practice and Theory*, ed. David Butler and Austin Ranney, 23–37. Washington, DC: American Enterprise Institute for Public Policy Research, 1978.

Butler, David, and Austin Ranney. "Theory." In *Referendums around the World: The Growing Use of Direct Democracy*, ed. David Butler and Austin Ranney, 11–23. Washington, DC: AEI Press, 1994.

Caldeira, Ken, and Michael E. Wickett. "Ocean Model Predictions of Chemistry Changes From Carbon Dioxide Emissions to the Atmosphere and Ocean." *Journal of Geophysical Research* 110 (2005): C09S04. DOI: 10.1029/2004JC002671.

Caldeira, Ken, and Lowell Wood. "Global and Arctic Climate Engineering: Numerical Model Studies." *Philosophical Transactions of the Royal Society A: Mathematical, Physical, and Engineering Sciences* 366 (2008): 4052–53. DOI: 10.1098/rsta.2008.0132.

Caldwell, Lynton Keith. *The National Environmental Policy Act: An Agenda for the Future*. Bloomington: Indiana University Press, 1998.

California Climate Change Center. *Our Changing Climate: Assessing the Risks to California*. Sacramento: California Climate Change Center, 2006.

Caplan, Arthur L. "Good, Better, or Best?" In *Human Enhancement*, ed. Julian Savulescu and Nick Bostrom, 199–209. Oxford: Oxford University Press, 2009.

Carrier, Martin. "Knowledge, Politics, and Commerce: Science under the Pressure of Practice." In *Science in the Context of Application*, ed. Martin Carrier and Alfred Nordmann, 11–30. Dordrecht: Springer, 2011.

Carson, Nancy. "Process, Prescience, and Pragmatism: The Office of Technology Assessment." In *Organizations for Policy Analysis: Helping Government Think*, ed. Carol H. Weiss, 236–51. Newbury Park, CA: Sage, 1992.

Carter, Colin A., and Guillaume P. Gruère. "Mandatory Labeling of Genetically Modified Foods: Does It Really Provide Consumer Choice?" *AgBioForum* 6 (2003). http://www.agbioforum.org/v6n12/v6n12a13-carter.htm.

Caswell, Julie A. "Labeling Policies for GMOs: To Each His Own?" *AgBioForum* 3 (2000): 53–57.

Caudill, David S. "Synthetic Science: A Response to Rabinow." *Law and Literature* 21 (2009): 431–44.

Center for Engineering, Ethics, and Society. *Ethics Education and Scientific and Engineering Research: What's Been Learned? What Should Be Done? Summary of a Workshop*. Washington, DC: National Academies Press, 2009.

Centers for Disease Control and Prevention. *Investigation of Human Health Effects Associated with Potential Exposure to Genetically Modified Corn*. 2001. http://www.cdc.gov/nceh/ehhe/cry9creport/pdfs/cry9creport.pdf.

Chan, Sarah, and John Harris. "In Support of Human Enhancement." *Studies in Ethics, Law, and Technology* 1 (2007). DOI: 10.2202/1941-6008.1007.

Chaudhry, Qasim, Hans Bouwmeester, and Rolf F. Hertel. "The Current Risk Assessment Paradigm in Relation to the Regulation of Nanotechnologies." In *International Handbook on Regulating Nanotechnologies*, ed. Graeme A. Hodge, Diana M. Bowman, and Andrew D. Maynard, 124–43. Cheltenham: Elgar, 2010.

Check, Erika. "Designs on Life." *Nature* 438 (2005): 417–18.

Chisholm, Sallie W., Paul G. Falkowski, and John J. Cullen. "Dis-Crediting Ocean Fertilization." *Science* 294 (2001): 309–10.

Cho, Mildred K., D. Magnus, A. L. Caplan, D. McGee, and the Ethics of Genomics Group. "Ethical Considerations in Synthesizing a Minimal Genome." *Science* 286 (1999): 2087–90.

Chubin, Daryl E. "Filling the Policy Vacuum Created by OTA's Demise." *Issues in Science and Technology*, Winter 2000–2001, 31–32.

CNA Corporation. *National Security and the Threat of Climate Change*. 2007. http://securityandclimate.cna.org/report/.

Coenen, Christopher. "Deliberating Visions: The Case of Human Enhancement in the Discourse on Nanotechnology and Convergence." In *Governing Future Technologies: Nanotechnology and the Rise of an Assessment Regime, Sociology of the Sciences Yearbook* 27, ed. Mario Kaiser, Monika Kurath, Sabine Maasen, and Christoph Rehmann-Sutter, 73–87. Dordrecht: Springer, 2010.

Coglianese, Cary. "Assessing Consensus: The Promise and Performance of Negotiated Rulemaking." *Duke Law Journal* 46 (1997): 1255–1349.

Cohen, Dara Kay, Mariano-Florentino Cuéllar, and Barry R. Weingast. "Crisis Bureaucracy: Homeland Security and the Political Design of Legal Mandates." *Stanford Law Review* 59 (2006): 673–759.

Collingridge, David. *The Social Control of Technology*. New York: St. Martin's, 1980.

Collins, Harry, and Robert Evans. *Rethinking Expertise*. Chicago: University of Chicago Press, 2007.

Collins, Jim. "Got Parts, Need Manual." *Nature* 465 (2010): 424.

Committee on Identifying and Assessing Unintended Effects of Genetically Engineered Foods on Human Health, Institute of Medicine and National Research Council of the National Academies. *Safety of Genetically Engineered Foods: Approaches to Assessing Unintended Health Effects*. Washington, DC: National Academies Press, 2004.

Cooper, Jon C. "Broad Programmatic, Policy, and Planning Assessments under the National Environmental Policy Act and Similar Devices: A Quiet Revolution in an Approach to Environmental Considerations." *Pace Environmental Law Review* 11 (1993): 89–156.

Coote, Anna, and Jo Lenaghan. *Citizens' Juries: Theory into Practice*. London: Institute for Public Policy Research, 1997.

Corley, Elizabeth A., and Dietram A. Scheufele. "Outreach Gone Wrong?: When We Talk Nano to the Public, We Are Leaving behind Key Audiences." *The Scientist*, January 2010, 22.

Corley, Elizabeth A., Dietram A. Scheufele, and Qian Hu. "Of Risks and Regulations: How Leading U.S. Nanoscientists Form Policy Stances about Nanotechnology." *Journal of Nanoparticle Research* 11 (2009): 1573–85.

Cormack, Christopher P. *CERCLA Nanotechnology Issues*. 2006. http://www.abanet.org/environ/nanotech/pdf/CERCLA.pdf.

Cornish, Edward. *Futuring: The Exploration of the Future*. Bethesda, MD: World Future Society, 2004.

Cornish, Edward. "Products of Genetic Engineering." *Dallas Morning News*, July 15, 1985, 11.

Cornwell, Laura, and Robert Costanza. "Environmental Bonds: Implementing the Precautionary Principle in Environmental Policy." In *Protecting Public Health and the Environment*, ed. Carolyn Raffensperger and Joel A. Tickner, 220–40. Washington, DC: Island, 1999.

Costanza, Robert, and Charles Perrings. "A Flexible Assurance Bonding System for Improved Environmental Management." *Ecological Economics* 2 (1990): 57–75.

Council on Environmental Quality. *The National Environmental Policy Act: A Study of Its Effectiveness after Twenty-Five Years*. 1997. http://ceq.hss.doe.gov/nepa/nepa25fn.pdf.

Cowan, Tadlock, and Geoffrey S. Becker. *Agricultural Biotechnology: Background and Recent Issues*. 2010. http://infousa.state.gov/economy/industry/docs/73949.pdf.

Crespi, John M., and Stéphan Marette. "'Does Contain' vs. 'Does Not Contain': Does It Matter Which GMO Label Is Used?" *European Journal of Law and Economics* 16 (2003): 327–44.

Cronin, Thomas E. *Direct Democracy: The Politics of Initiative, Referendum, and Recall.* Cambridge: Harvard University Press, 1989.

Crosby, Ned. "Citizens Juries: One Solution for Difficult Environmental Questions." In *Fairness and Competence in Citizen Participation: Evaluating Models for Environmental Discourse*, ed. Ortwin Renn, Thomas Webler, and Peter Wiedemann, 157–74. Boston: Kluwer Academic, 1995.

Crutzen, Paul. "Albedo Enhancement by Stratospheric Sulfur Injections: A Contribution to Resolve a Policy Dilemma?" *Climatic Change* 77 (2006): 211–19.

Dahl, Robert A. "Can International Organizations Be Democratic?: A Skeptic's View." In *Democracy's Edges*, ed. Ian Shapiro and Casiano Hacker-Cordon, 19–36. Cambridge: Cambridge University Press, 1999.

Dana, David A. "A Behavioral Economic Defense of the Precautionary Principle." *Northwestern University Law Review* 97 (2003): 1315–45.

Dana, David A. "When Less Liability May Mean More Precaution: The Case of Nanotechnology." *UCLA Journal of Environmental Law and Policy* 28 (2010): 153–200.

Daniels, Norman. "Can Anyone Really Be Talking about Ethically Modifying Human Nature?" In *Human Enhancement*, ed. Julian Savulescu and Nick Bostrom, 25–42. Oxford: Oxford University Press, 2009.

Darnall, Nicole, and Stephen Sides. "Assessing the Performance of Voluntary Environmental Programs: Does Certification Matter?" *Policy Studies Journal* 36 (2008): 95–117.

Davies, J. Clarence. *EPA and Nanotechnology: Oversight for the 21st Century.* Washington, DC: Woodrow Wilson International Center for Scholars, 2007.

Davies, J. Clarence. *Managing the Effects of Nanotechnology.* Washington, DC: Woodrow Wilson International Center for Scholars, 2006.

Davies, J. Clarence. *Nanotechnology Oversight: An Agenda for the New Administration.* Washington, DC: Woodrow Wilson International Center for Scholars, 2008.

Davies, J. Clarence. *Oversight of Next-Generation Nanotechnology.* Washington, DC: Woodrow Wilson International Center for Scholars, 2009.

Day, George S., and Paul J. J. Schoemaker. *Peripheral Vision: Detecting the Weak Signals That Will Make or Break Your Company.* Boston: Harvard Business School, 2006.

Degnan, Fred H. "Biotechnology and the Food Label." In *Labeling Genetically Modified Food: The Philosophical and Legal Debate*, ed. Paul Weirich, 17–31. Oxford: Oxford University Press, 2008.

Delli Carpini, Michael X., Fay Lomax Cook, and Lawrence R. Jacobs. "Public Deliberation, Discursive Participation, and Citizen Engagement." *Annual Review of Political Science* 7 (2004): 315–44.

Denison, Richard. "Ten Essential Elements in TSCA Reform." *Environmental Law Reporter* 39 (2009): 10,020–28.

Dewey, John. *The Public and Its Problems.* Chicago: Gateway, 1946.

Diamond, Jared. *Collapse: How Societies Choose to Fail or Succeed.* New York: Viking, 2005.

Dienel, Peter C., and Ortwin Renn. "Planning Cells: A Gate to 'Fractal' Mediation." In *Fairness and Competence in Citizen Participation: Evaluating Models for Environmental Discourse*, ed. Ortwin Renn, Thomas Webler, and Peter Wiedemann, 117–40. Boston: Kluwer Academic, 1995.

Dobbs, Dan B. *Law of Remedies: Damages, Equity, Restitution.* 2nd ed. St. Paul, MN: West, 1993.

Domingo, Jose L. "Toxicity Studies of Genetically Modified Plants: A Review of the Published Literature." *Critical Reviews in Food Science and Nutrition* 47 (2007): 721–33.

Dona, Artemis, and Ioannis S. Arvanitoyannis, "Health Risks of Genetically Modified Foods." *Critical Reviews in Food Science and Nutrition* 49 (2009): 164–75.

Doremus, Holly, Albert C. Lin, Ronald H. Rosenberg, and Thomas J. Schoenbaum. *Environmental Policy Law: Problems, Cases, and Readings.* 5th ed. New York: Foundation, 2008.

Douglas, Mary, and Aaron Wildavsky. *Risk and Culture: An Essay on the Selection of Technological and Environmental Dangers.* Berkeley: University of California Press, 1982.

Douglas, Thomas. "Enhancement in Sport, and Enhancement outside Sport." *Studies in Ethics, Law, and Technology* 1 (2007). DOI: 10.2202/1941-6008.1000.

Douglas, Thomas, and Julian Savulescu. "Synthetic Biology and the Ethics of Knowledge." *Journal of Medical Ethics* 36 (2010): 687–93.

Drexler, K. Eric. "Nanotechnology: From Feynman to Funding." *Bulletin of Science, Technology, and Society* 24 (2004): 24–25.

Drexler, K. Eric, and Jason Wejnert. "Nanotechnology and Policy." *Jurimetrics Journal* 45 (2004): 1–22.

Du, Wenchao, Yuanyuan Sun, Rong Ji, Jianguo Zhu, Jichun Wu, and Hongyan Guo. "TiO_2 and ZnO Nanoparticles Negatively Affect Wheat Growth and Soil Enzyme Activities in Agricultural Soil." *Journal of Environmental Monitoring* 13 (2011): 822–28.

Duinker, Peter N., and Lorne A. Greig. "Scenario Analysis in Environmental Impact Assessment: Improving Explorations of the Future." *Environmental Impact Assessment Review* 27 (2007): 206–19.

Dupuy, Jean-Pierre. "Some Pitfalls in the Philosophical Foundations of Nanoethics." *Journal of Medicine and Philosophy* 32 (2007): 237–61.

Dworkin, Roger B. "Science, Society, and the Expert Town Meeting: Some Comments on Asilomar." *Southern California Law Review* 51 (1978): 1471–82.

Dworkin, Ronald M. *Sovereign Virtue: The Theory and Practice of Equality.* Cambridge: Harvard University Press, 2000.

Ebbesen, Mette. "The Role of the Humanities and Social Sciences in Nanotechnology Research and Development." *Nanoethics* 2 (2008): 1–13.

Ebbesson, Jonas. "Public Participation." In *The Oxford Handbook of International Environmental Law*, ed. Daniel Bodansky, Jutta Brunnée, and Ellen Hey, 681–703. Oxford: Oxford University Press, 2007.

Edmond, Gary, and David Mercer. "Norms and Irony in the Biosciences: Ameliorating Critique in Synthetic Biology." *Law and Literature* 21 (2009): 445–66.

Ehrlich, Paul. *The Population Bomb.* New York: Ballantine, 1968.

Einsiedel, E. F. "Public Perceptions of Transgenic Animals." *Review of Science and Technology, International Office of Epizootics*, 24 (2005): 149–57.

Ellis, Jaye, and Alison FitzGerald. "The Precautionary Principle in International Law: Lessons from Fuller's Internal Morality." *McGill Law Journal* 49 (2004): 779–800.

Endy, Drew. "Foundations for Engineering Biology." *Nature* 438 (2005): 449–53.

Environmental Defense–DuPont Nano Partnership. *Nano Risk Framework.* 2007. http://www.nanoriskframework.com/files/2011/11/6496_Nano-Risk-Framework.pdf.

Environmental Law Institute. *Securing the Promise of Nanotechnology: Is U.S. Environmental Law up to the Job?* 2005. http://www.elistore.org/Data/products/d15_10.pdf.

Epstein, Gerald L., and Ashton B. Carter. "A Dedicated Organization in Congress." In *Science and Technology Advice for Congress*, ed. M. Granger Morgan and Jon M. Peha, 157–63. Washington, DC: Resources for the Future, 2003.

Erickson, Brent, Rina Singh, and Paul Winters. "Synthetic Biology: Regulating Industry Uses of New Biotechnologies." *Science* 333 (2011): 1254–56.

Esty, Daniel C. "The Case for a Global Environmental Organization." In *Managing the World Economy: Fifty Years after Bretton Woods*, ed. Peter B. Kenen, 287–309. Washington, DC: Institute for International Economics, 1994.

Esty, Daniel C. "On Portney's Complaint: Reconceptualizing Corporate Social Responsibility." In *Environmental Protection and the Social Responsibility of Firms: Perspectives from Law, Economics, and Business*, ed. Bruce L. Hay, Robert N. Stavins, and Richard H. K. Vietor, 137–44. Washington, DC: Resources for the Future, 2005.

Esty, Daniel C., and Andrew S. Winston. *Green to Gold: How Smart Companies Use Environmental Strategy to Innovate, Create Value, and Build Competitive Advantage.* New Haven: Yale University Press, 2006.

ETC Group. *Extreme Genetic Engineering: An Introduction to Synthetic Biology.* 2007. http://www.etcgroup.org/sites/www.etcgroup.org/files/publication/602/01/synbioreportweb.pdf.

European Commission. Scientific Committee on Emerging and Newly Identified Health Risks. Modified Opinion (after Public Consultation) on the Appropriateness of Existing Methodologies to Assess the Potential Risks Associated with Engineered and Adventitious Products of Nanotechnologies. 2006. http://ec.europa.eu/health/ph_risk/committees/04_scenihr/docs/scenihr_0_003b.pdf.

The European Group on Ethics in Science and New Technologies to the European Commission. *Ethics of Synthetic Biology.* 2009. http://ec.europa.eu/bepa/european-group-ethics/docs/opinion25_en.pdf.

European Technology Assessment Group. *Technology Assessment on Converging Technologies.* 2006. http://www.itas.kit.edu/downloads/etag_beua06a.pdf.

Farber, Daniel A. "Adaptation Planning and Climate Impact Assessments: Learning from NEPA's Flaws." *Environmental Law Reporter* 39 (2009): 10,605–614.

Farber, Daniel A. "Uncertainty." *Georgetown Law Journal* 99 (2011): 901–59.

Farina, Cynthia R., and Jeffrey J. Rachlinski. "Foreword: Post–Public Choice?" *Cornell Law Review* 87 (2002): 267–79.

Feenberg, Andrew. "Democratic Rationalization: Technology, Power, and Freedom." In *Philosophy of Technology: The Technological Condition: An Anthology*, ed. Robert C. Scharff and Val Dusek, 652–65. Malden, MA: Blackwell, 2003.

Feenberg, Andrew. *Transforming Technology: A Critical Theory Revisited.* New York: Oxford University Press, 2002.

Felcher, E. Marla. *The Consumer Product Safety Commission and Nanotechnology.* 2008. http://www.nanotechproject.org/process/assets/files/7033/pen14.pdf.

Fender, Jessica K. "The FDA and Nano: Big Problems with Tiny Technology." *Chicago-Kent Law Review* 83 (2008): 1063–95.

Ferester, Philip Michael. "Revitalizing the National Environmental Policy Act: Substantive Law Adaptations from NEPA's Progeny." *Harvard Environmental Law Review* 16 (1992): 207–70.

Finn, John E. "Sunset Clauses and Democratic Deliberation: Assessing the Significance of Sunset Provisions in Antiterrorism Legislation." *Columbia Journal of Transnational Law* 48 (2010): 442–502.

Fiorino, Daniel J. "Citizen Participation and Environmental Risk: A Survey of Institutional Mechanisms." *Science, Technology, and Human Values* 15 (1990): 226–43.

Fischer, David A., and Robert H. Jerry II. "Teaching Torts without Insurance: A Second-Best Solution." *St. Louis University Law Journal* 45 (2001): 857–96.

Fischer, Frank. "Are Scientists Irrational?: Risk Assessment in Practical Reason." In *Science and Citizens: Globalization and the Challenge of Engagement,* ed. Melissa Leach, Ian Scoones, and Brian Wynne, 54–65. London: Zed, 2005.

Fischer, Frank. *Citizens, Experts, and the Environment: The Politics of Local Knowledge.* Durham, NC: Duke University Press, 2000.

Fischer, Frank. "Technological Deliberation in a Democratic Society: The Case for Participatory Inquiry." *Science and Public Policy* 26 (1999): 294–302.

Fischhoff, Baruch. "Public Values in Risk Research." *Annals of the American Academy of Political and Social Science* 545 (1996): 75–83.

Fisher, Elizabeth, Pasky Pascual, and Wendy Wagner. "Understanding Environmental Models in Their Legal and Regulatory Context." *Journal of Environmental Law* 22 (2010): 251–83.

Fisher, Erik, and Roop L. Mahajan. "Contradictory Intent?: U.S. Federal Legislation on Integrating Societal Concerns into Nanotechnology Research and Development." *Science and Public Policy* 33 (2006): 5–16.

Fisher, Erik, Roop L. Mahajan, and Carl Mitcham. "Midstream Modulation of Technology: Governance from Within." *Bulletin of Science, Technology, and Society* 26 (2006): 485–96.

Fisher, Erik, and Clark A. Miller. "Collaborative Practices for Contextualizing the Engineering Laboratory." In *Engineering in Context,* ed. Steen Hyldgaard Christensen, Bernard Delahousse, and Martin Meganck, 369–81. Aarhus: Academica, 2009.

Flatt, Victor B. "The 'Worst Case' May Be the Best: Rethinking NEPA Law to Avoid Future Environmental Disasters." *Environmental and Energy Law and Policy Journal* 6, no. 2 (2011): 25–42.

Fleischer, Torsten, and Armin Grunwald. "Making Nanotechnology Developments Sustainable: A Role for Technology Assessment?" *Journal of Cleaner Production* 16 (2008): 889–98.

Fleming, James Rodger. *Historical Perspectives on Climate Change.* New York: Oxford University Press, 1998.

Fleming, James Rodger. "Weather as a Weapon." *Slate*, September 23, 2010. http://www.slate.com/id/2268232/.
Flournoy, Alyson C. "Legislating Inaction: Asking the Wrong Questions in Protective Environmental Decisionmaking." *Harvard Environmental Law Review* 15 (1991): 327–92.
Frey, Bruno S. "Efficiency and Democratic Political Organisation: The Case for the Referendum." *Journal of Public Policy* 12 (1992): 209–22.
Friends of the Earth, International Center for Technology Assessment, and ETC Group. *The Principles for the Oversight of Synthetic Biology*. 2012. http://www.foe.org/publications/reports.
Frodeman, Robert, and J. Britt Holbrook. "Science's Social Effects." *Issues in Science and Technology* 24 (2007): 28–30.
Frodeman, Robert, and Jonathan Parker. "Intellectual Merit and Broader Impact: The National Science Foundation's Broader Impacts Criterion and the Question of Peer Review." *Social Epistemology* 23 (2009): 337–45.
Fukuyama, Francis. *Our Posthuman Future: Consequences of the Biotechnology Revolution*. London: Profile, 2002.
Fukuyama, Francis. "Transhumanism." *Foreign Policy*, September–October 2004, 42–43.
Fuller, Steve. "Research Trajectories and Institutional Settings of New Converging Technologies." 2008. http://www.converging-technologies.org/docs/Knowledge%20NBIC%20D1.pdf.
Gardiner, Stephen M. "Is 'Arming the Future' with Geoengineering Really the Lesser Evil?: Some Doubts about the Ethics of Intentionally Manipulating the Climate System." In *Climate Ethics: Essential Readings*, ed. Stephen M. Gardiner, Simon Caney, Dale Jamieson, and Henry Shue, 284–312. New York: Oxford University Press, 2010.
Garfinkel, Michele S., Drew Endy, Gerald Epstein, and Ronald Friedman. *Synthetic Genomics: Options for Governance*. 2007. http://dspace.mit.edu/handle/1721.1/39141.
Garreau, Joel. *Radical Evolution: The Promise and Peril of Enhancing Our Minds, Our Bodies—And What It Means to Be Human*. New York: Doubleday, 2005.
Gaudet, Lyn M., and Gary E. Marchant. "Administrative Law Tools for More Adaptive and Responsive Regulation." In *The Growing Gap between Emerging Technologies and Legal-Ethical Oversight*, ed. Gary E. Marchant, Braden R. Allenby, and Joseph R. Herkert, 167–82. New York: Springer Science and Business Media, 2011.
Gavelin, Karin, and Richard Wilson. *Democratic Technologies?: The Final Report of the Nanotechnology Engagement Group*. 2007. http://www.involve.org.uk/wp-content/uploads/2011/03/Democratic-Technologies.pdf.
Gee, David, and Morris Greenberg. "Asbestos: From 'Magic' to Malevolent Mineral." In *The Precautionary Principle in the 20th Century: Late Lessons from Early Warnings*, ed. Poul Harremoës, David Gee, Malcolm MacGarvin, Andy Stirling, Jane Keys, Brian Wynne, and Sofia Guedes Vaz, 49–63. London: Earthscan, 2002.
Geerlings, Hans, and Kenneth David. "Engagement and Translation: Perspective of a Natural Scientist." In *What Can Nanotechnology Learn from Biotechnology?:*

Social and Ethical Lessons for Nanoscience from the Debate over Agrifood Biotechnology and GMOs, ed. Kenneth David and Paul B. Thompson, 189–220. Burlington, MA: Academic, 2008.

Genus, Audley. "Rethinking Constructive Technology Assessment as Democratic, Reflective Discourse." *Technological Forecasting and Social Change* 73 (2006): 13–26.

Gerstein, Mark, et al. "What Is a Gene, Post-ENCODE?: History and Updated Definition." *Genome Research* 17 (2007): 669–81.

Gibbs, W. Wayt. "Synthetic Life." *Scientific American*, May 2004, 75–81.

Giffin, Craig B. "West Virginia's Seemingly Eternal Struggle for a Fiscally and Environmentally Adequate Coal Mining Reclamation Bonding Program." *West Virginia Law Review* 107 (2004): 105–86.

Gifford, Donald G. "The Peculiar Challenges Posed by Latent Diseases Resulting from Mass Products." *Maryland Law Review* 64 (2005): 613–98.

Glendenning, Chellis. "Notes toward a Neo-Luddite Manifesto." In *Philosophy of Technology: The Technological Condition: An Anthology*, ed. Robert C. Scharff and Val Dusek, 603–5. Malden, MA: Blackwell, 2003.

Glicksman, Robert L., David L. Markell, and Daniel R. Mandelker. *Environmental Protection Law and Policy*. 4th ed. New York: Aspen, 2003.

Golan, Elise, Fred Kuchler, and Lorraine Mitchell, with contributions by Cathy Greene and Amber Jessup. "Economics of Food Labeling." *Journal of Consumer Policy* 24 (2001): 117–84.

Goodell, Jeff. "A Hard Look at the Perils and Potential of Geoengineering." *Yale Environment 360*. 2010. http://e360.yale.edu/content/feature.msp?id=2260.

Gordon, Bart. *Engineering the Climate: Research Needs and Strategies for International Coordination*. 2010. http://www.whoi.edu/fileserver.do?id=74967&pt=2&p=81828.

Greely, Hank. "Remarks on Human Biological Enhancement." *University of Kansas Law Review* 56 (2008): 1139–57.

Grossman, Margaret Rosso. "European Community Legislation for Traceability and Labeling of Genetically Modified Crops, Food, and Feed." In *Labeling Genetically Modified Food: The Philosophical and Legal Debate*, ed. Paul Weirich, 32–62. Oxford: Oxford University Press, 2008.

Gruère, Guillaume P., and S. R. Rao. "A Review of International Labeling Policies of Genetically Modified Food to Evaluate India's Proposed Rule." *AgBioForum* 10 (2007): 51–64.

Grushcow, Jeremy M. "Measuring Secrecy: A Cost of the Patent System Revealed." *Journal of Legal Studies* 33 (2004): 59–84.

Guston, David H. "The Center for Nanotechnology in Society at Arizona State University and the Prospects for Anticipatory Governance." In *Nanoscale: Issues and Perspectives for the Nano Century*, ed. Nigel M. de S. Cameron and M. Ellen Mitchell, 377–92. Hoboken, NJ: Wiley, 2007.

Guston, David H. "Insights from the Office of Technology Assessment and Other Assessment Experiences." In *Science and Technology Advice for Congress*, ed. M. Granger Morgan and Jon M. Peha, 77–89. Washington, DC: Resources for the Future, 2003.

Guston, David H., and Daniel Sarewitz. "Real-Time Technology Assessment." *Technology in Society* 24 (2002): 93–109.

Habermas, Jürgen. *Between Facts and Norms*. Cambridge: MIT Press, 1996.
Habermas, Jürgen. *Legitimation Crisis*. Trans. Thomas McCarthy. Boston: Beacon, 1975.
Habermas, Jürgen. *The Postnational Constellation: Political Essays*. Trans. and ed. Max Pensky. 1998; Cambridge: MIT Press, 2001.
Habermas, Jürgen. *The Theory of Communicative Action*. Trans. Thomas McCarthy. Boston: Beacon, 1984.
Haemer, Robert B. "Reform of the Toxic Substances Control Act: Achieving Balance in the Regulation of Toxic Substances." *Environmental Lawyer* 6 (1999–2000): 99–134.
Hamlett, Patrick, Michael D. Cobb, and David H. Guston. *National Citizens' Technology Forum: Nanotechnologies and Human Enhancement*. CNS-ASU Report R08-0003. Glendale: Center for the Study of Nanotechnology in Society, Arizona State University, 2008.
Hansen, Steffen Foss, and Joel A. Tickner. "The Challenges of Adopting Voluntary Health, Safety, and Environment Measures for Manufactured Nanomaterials: Lessons from the Past for More Effective Adoption in the Future." *Nanotechnology Law and Business* 4 (2007): 341–59.
Hartens, Fred, and Harm Kuipers. "Effects of Androgenic-Anabolic Steroids in Athletes." *Sports Medicine* 34 (2004): 513–54.
Harvey, L. D. D. "Mitigating the Atmospheric CO_2 Increase and Ocean Acidification by Adding Limestone Powder to Upwelling Regions." *Journal of Geophysical Research-Oceans* 113 (2008). C04028. DOI: 10.1029/2007/JC004383.
Haseloff, Jim, and Jim Ajioka. "Synthetic Biology: History, Challenges, and Prospects." *Journal of Royal Society Interface*. 2009. http://rsif.royalsocietypublishing.org/site/misc/syntheticbiology_focus.xhtml.
Hathaway, Oona. "Path Dependence in the Law: The Course and Pattern of Legal Change in a Common Law System." *Iowa Law Review* 86 (2001): 601–66.
Held, David. *Democracy and the Global Order: From the Modern State to Cosmopolitan Power*. Cambridge: Cambridge University Press, 1995.
Held, David. "Democratic Accountability and Political Effectiveness from a Cosmopolitan Perspective." *Government and Opposition* 39 (2004): 364–91.
Helland, Aasgier, Peter Wick, Andreas Koehler, Kaspar Schmid, and Claudia Som. "Reviewing the Environmental and Human Health Knowledge Base of Carbon Nanotubes." *Environmental Health Perspectives* 115 (2007): 1125–31.
Hennen, Leonhard. "Impacts of Participatory TA on Its Societal Environment." In *European Participatory Technology Assessment*, ed. Lars Klüver et al., 154–68. Copenhagen: Danish Board of Technology, 2000. http://www.tekno.dk/pdf/projekter/europta_Report.pdf.
Heslop, Louise A. "If We Label It, Will They Care?: The Effect of GM-Ingredient Labelling on Consumer Responses." *Journal of Consumer Policy* 29 (2006): 203–28.
Hester, Tracy. *Remaking the World to Save It: Applying U.S. Environmental Laws to Climate Engineering Projects*. 2011. http://papers.ssrn.com/sol3/papers.cfm?abstract_id=1755203.
Hill, Christopher T. "An Expanded Analytical Capability in the Congressional Research Service, the General Accounting Office, or the Congressional Budget

Office." In *Science and Technology Advice for Congress*, ed. M. Granger Morgan and Jon M. Peha, 106–17. Washington, DC: Resources for the Future, 2003.

Holbrook, J. Britt. "Assessing the Science-Society Relation: The Case of the U.S. National Science Foundation's Second Merit Review Criterion." *Technology in Society* 27 (2005): 437–51.

Hörning, Georg. "Citizens' Panels as a Form of Deliberative Technology Assessment." *Science and Public Policy* 26 (1999): 351–59.

Houck, Oliver A. "How'd We Get Divorced?: The Curious Case of NEPA and Planning." *Environmental Law Reporter* 39 (2009): 10,645–50.

Huebner, Jonathan. "A Possible Declining Trend for Worldwide Innovation." *Technological Forecasting and Social Change* 72 (2005): 980–86.

Huffman, Wallace E., and Matt Rousu. "Consumer Attitudes and Market Resistance to Biotech Products." In *Regulating Agricultural Biotechnology: Economics and Policy*, ed. Richard E. Just, Julian M. Alston, and David Zilberman, 201–25. New York: Springer, 2006.

Hughes, James. "Beyond Human Nature." In *Unnatural Selection: The Challenges of Engineering Tomorrow's People*, ed. Peter Healey and Steve Rayner, 51–59. London: Earthscan, 2009.

Hughes, James. *Citizen Cyborg: Why Democratic Societies Must Respond to the Redesigned Human of the Future*. Cambridge, MA: Westview, 2004.

Hughes, James. "Contradictions from the Enlightenment Roots of Transhumanism." *Journal of Medicine and Philosophy* 35 (2010): 622–40.

Hunter, David, James Salzman, and Durwood Zaelke. *International Environmental Law and Policy*. 4th ed. New York: Foundation, 2011.

Hunter, David J., S. E. Hankinson, F. Laden, G. A. Colditz, J. E. Manson, W. C. Willett, F. E. Speizer, and M. S. Wolff. "Plasma Organochlorine Levels and the Risk of Breast Cancer." *New England Journal of Medicine* 337 (1997): 1253–58.

Intergovernmental Panel on Climate Change. *Carbon Dioxide Capture and Storage*. Cambridge: Cambridge University Press, 2005.

Intergovernmental Panel on Climate Change. *Climate Change 2007: Impacts, Adaptation, and Vulnerability*. Cambridge: Cambridge University Press, 2007.

Intergovernmental Panel on Climate Change. *Climate Change 2007: The Physical Science Basis*. Cambridge: Cambridge University Press, 2007. http://www.ipcc.ch/ipccreports/ar4-wg1.htm.

Intergovernmental Panel on Climate Change. *Fourth Assessment Report, Climate Change 2007: Synthesis Report*. Geneva, Switzerland: Intergovernmental Panel on Climate Change, 2007.

International Maritime Organization. *Report of the Twenty-Ninth Consultative Meeting and the Second Meeting of Contracting Parties*. 2007. http://www.imo.org/includes/blastDataOnly.asp/data_id%3D20797/17.pdf.

International Risk Governance Council. *Guidelines for the Appropriate Risk Governance of Synthetic Biology*. 2010. http://www.irgc.org/IMG/pdf/irgc_SB_final_07jan_web.pdf.

Irwin, Alan. *Citizen Science: A Study of People, Expertise, and Sustainable Development*. London: Routledge, 1995.

Irwin, Alan. "Constructing the Scientific Citizen: Science and Democracy in the Biosciences." *Public Understanding of Science* 10 (2001): 1–18.

Jacobstein, Neil. "Foresight Guidelines for Responsible Nanotechnology Development." Draft Version 6. 2006. http://www.foresight.org/guidelines/Foresight GuidelinesV6.pdf.

Jaffe, Gregory. "The Next Generation." *Environmental Forum*, March–April 2009, 38–42.

Jamieson, Dale. "Can Space Reflectors Save Us?" *Slate*, September 23, 2010. http://www.slate.com/id/2268034/.

Jamieson, Dale. "Ethics and Intentional Climate Change." *Climatic Change* 33 (1996): 323–36.

Jasanoff, Sheila. *Designs on Nature: Science and Democracy in Europe and the United States*. Princeton: Princeton University Press, 2005.

Jasanoff, Sheila. *The Fifth Branch: Science Advisors as Policymakers*. Cambridge: Harvard University Press, 1990.

Jasanoff, Sheila. "Introduction: Rewriting Life, Reframing Rights." In *Reframing Rights: Bioconstitutionalism in the Genetic Age*, ed. Sheila Jasanoff, 1–27. Cambridge: MIT Press, 2011.

Jasanoff, Sheila. *Science at the Bar: Law, Science, and Technology in America*. Cambridge: Harvard University Press, 1995.

Jasanoff, Sheila. "Technologies of Humility: Citizen Participation in Governing Science." *Minerva* 41 (2003): 223–44.

Jelsma, Jaap. "Learning about Learning in the Development of Biotechnology." In *Managing Technology in Society: The Approach of Constructive Technology Assessment*, ed. Arie Rip, Thomas J. Misa, and Johan Schot, 141–65. London and New York: Pinter, 1995.

Joss, Simon. "Participation in Parliamentary Technology Assessment: From Theory to Practice." In *Parliaments and Technology: The Development of Technology Assessment in Europe*, ed. Norman J. Vig and Herbert Paschen, 325–62. Albany: State University of New York Press, 2000.

Kaebnick, Gregory E. "Should Moral Objections to Synthetic Biology Affect Public Policy?" *Nature Biotechnology* 27 (2009): 1106–8.

Kahan, Dan M., Donald Braman, and Gregory N. Mandel. *Risk and Culture: Is Synthetic Biology Different?* 2009. SSRN: http://ssrn.com /abstract=1347165.

Kahan, Dan M., Donald Braman, Paul Slovic, John Gastil, and Geoffrey Cohen. "Cultural Cognition of the Risks and Benefits of Nanotechnology." *Nature Nanotechnology* 4 (2009): 87–90.

Karkkainen, Bradley C. "Toward a Smarter NEPA: Monitoring and Managing Government's Environmental Performance." *Columbia Law Review* 102 (2002): 903–72.

Karkkainen, Bradley C. "Whither NEPA?" *New York University Environmental Law Journal* 12 (2004): 333–63.

Karn, Barbara P., and Lynn L. Bergeson. "Green Nanotechnology: Straddling Promise and Uncertainty." *Natural Resources and Environment* 24, no. 2 (2009): 9–13, 23.

Kaufmann, Alain, Claude Joseph, Catherine El-Bez, and Marc Audétat. "Why Enrol Citizens in the Governance of Nanotechnology?" In *Governing Future Technologies: Nanotechnology and the Rise of an Assessment Regime, Sociology of the Sciences Yearbook* 27, ed. Mario Kaiser, Monika Kurath, Sabine Maasen, and Christoph Rehmann-Sutter, 201–15. Dordrecht: Springer, 2010.

Kearns, Peter. "Nanomaterials: Getting the Measure." *OECD Observer*, May 2010.

http://www.oecdobserver.org/news/fullstory.php/aid/3291/Nanomaterials:_Getting_the_measure_.html.
Keim, Brandon. "Gene Therapy: Is Death an Acceptable Risk?" *Wired*, August 30, 2007. http://www.wired.com/medtech/genetics/news/2007/08/gene_therapy?currentPage=2.
Keith, David W. "Geoengineering." In *Encyclopedia of Global Change: Environmental Change and Human Society*, ed. Andrew S. Goudie, 495–502. Oxford: Oxford University Press, 2002.
Keith, David W. "Geoengineering." *Nature* 409 (2001): 420.
Keith, David W. "Geoengineering the Climate: History and Prospect." *Annual Review of Energy and the Environment* 25 (2000): 245–84.
Keith, David W. "Why Capture CO_2 from the Atmosphere?" *Science* 325 (2009): 1654–55.
Keith, David W., Kenton Heidel, and Robert Cherry. "Capturing CO_2 from the Atmosphere: Rationale and Process Design Considerations." In *Geo-Engineering Climate Change: Environmental Necessity or Pandora's Box?*, ed. Brian Launder and J. Michael Thompson, 107–26. Cambridge: Cambridge University Press, 2010.
Keith, David W., Edward Parson, and M. Granger Morgan. "Research on Global Sun Block Needed Now." *Nature* 463 (2010): 426–27.
Kelle, Alexander. "Security Issues Related to Synthetic Biology: Between Threat Perceptions and Governance Options." In *Synthetic Biology: The Technoscience and Its Societal Consequences*, ed. Markus Schmidt, Alexander Kelle, Agomoni Ganguli-Mitra, and Huib de Vriend, 101–19. Dordrecht: Springer, 2009.
Kenney, Martin. *Biotechnology: The University-Industrial Complex*. New Haven: Yale University Press, 1986.
Khalil, Ahmad, and James J. Collins. "Synthetic Biology: Applications Come of Age." *Nature Reviews* 11 (2010): 367–79.
Khushf, George. "An Ethic for Enhancing Human Performance through Integrative Technologies." In *Managing Nano-Bio-Info-Cogno Innovations: Converging Technologies in Society*, ed. William Sims Bainbridge and Mihail C. Roco, 255–78. Dordrecht: Springer, 2006.
Kimbrell, George A. "Governance of Nanotechnology and Nanomaterials: Principles, Regulation, and Renegotiating the Social Contract." *Journal of Law, Medicine, and Ethics* 37 (2009): 706–23.
Kintisch, Eli. *Hack the Planet: Science's Best Hope—or Worst Nightmare—for Averting Climate Catastrophe*. Hoboken, NJ: Wiley, 2010.
Kintisch, Eli. "March Geoengineering Confab Draws Praise, Criticism." *ScienceInsider*, November 6, 2009. http://news.sciencemag.org/scienceinsider/2009/11/march-geoengine.html.
Klaine, Stephen J., Pedro J. J. Alvarez, Graeme E. Batley, Teresa F. Fernandes, Richard D. Handy, Delina Y. Lyon, Shaily Mahendra, Michael J. Mclaughlin, and Jamie R. Lead. "Nanomaterials in the Environment: Behavior, Fate, Bioavailability, and Effects." *Environmental Toxicology and Chemistry* 27 (2008): 1825–51.
Kleinman, Daniel Lee. "Beyond the Science Wars: Contemplating the Democratization of Science." *Politics and the Life Sciences* 17 (1998): 133–45.
Kline, Stephen J. "What Is Technology." In *Philosophy of Technology—The Techno-*

logical Condition: An Anthology, ed. Robert C. Scharff and Val Dusek, 210–12. Malden, MA: Blackwell, 2003.

Klüver, Lars. "The Danish Board of Technology." In *Parliaments and Technology: The Development of Technology Assessment in Europe*, ed. Norman J. Vig and Herbert Paschen, 173–97. Albany: State University of New York Press, 2000.

Klüver, Lars, et al. *European Participatory Technology Assessment.* 2000. http://www.tekno.dk/pdf/projekter/europta_Report.pdf.

Knezo, Genevieve J. *Technology Assessment in Congress: History and Legislative Options.* Congressional Research Service Report RS21586. 2005.

Kniesner, Thomas J., and John D. Leeth. "Abolishing OSHA." *Regulation*, Fall 1995, 46.

Knight, Andrew J. "Perceptions, Knowledge, and Ethical Concerns with GM Foods and the GM Process." *Public Understanding of Science* 18 (2009): 177–88.

Koch, Tom. "Enhancing Who? Enhancing What?: Ethics, Bioethics, and Transhumanism." *Journal of Medicine and Philosophy* 35 (2010): 685–99.

Kolbert, Elizabeth. "The Darkening Sea." *New Yorker*, November 20, 2006, 66–75.

Korobkin, Russell. "Embryonic Histrionics: A Critical Evaluation of the Bush Stem Cell Funding Policy and the Congressional Alternative." *Jurimetrics* 47 (2006): 1–29.

Kourany, Janet A. "Integrating the Ethical into Scientific Rationality." In *Science in the Context of Application*, ed. Martin Carrier and Alfred Nordmann, 371–86. Dordrecht: Springer, 2011.

Kraft, Michael E., and Norman J. Vig. "Environmental Policy from the 1970s to the 1990s: Continuity and Change." In *Environmental Policy in the 1990s: Toward a New Agenda*, ed. Norman J. Vig and Michael E. Kraft, 2nd ed., 3–29. Washington, DC: CQ Press, 1994.

Krimsky, Sheldon. "From Asilomar to Industrial Biotechnology: Risks, Reductionism, and Regulation." *Science as Culture* 14 (2005): 309–23.

Krimsky, Sheldon. *Genetic Alchemy: The Social History of the Recombinant DNA Controversy.* Cambridge: MIT Press, 1982.

Krimsky, Sheldon. *Science in the Public Interest: Has the Lure of Profits Corrupted Biomedical Research?* Lanham, MD: Rowman and Littlefield, 2003.

Krimsky, Sheldon, and Roger P. Wrubel. *Agricultural Biotechnology and the Environment: Science, Policy, and Social Issues.* Urbana: University of Illinois Press, 1996.

Krueger, Roger W. "The Public Debate on Agrobiotechnology: A Biotech Company's Perspective." *AgBioForum* 4 (2001). http://www.agbioforum.org/v4n34/v4n34a09-krueger.htm.

Kulinowski, Kristen M., and Vicki L. Colvin. "The Environmental Impact of Engineered Nanomaterials." In *Nanotechnology and the Environment: Applications and Implications*, ed. Barbara Karn, Tina Masciangioli, Wei-xian Zhang, Vicki Colvin, and Paul Alivisatos, 21–26. Washington, DC: American Chemical Society, 2005.

Kunich, John Charles. "Mother Frankenstein, Doctor Nature, and the Environmental Law of Genetic Engineering." *Southern California Law Review* 74 (2001): 807–72.

Kurath, Monika, and Priska Gisler. "Informing, Involving or Engaging?: Science

Communication, in the Ages of Atom-, Bio- and Nanotechnology." *Public Understanding of Science* 18 (2009): 559–73.

Kurzweil, Ray. *The Singularity Is Near: When Humans Transcend Biology.* New York: Viking, 2005.

Kuzma, Jennifer, ed. *The Nanotechnology-Biology Interface: Exploring Models for Oversight.* 2005. http://www.nanotechproject.org/process/assets/files/2721/53_nanotechnologybiologyinterfaceexploringmodelsoversight.pdf.

Kuzma, Jennifer, P. Najmaie, and J. Larson. "Evaluating Oversight Systems for Emerging Technologies: A Case Study of Genetically Engineered Organisms." *Journal of Law, Medicine, and Ethics* 37 (2009): 546–86.

Kwok, Roberta. "Five Hard Truths for Synthetic Biology." *Nature* 463 (2010): 288–90.

Kysar, Douglas A. "Climate Change, Cultural Transformation, and Comprehensive Rationality." *Boston College Environmental Affairs Law Review* 31 (2004): 555–90.

Kysar, Douglas A. "Ecologic: Nanotechnology, Environmental Assurance Bonding, and Symmetric Humility." *UCLA Journal of Environmental Law and Policy* 28 (2010): 201–50.

Kysar, Douglas A. "Preference for Processes: The Process/Product Distinction and the Regulation of Consumer Choice." *Harvard Law Review* 118 (2004): 526–642.

Lackner, Klaus S. "Capture of Carbon Dioxide from Ambient Air." *European Physical Journal of Special Topics* 176 (2009): 93–106.

Lambrecht, Bill. *Dinner at the New Gene Café: How Genetic Engineering Is Changing What We Eat, How We Live, and the Global Politics of Food.* New York: St. Martin's, 2001.

Latham, John, Philip Rasch, Chih-Chieh Chen, Laura Kettles, Alan Gadian, Andrew Gettelman, Hugh Morrison, Keith Bower, and Tom Choularton. "Global Temperature Stabilization via Controlled Albedo Enhancement of Low-Level Maritime Clouds." *Philosophical Transactions of the Royal Society A: Mathematical, Physical, and Engineering Sciences* 366 (2008): 3969–87.

Laurent, Brice. *Replicating Participatory Devices: The Consensus Conference Confronts Nanotechnology.*" Centre de Sociologie de l'Innovation Working Paper 18. 2009. http://www.csi.ensmp.fr/Items/WorkingPapers/Download/WP_CSI_018.pdf.

"The Law of the Sea" (editorial). *Geoscience* 2 (2009): 153.

Lazarus, Richard. "Environmental Law after Katrina: Reforming Environmental Law by Reforming Environmental Lawmaking." *Tulane Law Review* 81 (2007): 1–39.

Leach, Melissa, and Ian Scoones. "Science and Citizenship in Global Context." In *Science and Citizens: Globalization and the Challenge of Engagement*, ed. Melissa Leach, Ian Scoones, and Brian Wynne, 15–38. London: Zed, 2005.

Leinen, Margaret. "The Asilomar International Conference on Climate Intervention Technologies: Background and Overview." *Stanford Journal of Law, Science and Policy* 4 (2011): 1–5.

Leiserowitz, Anthony. *International Public Opinion, Perception, and Understanding of Climate Change.* 2009. http://environment.yale.edu/climate/publications/international-public-opinion-perception-and-understanding-of-global-cl/.

Lemaux, Peggy G. "Genetically Engineered Plants and Foods: A Scientist's Analysis of the Issues (Part I)." *Annual Review of Plant Biology* 59 (2008): 771–812.

Lempert, Richard. "Low Probability/High Consequence Events: Dilemmas of Damage Compensation." *DePaul Law Review* 58 (2009): 357–91.

Li, Jasmine, S. Muralikrishnan, C. T. Ng, L. Y. Yung, and B. H. Bay. "Nanoparticle-Induced Pulmonary Toxicity." *Experimental Biology and Medicine* 235 (2010): 1025–33.

Li, Ning, T. Xia, and A. E. Nel. "The Role of Oxidative Stress in Ambient Particulate Matter-Induced Lung Diseases and Its Implications in the Toxicity of Engineered Nanoparticles." *Free Radical Biology and Medicine* 44 (2008): 1689–99.

Light, Paul C. "The Homeland Security Hash." *Wilson Quarterly*, Spring 2007, 36–44.

Lin, Albert C. "Beyond Tort: Compensating Victims of Environmental Toxic Injury." *Southern California Law Review* 78 (2005): 1439–1528.

Lin, Albert C. "Deciphering the Chemical Soup: Using Public Nuisance to Compel Chemical Testing." *Notre Dame Law Review* 85 (2010): 955–1013.

Lin, Albert C. "Geoengineering Governance." *Issues in Legal Scholarship* 8 (2009). http://www.degruyter.com/view/j/ils.2009.8.issue-1/ils.2009.8.1.1112/ils.2009.8.1.1112.xml.

Lin, Albert C. "Size Matters: Regulating Nanotechnology." *Harvard Environmental Law Review* 31 (2007): 349–408.

Lin, Albert C. "The Unifying Role of Harm in Environmental Law." *Wisconsin Law Review* 2006 (2006): 897–985.

Lin, Patrick. "Nanotechnology Bound: Evaluating the Case for More Regulation." *Nanoethics* 1 (2007): 105–22.

Lindstrom, Matthew J., and Zachary A. Smith. *The National Environmental Policy Act: Judicial Misconstruction, Legislative Indifference, and Executive Neglect.* College Station: Texas A&M University Press, 2001.

Linkov, Igor, F. Kyle Satterstrom, John C. Monica Jr., Steffen Foss Hansen, and Thomas A. Davis. "Nano Risk Governance: Current Developments and Future Perspectives." *Nanotechnology Law and Business* 6 (2009): 203–20.

Linstone, Harold A. "Technological Slowdown or Societal Speedup—The Price of System Complexity?" *Technological Forecasting and Social Change* 51 (1996): 195–205.

Lobel, Orly. "The Renew Deal: The Fall of Regulation and the Rise of Governance in Contemporary Legal Thought." *Minnesota Law Review* 89 (2004): 342–470.

Locke, John. *Two Treatises of Government.* Ed. Peter Laslett. 1689; Cambridge: Cambridge University Press, 1968.

Lyndon, Mary L. "Information Economics and Chemical Toxicity; Designing Laws to Produce and Use Data." *Michigan Law Review* 87 (1989): 1795–1861.

Lyndon, Mary L. "Tort Law and Technology." *Yale Journal on Regulation* 12 (1995): 137–76.

MacCracken, Michael C. "On the Possible Use of Geoengineering to Moderate Specific Climate Change Impacts." *Environmental Research Letters* 4 (2009): 045107. DOI: 10.1088/1748-9326/4/4/045107.

Malthus, Thomas. *An Essay on the Principle of Population.* Ed. Philip Appleman. 2nd ed. New York: Norton, 2004.

Mandel, Gregory N. "Confidence-Building Measures for Genetically Modified Products: Stakeholder Teamwork on Regulatory Proposals." *Jurimetrics* 44 (2003): 41–61.

Mandel, Gregory N. "Gaps, Inexperience, Inconsistencies, and Overlaps: Crisis in the Regulation of Genetically Modified Plants and Animals." 45 *William and Mary Law Review* (2004): 2167–2259.
Mandel, Gregory N. "Nanotechnology Governance." *Alabama Law Review* 59 (2008): 1323–84.
Mandel, Gregory N. "Regulating Emerging Technologies." *Law, Innovation, and Technology* 1 (2009): 75–92.
Mandelker, Daniel R. *NEPA Law and Litigation.* 2nd ed. Deerfield, IL: Clark Boardman Callaghan, 1992.
Mandelker, Daniel R. "Thoughts on NEPA at 40." *Environmental Law Reporter* 39 (2009): 10,640–41.
Manning, John F. "Competing Presumptions about Statutory Coherence." *Fordham Law Review* 74 (2006): 2009–50.
Marchant, Gary E., and Douglas J. Sylvester. "Transnational Models for Regulation of Nanotechnology." *Journal of Law, Medicine, and Ethics* 34 (2006): 714–25.
Marchant, Gary E., Douglas J. Sylvester, and Kenneth W. Abbott. "A New Soft Law Approach to Nanotechnology Oversight: A Voluntary Product Certification Scheme." *UCLA Journal of Environmental Law and Policy* 28 (2010): 123–52.
Marchant, Gary E., Douglas J. Sylvester, and Kenneth W. Abbott. "Risk Management Principles for Nanotechnology." *Nanoethics* 2 (2008): 43–60.
Margolis, Robert M., and David H. Guston. "The Origins, Accomplishments, and Demise of the Office of Technology Assessment." In *Science and Technology Advice for Congress*, ed. M. Granger Morgan and Jon M. Peha, 53–76. Washington, DC: Resources for the Future, 2003.
Matsuura, Jeffrey H. *Nanotechnology Regulation and Policy Worldwide.* Boston: Artech, 2006.
Matthews, H. Damon, and Ken Caldeira. "Transient Climate-Carbon Simulations of Planetary Geoengineering." *Proceedings of the National Academy of Sciences* 104 (2007): 9949–54.
Maynard, Andrew D., and Eileen D. Kuempel. "Airborne Nanostructured Particles and Occupational Health." *Journal of Nanoparticle Research* 7 (2005): 587–614.
Maynard, Andrew D., and David Rejeski. "Too Small to Overlook." *Nature* 460 (2009): 174.
McCarthy, Paul. "Liberal Freedoms: Enhancement Is/Nt Eugenics?" *Studies in Ethics, Law, and Technology* 1 (2007). DOI: 10.2202/1941-6008.1002.
McClellan, Justin, David W. Keith, and Jay Apt. "Cost Analysis of Stratospheric Albedo Modification Delivery Systems." *Environmental Research Letters* 7 (2012): 034019. DOI: 10.1088/1748-9326/7/3/034019.
McDonald, Barry P. "Government Regulation or Other 'Abridgements' of Scientific Research: The Proper Scope of Judicial Review under the First Amendment." *Emory Law Journal* 54 (2005): 979–1091.
McElfish, James, and Elissa Parker. *Rediscovering the National Environmental Policy Act: Back to the Future.* Washington, DC: Environmental Law Institute, 1995.
McGarity, Thomas O. "Frankenfood Free: Consumer Sovereignty, Federal Regulation, and Industry Control in Marketing and Choosing Food in the United States." In *Labeling Genetically Modified Food: The Philosophical and Legal Debate*, ed. Paul Weirich, 128–50. Oxford: Oxford University Press, 2008.

McGarity, Thomas O. "Seeds of Distrust: Federal Regulation of Genetically Modified Foods." *University of Michigan Journal of Law Reform* 35 (2002): 403–510.

McGarity, Thomas O., and Wendy E. Wagner. *Bending Science: How Special Interests Corrupt Public Health Research*. Cambridge: Harvard University Press, 2008.

McKibben, Bill. *Enough: Staying Human in an Engineered Age*. New York: Holt, 2003.

McMahon, Joseph. "The *EC-Biotech* Decision: Another Missed Opportunity?" In *The Regulation of Genetically Modified Organisms: Comparative Approaches*, ed. Luc Bodiguel and Michael Cardwell, 337–55. Oxford: Oxford University Press, 2010.

Meadows, Donella H., Jørgen Randers, Dennis L. Meadows, and William W. Behrens. *Limits to Growth: A Report for the Club of Rome's Project on the Predicament of Mankind*. New York: Universe, 1972.

Meili, Christoph, and Markus Widmer. "Voluntary Measures in Nanotechnology Risk Governance: The Difficulty of Holding the Wolf by the Ears." In *International Handbook on Regulating Nanotechnologies*, ed. Graeme A. Hodge, Diana M. Bowman, and Andrew D. Maynard, 446–61. Cheltenham: Elgar, 2010.

Mellon, Margaret. "A View from the Advocacy Community." In *What Can Nanotechnology Learn from Biotechnology?: Social and Ethical Lessons for Nanoscience from the Debate over Agrifood Biotechnology and GMOs*, ed. Kenneth David and Paul B. Thompson, 81–88. Burlington, MA: Academic, 2008.

Mendeloff, John M. *The Dilemma of Toxic Substance Regulation*. Cambridge: MIT Press, 1988.

Metcalfe, Dean D. "What Are the Issues in Addressing the Allergenic Potential of Genetically Modified Foods?" *Environmental Health Perspectives* 111 (2003): 1110–13.

Michaels, David. *Doubt Is Their Product: How Industry's Assault on Science Threatens Your Health*. New York: Oxford University Press, 2008.

Miles, John. "Nanotechnology Captured." In *International Handbook on Regulating Nanotechnologies*, ed. Graeme A. Hodge, Diana M. Bowman, and Andrew D. Maynard, 83–106. Cheltenham: Elgar, 2010.

Miller, Henry I., and Gregory Conko. *The Frankenfood Myth: How Protest and Politics Threaten the Biotech Revolution*. Westport, CT: Praeger, 2004.

Modis, Theodore. "Discussion of Huebner Article." *Technological Forecasting and Social Change* 72 (2005): 987–88.

Modis, Theodore. "Forecasting the Growth of Complexity and Change." *Technological Forecasting and Social Change* 69 (2002): 377–404.

Monica, John C., Jr., and John C. Monica. "Examples of Recent EPA Regulation of Nanoscale Materials under the Toxic Substances Control Act." *Nanotechnology Law and Business* 6 (2009): 388–406.

Mooallem, Jon. "Do-It-Yourself Genetic Engineering." *New York Times Magazine*, February 14, 2010, 40.

Mooney, Chris. "A Short History of Sunsets." *Legal Affairs*, January–February 2004, 67–71.

Morgan, M. Granger, and Jon M. Peha. "Analysis, Governance, and the Need for Better Institutional Arrangements." In *Science and Technology Advice for Congress*, ed. M. Granger Morgan and Jon M. Peha, 3–20. Washington, DC: Resources for the Future, 2003.

Morrison, James L. "Environmental Scanning." In *A Primer for New Institutional Researchers*, ed. M. A. Whitely, J. D. Porter, and R. H. Fenske, 86–99. Tallahassee, FL: Association for Institutional Research, 1992. http://horizon.unc.edu/courses/papers/enviroscan/default.html.

Morrow, David R., Robert E. Kopp, and Michael Oppenheimer. "Toward Ethical Norms and Institutions for Climate Engineering Research." *Environmental Research Letters* 4 (2009): 045106. DOI: 10.1088/1748-9326/4/4/045106.

Mutlu, Gokhan M., et al. "Biocompatible Nanoscale Dispersion of Single-Walled Carbon Nanotubes Minimizes in vivo Pulmonary Toxicity." *Nano Letters* 10 (2010): 1664–70.

Nanz, Patrizia, and Jens Steffek. "Global Governance, Participation, and the Public Sphere." *Government and Opposition* 39 (2004): 314–35.

National Research Council. *Environmental Effects of Transgenic Plants: The Scope and Adequacy of Regulation*. Washington, DC: National Academies Press, 2002.

National Research Council. *Genetically Modified Pest-Protected Plants: Science and Regulation*. Washington, DC: National Academies Press, 2000.

National Research Council. *Human Performance Modification: Review of Worldwide Research with a View to the Future*. Washington, DC: National Academies Press, 2012.

National Research Council. *The Impact of Genetically Engineered Crops on Farm Sustainability in the United States*. Washington, DC: National Academies Press, 2010.

National Research Council. *A Research Strategy for Environmental, Health, and Safety Aspects of Engineered Nanomaterials*. Washington, DC: National Academies Press, 2012.

National Research Council. *Review of Federal Strategy for Nanotechnology-Related Environmental, Health, and Safety Research*. Washington, DC: National Academies Press, 2009.

National Research Council. *Science and Judgment in Risk Assessment*. Washington, DC: National Academy Press, 1994.

National Research Council. *Understanding Risk: Informing Decisions in a Democratic Society*. Ed. Paul C. Stern and Harvey V. Fineberg. Washington, DC: National Academy Press, 1996.

National Science Board. *Science and Engineering Indicators*. 2008. http://www.nsf.gov/statistics/seind08/.

National Science Board. *Science and Engineering Indicators*. 2010. http://www.nsf.gov/statistics/seind10/.

National Science Foundation. *Merit Review Broader Impacts Criterion: Representative Activities*. 2007. http://www.nsf.gov/pubs/gpg/broaderimpacts.pdf.

National Science Foundation. *Proposal and Award Policies and Procedures Guide*. 2009. http://www.nsf.gov/publications/pub_summ.jsp?ods_key=nsf0929&org=NSF.

Nel, Andre E., Lutz Mädler, Darrell Velegol, Tian Xia, Eric M. V. Hoek, Ponisseril Somasundaran, Fred Klaessig, Vince Castranova, and Mike Thompson. "Understanding Biophysicochemical Interactions at the Nano-Bio Interface." *Nature Materials* 8 (2009): 543–57.

Nel, Andre, T. Xia, L. Madler, and N. Li. "Toxic Potential of Materials at the Nanolevel." *Science* 311 (2006): 622–27.

Newberry, Byron. "Are Engineers Instrumentalists?" *Technology in Society* 29 (2007): 107–19.

Nisbet, Matthew C., and Dietram A. Scheufele. "What's Next for Science Communication?: Promising Directions and Lingering Distractions." *American Journal of Botany* 96 (2009): 1767–78.

Noah, Lars. "The Imperative to Warn: Disentangling the 'Right to Know' from the 'Need to Know' about Consumer Product Hazards." *Yale Journal on Regulation* 11 (1994): 293–400.

Nordmann, Alfred. *Converging Technologies—Shaping the Future of European Societies.* 2004. http://www.ntnu.no/2020/final_report_en.pdf.

Nordmann, Alfred, and Astrid Schwarz. "Lure of the 'Yes': The Seductive Power of Technoscience." In *Governing Future Technologies: Nanotechnology and the Rise of an Assessment Regime, Sociology of the Sciences Yearbook* 27, ed. Mario Kaiser, Monika Kurath, Sabine Maasen, and Christoph Rehmann-Sutter, 255–77. Dordrecht: Springer, 2010.

Nye, Joseph S., Jr. "Globalization's Democratic Deficit: How to Make International Institutions More Accountable." *Foreign Affairs*, July–August 2001, 2–6.

Oberdörster, Gunter. "Safety Assessment for Nanotechnology and Nanomedicine: Concepts of Nanotoxicology." *Journal of Internal Medicine* 267 (2009): 89–105.

Oberdörster, Gunter, E. Oberdörster, and J. Oberdörster. "Nanotoxicology: An Emerging Discipline Evolving from Studies of Ultrafine Particles." *Environmental Health Perspectives* 113 (2005): 823–39.

Olson, Mancur. *The Logic of Collective Action: Public Goods and the Theory of Groups.* Cambridge: Harvard University Press, 1971.

Organisation for Economic Co-Operation and Development. *OECD Programme on the Safety of Manufactured Nanomaterials, 2009–2012: Operational Plans of the Projects.* 2010. http://search.oecd.org/officialdocuments/displaydocumentpdf/?doclanguage=en&cote=ENV/JM/MONO(2010)11.

Owen, David G. "Bending Nature, Bending Law." *Florida Law Review* 62 (2010): 569–616.

Page, Benjamin I., and Robert Y. Shapiro. "Effects of Public Opinion on Policy." *American Political Science Review* 77 (1983): 175–90.

Parens, Erik, Josephine Johnston, and Jacob Moses. *Ethical Issues in Synthetic Biology: An Overview of the Debates.* 2009. http://www.synbioproject.org/process/assets/files/6334/synbio3.pdf.

Park, Eun-Jung, W. S. Cho, J. Jeong, J. Yi, K. Choi, K. Park. "Pro-Inflammatory and Potential Allergic Responses Resulting from B Cell Activation in Mice Treated with Multi-Walled Carbon Nanotubes by Intratracheal Instillation." *Toxicology* 259 (2009): 113–21.

Patterson, Thomas E. *The Vanishing Voter: Public Involvement in an Age of Uncertainty.* New York: Knopf, 2002.

Peha, Jon M. "Science and Technology Advice for Congress: Past, Present, and Future." *Renewable Resources Journal* 24 (2006): 19–23.

Pei, Lei, et al. "Regulatory Frameworks for Synthetic Biology." In *Synthetic Biology: Industrial and Environmental Applications*, ed. Markus Schmidt, 157–226. Weinheim, Germany: Wiley-Blackwell, 2012.

Pelletier, David L. "FDA's Regulation of Genetically Engineered Foods: Scientific, Legal and Political Dimensions." *Food Policy* 31 (2006): 570–91.

Pelletier, David L. "Science, Law, and Politics in FDA's Genetically Engineered Foods Policy: Scientific Concerns and Uncertainties." *Nutrition Reviews* 63 (2005): 210–23.

Pennisi, Elizabeth. "Synthetic Genome Brings New Life to Bacterium." *Science* 328 (2010): 958–59.

Percival, Robert V., Christopher H. Schroeder, Alan S. Miller, and James P. Leape. *Environmental Regulation: Law, Science, and Policy*. 4th ed. New York: Aspen, 2003.

Perez, Oren. "Precautionary Governance and the Limits of Scientific Knowledge: A Democratic Framework for Regulating Nanotechnology." *UCLA Journal of Environmental Law and Policy* 28 (2010): 29–76.

Perrings, Charles. "Environmental Bonds and Environmental Research in Innovative Activities." *Ecological Economics* 1 (1989): 95–110.

Petermann, Thomas. "Technology Assessment Units in the European Parliamentary Systems." In *Parliaments and Technology: The Development of Technology Assessment in Europe*, ed. Norman J. Vig and Herbert Paschen, 37–61. Albany: State University of New York Press, 2000.

Peters, Philip G., and Thomas A. Lambert. "Regulatory Barriers to Consumer Information about Genetically Modified Foods." In *Labeling Genetically Modified Food: The Philosophical and Legal Debate*, ed. Paul Weirich, 151–77. Oxford: Oxford University Press, 2008.

Peterson, James Edward. "Can Algae Save Civilization?: A Look at Technology, Law, and Policy Regarding Iron Fertilization of the Ocean to Counteract the Greenhouse Effect." *Colorado Journal of International Environmental Law and Policy* 6 (1995): 61–108.

Phoenix, Chris, and Eric Drexler. "Safe Exponential Manufacturing." *Nanotechnology* 15 (2004): 869–72.

Pidgeon, Nick, Barbara Herr Harthorn, Karl Bryant, and Tee Rogers-Hayden. "Deliberating the Risks of Nanotechnologies for Energy and Health Applications in the United States and United Kingdom." *Nature Nanotechnology* 4 (2009): 95–98.

Pierson, Paul. "Increasing Returns, Path Dependence, and the Study of Politics." *American Political Science Review* 94 (2000): 251–67.

Pitt, Joseph. *Thinking about Technology*. New York: Seven Bridges, 2000.

Plater, Zygmunt, J. B., Robert H. Abrams, William Goldfarb, Lisa Heinzerling, David L. Wirth, and Robert L. Graham. *Environmental Law and Policy: Nature, Law, and Society*. 3rd ed. New York: Aspen, 2004.

Plein, L. Christopher, and David J. Webber. "The Role of Technology Assessment in Congressional Consideration of Biotechnology." In *Science, Technology, and Politics: Policy Analysis in Congress*, ed. Gary C. Bryner, 123–51. Boulder, CO: Westview, 1992.

Plough, Alonzo, and Sheldon Krimsky. "The Emergence of Risk Communication Studies: Social and Political Context." *Science, Technology, and Human Values* 12 (1987): 4–10.

Poland, Craig A., et al. "Carbon Nanotubes Introduced into the Abdominal Cavity of Mice Show Asbestos-Like Pathogenicity in a Pilot Study." *Nature Nanotechnology* 3 (2008): 423–28.

Pollack, Mark A., and Gregory C. Shaffer. *When Cooperation Fails: The International Law and Politics of Genetically Modified Foods.* Oxford: Oxford University Press, 2009.

Pollan, Michael. *The Omnivore's Dilemma: A Natural History of Four Meals.* New York: Penguin, 2006.

Popovsky, Mark. "Nanotechnology and Environmental Insurance." *Columbia Journal of Environmental Law* 36 (2011): 125–62.

Portney, Paul R. "Corporate Social Responsibility: An Economic and Public Policy Perspective." In *Environmental Protection and the Social Responsibility of Firms: Perspectives from Law, Economics, and Business*, ed. Bruce L. Hay, Robert N. Stavins, and Richard H. K. Vietor, 107–31. Washington, DC: Resources for the Future, 2005.

Posner, Richard A. *Catastrophe: Risk and Response.* Oxford: Oxford University Press, 2004.

Postel, Sandra. *Last Oasis: Facing Water Scarcity.* New York: Norton, 1997.

Powell, Maria, and Daniel Lee Kleinman. "Building Citizen Capacities for Participation in Nanotechnology Decision-Making: The Democratic Virtues of the Consensus Conference Model." *Public Understanding of Science* 17 (2008): 329–48.

Powles, Stephen B. "Evolved Glyphosate-Resistant Weeds around the World: Lessons to Be Learnt." *Pest Management Science* 64 (2008): 360–65.

Preston, Christopher J. "Synthetic Biology: Drawing a Line in Darwin's Sand." *Environmental Values* 17 (2008): 23–39.

Priest, Susanna. "Biotechnology, Nanotechnology, Media, and Public Opinion." In *What Can Nanotechnology Learn from Biotechnology?: Social and Ethical Lessons for Nanoscience from the Debate over Agrifood Biotechnology and GMOs*, ed. Kenneth David and Paul B. Thompson, 221–34. Burlington, MA: Academic, 2008.

Rabinow, Paul. "Prosperity, Amelioration, Flourishing: From a Logic of Practical Judgment to Reconstruction." *Law and Literature* 21 (2009): 301–20.

Rabinow, Paul, and Gaymon Bennett. *Designing Human Practices: An Experiment with Synthetic Biology*. Chicago: University of Chicago Press, 2012.

Rakhlin, Maksim. "Regulating Nanotechnology: A Private-Public Solution." *Duke Law and Technology Review* 2008 (2008): article 2.

Ramey, James T. "The Promise of Nuclear Energy." *Annals of the American Academy of Political and Social Science* 410 (1973): 11–23.

Rappert, Brian. "Pacing Science and Technology with Codes of Conduct: Rethinking What Works." In *The Growing Gap between Emerging Technologies and Legal-Ethical Oversight*, ed. Gary E. Marchant, Braden R. Allenby, and Joseph R. Herkert, 109–26. New York: Springer Science and Business Media, 2011.

Rasch, Philip J., John Latham, and Chih-Chieh Chen. "Geoengineering by Cloud Seeding: Influence on Sea Ice and Climate System." *Environmental Research Letters* 4 (2009): 045112. DOI: 10.1088/1748-9326/4/4/045112.

Rasch, Philip J., S. Tilmes, R. P. Turco, A. Robock, L. Oman, C. C. Chen, G. L. Stenchikov, and R. R. Garcia. "An Overview of Geoengineering of Climate Using Stratospheric Sulphate Aerosols." *Philosophical Transactions of the Royal Society A: Mathematical, Physical and Engineering Sciences* 366 (2008): 4007–37.

Raustiala, Kal. "States, NGOs, and International Environmental Institutions." *International Studies Quarterly* 41 (1997): 719–40.
Rawls, John. "The Idea of Public Reason Revisited." *University of Chicago Law Review* 64 (1997): 765–808.
Rawls, John. *Political Liberalism.* New York: Columbia University Press, 1996.
Rayner, S., C. Redgwell, J. Savulescu, N. Pidgeon, and T. Kruger. *Memorandum on Draft Principles for the Conduct of Geoengineering Research: House of Commons Science and Technology Committee Enquiry into the Regulation of Geoengineering.* 2009. http://www.sbs.ox.ac.uk/research/sts/Documents/regulation-of-geoengineering.pdf.
Rechtschaffen, Clifford. "The Warning Game: Evaluating Warnings under California's Prop 65." *Ecology Law Quarterly* 23 (1996): 303–68.
Reichhardt, Tony. "Will Souped Up Salmon Sink or Swim?" *Nature* 406 (2000): 10–12.
Reinhardt, Forest L. "Environmental Protection and the Social Responsibility of Firms: Perspectives from the Business Literature." In *Environmental Protection and the Social Responsibility of Firms: Perspectives from Law, Economics, and Business,* ed. Bruce L. Hay, Robert N. Stavins, and Richard H. K. Vietor, 159–68. Washington, DC: Resources for the Future, 2005.
Rejeski, David. "Comment on a Framework Convention for Nanotechnology?" *Environmental Law Reporter* 38 (2008): 10,518–19.
Rejeski, David. "The Molecular Economy." *Environmental Forum,* January–February 2010, 36–41.
Rejeski, David. "Public Policy on the Technological Frontier." In *The Growing Gap between Emerging Technologies and Legal-Ethical Oversight,* ed. Gary E. Marchant, Braden R. Allenby, and Joseph R. Herkert, 47–59. New York: Springer Science and Business Media, 2011.
Remmen, Arne. "Pollution Prevention, Cleaner Technologies, and Industry." In *Managing Technology in Society: The Approach of Constructive Technology Assessment,* ed. Arie Rip, Thomas J. Misa, and Johan Schot, 199–222. London and New York: Pinter, 1995.
Renn, Ortwin, Thomas Webler, and Peter Wiedemann. "The Pursuit of Fair and Competent Citizen Participation." In *Fairness and Competence in Citizen Participation: Evaluating Models for Environmental Discourse,* ed. Ortwin Renn, Thomas Webler, and Peter Wiedemann, 339–67. Boston: Kluwer Academic, 1995.
Resnik, David B. *Playing Politics with Science: Balancing Scientific Independence and Government Oversight.* Oxford: Oxford University Press, 2009.
Rice, Todd W. "The Historical, Ethical, and Legal Background of Human-Subjects Research." *Respiratory Care* 53 (2008): 1325–29.
Rip, Arie. "Assessing the Impacts of Innovation: New Developments in Technology Assessment." In *Social Sciences and Innovation,* 197–213. Paris: OECD, 2001.
Rip, Arie. "Nanoscience and Nanotechnologies: Bridging Gaps through Constructive Technology Assessment." In *Handbook of Transdisciplinary Research,* ed. Gertrude Hirsch Hadorn, Holger Hoffmann-Riem, Susette Biber-Klemm, Walter Grossenbacher-Mansuy, Dominique Joye, Christian Pohl, Urs Wiesmann, and Elisabeth Zemp, 145–57. Dordrecht: Springer, 2008.
Rip, Arie, Johan Schot, and Thomas J. Misa. "Constructive Technology Assessment: A New Paradigm for Managing Technology in Society." In *Managing Technol-*

ogy in Society: The Approach of Constructive Technology Assessment, ed. Arie Rip, Thomas J. Misa, and Johan Schot, 1–12. London and New York: Pinter, 1995.

Rip, Arie, and Marloes Van Amerom. "Emerging De Facto Agendas Surrounding Nanotechnology: Two Cases Full of Contingencies, Lock-Outs, and Lock-Ins." In *Governing Future Technologies, Sociology of the Sciences Yearbook*, ed. Mario Kaiser, Monika Kurath, Sabine Maasen, and Christoph Rehmann-Sutter, 131–55. Dordrecht: Springer, 2010.

Robock, Alan. "20 Reasons Why Geoengineering May be a Bad Idea." *Bulletin of the Atomic Scientists*, May–June 2008, 14–18. http://www.thebulletin.org/files/064002006_0.pdf.

Robock, Alan. "Whither Geoengineering?" *Science* 320 (2008): 1166–67.

Robock, Alan, Martin Bunzl, Ben Kravitz, and Georgiy L. Stenchikov. "A Test for Geoengineering?" *Science* 327 (2010): 530–31.

Robock, Alan, Allison Marquardt, Ben Kravitz, and Georgiy Stenchikov. "Benefits, Risks, and Costs of Stratospheric Geoengineering." *Geophysical Research Letters* 36 (2009): L19703. DOI:10.1029/2009GL039209.

Robock, Alan, Luke Oman, and Georgiy L. Stenchikov. "Regional Climate Responses to Geoengineering with Tropical and Arctic SO_2 Injections." *Journal of Geophysical Research* 113 (2008): article D16101. DOI: 10.1029/2008JD010050.

Rockström, Johan, et al. "A Safe Operating Space for Humanity." *Nature* 461 (2009): 472–75.

Roco, Mihail C. "Introduction." In *Societal Implications of Nanoscience and Nanotechnology*, ed. Mihail C. Roco and William Sims Bainbridge, 1–3. Dordrecht: Kluwer Academic, 2001.

Roco, Mihail C. "Possibilities for Global Governance of Converging Technologies." *Journal of Nanoparticle Research* 10 (2008): 11–29.

Roco, Mihail C., and William Bainbridge, eds. *Converging Technologies for Improving Human Performance: Nanotechnology, Biotechnology, Information Technology, and Cognitive Science*. Dordrecht: Kluwer Academic, 2002.

Rodemeyer, Michael. "Back to the Future: Revisiting OTA Ten Years Later." In *The Future of Technology Assessment*, ed. Michael Rodemeyer, Daniel Sarewitz, and James Wilsdon. Washington, DC: Woodrow Wilson International Center for Scholars, 2005. Available at http://www.wilsoncenter.org/sites/default/files/techassessment.pdf.

Rodemeyer, Michael. *New Life, Old Bottles: Regulating First-Generation Products of Synthetic Biology*. 2009. http://www.synbioproject.org/library/publications/archive/synbio2/.

Rogers-Hayden, Tee, and Nick Pidgeon. "Moving Engagement 'Upstream'?: Nanotechnologies and the Royal Society and Royal Academy of Engineering's Inquiry." *Public Understanding of Science* 16 (2007): 345–64.

Rosa, Hartmut. "Social Acceleration: Ethical and Political Consequences of a Desynchronized High-Speed Society." In *High-Speed Society: Social Acceleration, Power, and Modernity*, ed. Hartmut Rosa and William E. Scheuerman, 77–112. University Park: Pennsylvania State University Press, 2009.

The Royal Academy of Engineering. *Synthetic Biology: Scope, Applications and Implications*. 2009. http://www.raeng.org.uk/news/publications/list/reports/Synthetic_biology.pdf.

The Royal Society. *Geoengineering the Climate: Science, Governance, and Uncer-*

tainty. 2009. http://royalsociety.org/policy/publications/2009/geoengineering-climate/.

The Royal Society and the Royal Academy of Engineering. *Nanoscience and Nanotechnologies: Opportunities and Uncertainties*. 2004. http://www.nanotec.org.uk/finalReport.htm.

Russell, Edmund P., III. "The Strange Career of DDT: Experts, Federal Capacity, and Environmentalism in World War II." *Technology and Culture* 40 (1999): 770–96.

Sabel, Charles F., and William H. Simon. "Contextualizing Regimes: Institutionalization as a Response to the Limits of Interpretation and Policy Engineering." *Michigan Law Review* 110 (2012): 1265–1308.

Sandberg, Anders, and Nick Bostrom. "Cognitive Enhancement: A Review of Technology." *EU ENHANCE Project Publication*. 2007. http://diyhpl.us/~bryan/papers2/neuro/implants/Anders%20Sandberg,%20Nick%20Bostrom%20%20Cognitive%20Enhancement%20Tech%20Review.pdf.

Sandel, Michael J. "The Case against Perfection: What's Wrong with Designer Children, Bionic Athletes, and Genetic Engineering." *Atlantic Monthly*, April 2004, 51–62.

Sarahan, Paul C. "Nanotechnology Safety: A Framework for Identifying and Complying with Workplace Safety Requirements." *Nanotechnology Law and Business* 5 (2008): 191–205.

Sarewitz, Daniel. "Anticipatory Governance of Emerging Technologies." In *The Growing Gap between Emerging Technologies and Legal-Ethical Oversight*, ed. Gary E. Marchant, Braden R. Allenby, and Joseph R. Herkert, 95–105. New York: Springer Science and Business Media, 2011.

Sarewitz, Daniel. "Governing Our Future Selves." In *Unnatural Selection: The Challenges of Engineering Tomorrow's People*, ed. Peter Healey and Steve Rayner, 215–20. London: Earthscan, 2009.

Sarewitz, Daniel. "This Won't Hurt a Bit: Assessing and Governing Rapidly Advancing Technologies in a Democracy." In *The Future of Technology Assessment*, ed. Michael Rodemeyer, Daniel Sarewitz, and James Wilsdon. Washington, DC: Woodrow Wilson International Center for Scholars, 2005. Available at http://www.wilsoncenter.org/sites/default/files/techassessment.pdf.

Satterfield, T., M. Kandlikar, C. E. Beaudrie, J. Conti, and B. Herr Harthorn. "Anticipating the Perceived Risk of Nanotechnologies." *Nature Nanotechnology* 4 (2009): 752–58.

Schelling, Thomas C. "The Economic Diplomacy of Geoengineering." *Climatic Change* 33 (1996): 303–7.

Scheuerman, William E. "Constitutionalism in an Age of Speed." *Constitutional Commentary* 19 (2002): 353–90.

Schmidt, Markus. "Do I Understand What I Create?: Biosafety Issues in Synthetic Biology." In *Synthetic Biology: The Technoscience and Its Societal Consequences*, ed. Markus Schmidt, Alexander Kelle, Agomoni Ganguli-Mitra, and Huib de Vriend, 81–100. Dordrecht: Springer, 2009.

Schmidt, Markus, ed. *Synthetic Biology: Industrial and Environmental Applications*. Weinheim, Germany: Wiley-Blackwell, 2012.

Schnaiberg, Allan. *The Environment: From Surplus to Scarcity*. New York: Oxford University Press, 1980.

Schneider, Stephen H. "Geoengineering: Could—or Should—We Do It?" *Climatic Change* 33 (1996): 291–302.
Scholte, Jan Aart. "Civil Society and Democratically Accountable Global Governance." *Government and Opposition* 39 (2004): 211–33.
Schot, Johan. "Towards New Forms of Participatory Technology Development." *Technology Analysis and Strategic Management* 13 (2001): 39–52.
Schot, Johan, and Arie Rip. "The Past and Future of Constructive Technology Assessment." *Technology Forecasting and Social Change* 54 (1996): 251–68.
Schroeder, Christopher H. "Deliberative Democracy's Attempt to Turn Politics into Law." *Law and Contemporary Problems* 65 (2002): 95–132.
Schuiling, R. D., and P. Krijgsman. "Enhanced Weathering: An Effective and Cheap Tool to Sequester CO_2." *Climate Change* 74 (2006): 349–54.
Schummer, Joachim. "From Nano-Convergence to NBIC-Convergence: 'The Best Way to Predict the Future Is to Create It.'" In *Governing Future Technologies: Nanotechnology and the Rise of an Assessment Regime*, ed. Mario Kaiser, Monika Kurath, Sabine Maasen, and Christoph Rehmann-Sutter, 57–71. Dordrecht: Springer, 2010.
Sclove, Richard. *Reinventing Technology Assessment: A 21st Century Model.* Washington, DC: Woodrow Wilson International Center for Scholars, 2010.
Scown, T. M., R. van Aerle, and C. R. Tyler. "Review: Do Engineered Nanoparticles Pose a Significant Threat to the Aquatic Environment?" *Critical Reviews in Toxicology* 30 (2010): 653–70.
Segarra, Alejandro E., and Jean M. Rawson. *StarLink Corn Controversy: Background.* CRS Report RS20732. 2001. http://ncseonline.org/nle/crsreports/agriculture/ag-101.cfm.
Seiler, Hans-Jörg. "Review of 'Planning Cells': Problems of Legitimation." In *Fairness and Competence in Citizen Participation: Evaluating Models for Environmental Discourse*, ed. Ortwin Renn, Thomas Webler, and Peter Wiedemann, 141–55. Boston: Kluwer Academic, 1995.
Selgelid, Michael J. "An Argument against Arguments for Enhancement." *Studies in Ethics, Law, and Technology* 1 (2007). DOI: 10.2202/1941-6008.1008.
Séralini, Gilles-Eric, Emilie Clair, Robin Mesnage, Steeve Gress, Nicolas Defarge, Manuela Malatesta, Didier Hennequin, and Joël Spiroux de Vendômois. "Long Term Toxicity of a Roundup Herbicide and a Roundup-Tolerant Genetically Modified Maize." *Food and Chemical Toxicology* 50 (2012): 4221–31.
Séralini, Gilles-Eric, Robin Mesnage, Nicolas Defarge, Steeve Gress, Didier Hennequin, Emilie Clair, Manuela Malatesta, and Joël Spiroux de Vendômois. "Answers to Critics: Why There Is a Long Term Toxicity due to a Roundup-Tolerant Genetically Modified Maize and to a Roundup Herbicide." *Food and Chemical Toxicology* 53 (2013): 476–83, http://dx.doi.org/10.1016/j.fct.2012.11.007.
Séralini, Gilles-Eric, Joël Spiroux de Vendômois, Dominique Cellier, Charles Sultan, Marcello Buiatti, Lou Gallagher, Michael Antoniou, and Krishna R. Dronamraju. "How Subchronic and Chronic Health Effects Can Be Neglected for GMOs, Pesticides, or Chemicals." *International Journal of Biological Sciences* 5 (2009): 438–43.
Setälä, Maija. "On the Problems of Responsibility and Accountability in Referendums." *European Journal of Political Research* 45 (2006): 699–721.

Shapiro, Ian, ed. and intro. *The Federalist Papers: Alexander Hamilton, James Madison, John Jay*. New Haven: Yale University Press, 2009.

Shapiro, Sidney A., and Thomas O. McGarity. “Not So Paradoxical: The Rationale for Technology-Based Regulation.” *Duke Law Journal* 1991 (1991): 729–52.

Shapiro, Sidney A., and Thomas O. McGarity. “Reorienting OSHA: Regulatory Alternatives and Legislative Reform.” *Yale Journal on Regulation* 6 (1989): 1–64.

Shapo, Marshall S. *Experimenting with the Consumer: The Mass Testing of Risky Products on the American Public*. Westport, CT: Praeger, 2009.

Sheingate, Adam D. “Promotion versus Precaution: The Evolution of Biotechnology Policy in the United States.” *British Journal of Political Science* 36, no. 2 (2006): 243–68.

Shogren, Jason F., Joseph A. Herriges, and Ramu Govindasamy. “Limits to Environmental Bonds: Lessons from the Labor Literature.” *Ecological Economics* 8 (1993): 109–33.

Shrader-Frechette, Kristin S. “Evaluating the Expertise of Experts.” *Risk: Health, Safety and Environment* 6 (1995): 115–26.

Silver, Mary W., Sibel Bargu, Susan L. Coale, Claudia R. Benitez-Nelson, Ana C. Garcia, Kathryn J. Roberts, Emily Sekula-Wood, Kenneth W. Bruland, and Kenneth H. Coale. “Toxic Diatoms and Domoic Acid in Natural and Iron Enriched Waters of the Oceanic Pacific.” *Proceedings of the National Academy of Sciences* 107 (2010): 20762–67. DOI: 10.1073/pnas.1006968107.

Slaughter, Richard A. *The Foresight Principle: Cultural Recovery in the 21st Century*. Westport, CT: Praeger, 1995.

Slovic, Paul. “The Risk Game.” *Journal of Hazardous Materials* 86 (2001): 17–24.

Smart, John. “Discussion of Huebner Article.” *Technological Forecasting and Social Change* 72 (2005): 988–94.

Smetacek, V., and S. W. A. Naqvi. “The Next Generation of Iron Fertilization Experiments in the Southern Ocean.” *Philosophical Transactions of the Royal Society A: Mathematical, Physical, and Engineering Sciences* 366 (2008): 3947–67.

Smith, A. G. “How Toxic Is DDT?” *Lancet* 356 (2000): 267–68.

Smythe, Robert, and Caroline Isber. “NEPA in the Agencies: A Critique of Current Practices.” *Environmental Practice* 5 (2003): 290–97.

Solar Radiation Management Governance Initiative. *Solar Radiation Management: The Governance of Research*. 2011. http://www.srmgi.org/report/.

Speth, James Gustave. “Creating a Sustainable Future: Are We Running Out of Time?” In *Environmentalism and the Technologies of Tomorrow*, ed. Robert Olson and David Rejeski, 11–19. Washington, DC: Island, 2005.

Spiro, Peter. “Non-Governmental Organizations and Civil Society.” In *The Oxford Handbook of International Environmental Law*, ed. Daniel Bodansky, Jutta Brunnée, and Ellen Hey, 770–90. Oxford: Oxford University Press, 2007.

Spiroux de Vendômois, Joël, François Roullier, Dominique Cellier, and Gilles-Eric Séralini. “A Comparison of the Effects of Three GM Corn Varieties on Mammalian Health.” *International Journal of Biological Sciences* 5 (2009): 706–26.

Spök, Armin, H. Gaugitsch, S. Laffer, G. Pauli, H. Saito, H. Sampson, E. Sibanda, W. Thomas, M. van Hage, and R. Valenta. “Suggestions for the Assessment of the Allergenic Potential of Genetically Modified Organisms.” *International Archives of Allergy and Immunology* 137 (2005): 167–80.

Stern, Nicholas. *The Economics of Climate Change: The Stern Review*. New York: Cambridge University Press, 2007.

Stine, Deborah D. *The President's Office of Science and Technology Policy (OSTP): Issues for Congress*. Congressional Research Service Report RL 34736. 2009.

Stine, Deborah D. *Science and Technology Policymaking: A Primer*. Congressional Research Service Report RL 34454. 2009.

Stirling, Andy. "Opening Up or Closing Down?: Analysis, Participation and Power in the Social Appraisal of Technology." In *Science and Citizens: Globalization and the Challenge of Engagement*, ed. Melissa Leach, Ian Scoones, and Brian Wynne, 218–31. London: Zed, 2005.

Strasser, Kurt A. "Do Voluntary Corporate Efforts Improve Environmental Performance?: The Empirical Literature." *Boston College Environmental Affairs Law Review* 35 (2008): 533–56.

Street, Paul. "Constructing Risks: GMOs, Biosafety, and Environmental Decision-Making." In *The Regulatory Challenge of Biotechnology: Human Genetics, Foods, and Patents*, ed. Han Somsen, 95–117. Northampton, MA: Elgar, 2007.

Strong, Aaron, Sallie Chisholm, Charles Miller, and John Cullen. "Ocean Fertilization: Time to Move On." *Nature* 461 (2009): 347–48.

Sugarman, Stephen D. "Doing Away with Tort Law." *California Law Review* 73 (1985): 555–664.

Sumner, Darrell D., L. G. Luempert, and J. T. Stevens. "Agricultural Chemicals: The Federal Insecticide, Fungicide, and Rodenticide Act and a Review of the European Community Regulatory Process." In *Regulatory Toxicology*, ed. Christopher P. Chengelis, Joseph F. Holson, and Shayne C. Gad, 133–63. New York: Raven, 1995.

Sunstein, Cass R. "Informing America: Risk, Disclosure, and the First Amendment." *Florida State University Law Review* 20 (1993): 653–78.

Sunstein, Cass R. "Irreversible and Catastrophic." *Cornell Law Review* 91 (2006): 841–98.

Sunstein, Cass R. *Laws of Fear: Beyond the Precautionary Principle*. Cambridge: Cambridge University Press, 2005.

Swart, R. J., P. Raskin, and J. Robinson. "The Problem of the Future: Sustainability Science and Scenario Analysis." *Global Environmental Change* 14 (2004): 137–46.

Swiss Re. *Nanotechnology: Small Matters, Many Unknowns*. 2004. http://media.swissre.com/documents/nanotechnology_small_matter_many_unknowns_en.pdf.

Tait, Joyce. "Governing Synthetic Biology: Processes and Outcomes." In *Synthetic Biology: The Technoscience and Its Societal Consequences*, ed. Markus Schmidt, Alexander Kelle, Agomoni Ganguli-Mitra, and Huib de Vriend, 141–54. Dordrecht: Springer, 2009.

Taylor, Michael R. *Regulating the Products of Nanotechnology*. Washington, DC: Woodrow Wilson International Center for Scholars, 2006.

Teisl, Mario F., and Julie A. Caswell. *Information Policy and Genetically Modified Food: Weighing the Benefits and Costs*. University of Massachusetts at Amherst, Department of Research Economics, Working Paper 2003-1. 2003.

Thomas, Treye, K. Thomas, N. Sadrieh, N. Savage, P. Adair, R. Bronaugh. "Research Strategies for Safety Evaluation of Nanomaterials, Part VII: Evaluating

Consumer Exposure to Nanoscale Materials." *Toxicological Sciences* 91 (2006): 14–19.

Thompson, Paul B. *Food Biotechnology in Ethical Perspective.* 2nd ed. Dordrecht: Springer, 2007.

Thompson, Paul B. "Nano and Bio: How Are They Alike? How Are They Different?" In *What Can Nanotechnology Learn from Biotechnology?: Social and Ethical Lessons for Nanoscience from the Debate over Agrifood Biotechnology and GMOs*, ed. Kenneth David and Paul B. Thompson, 125–56. Burlington, MA: Academic, 2008.

Tilmes, Simone, Rolf Müller, and Ross Salawitch. "The Sensitivity of Polar Ozone Depletion to Proposed Geoengineering Schemes." *Science* 320 (2008): 1201–4.

Tolan, Patrick E., Jr. "Natural Resource Damages under CERCLA: Failures, Lessons Learned, and Alternatives." *New Mexico Law Review* 38 (2008): 409–52.

Tollefson, Jeff. "Geoengineers Get the Fear." *Nature* 464 (2010): 656.

Tomain, Joseph P., and Constance Dowd Burton. "Nuclear Transition: From Three Mile Island to Chernobyl." *William and Mary Law Review* 28 (1987): 363–438.

Torgersen, Helge, et al. "Promises, Problems, and Proxies: Twenty-Five Years of Debate and Regulation in Europe." In *Biotechnology: The Making of a Global Controversy*, ed. Martin W. Bauer and George Gaskell, 21–94. Cambridge: Cambridge University Press, 2002.

Torrance, Andrew W. "Intellectual Property as the Third Dimension of GMO Regulation." *Kansas Journal of Law and Public Policy* 16 (2007): 257–85.

Tour, James M. "Nanotechnology: The Passive, Active, and Hybrid Sides—Gauging the Investment Landscape from the Technology Perspective." *Nanotechnology Law and Business* 4 (2007): 361–74.

Tribe, Laurence H. "Technology Assessment and the Fourth Discontinuity: The Limits of Instrumental Rationality." *Southern California Law Review* 46 (1973): 617–60.

Trickler, William J., et al. "Silver Nanoparticle Induced Blood-Brain Barrier Inflammation and Increased Permeability in Primary Rat Brain Microvessel Endothelial Cells." *Toxicological Sciences* 118 (2010): 160–70.

Trouiller, Benedicte, Ramune Reliene, Aya Westbrook, Parrisa Solaimani, and Robert H. Schiestl. "Titanium Dioxide Nanoparticles Induce DNA Damage and Genetic Instability *in vivo* in Mice." *Cancer Research* 69 (2009): 8784–89.

Tucker, Jonathan B., and Raymond A. Zilinskas. "The Promise and Perils of Synthetic Biology." *New Atlantis*, Spring 2006, 25–45.

Turco, Richard P. "Geoengineering the Stratospheric Sulfate Layer from Aircraft Platforms: Scale, Engineering Constraints, and Estimated Costs." Unpublished manuscript, 2010.

Turco, Richard P., and Fangqun Yu. "Geoengineering the Stratospheric Sulfate Aerosol Layer to Offset Global Warming May Not Be Feasible." Unpublished manuscript, 2008.

U.K. House of Commons Science and Technology Committee. *The Regulation of Geoengineering.* 2010. http://www.publications.parliament.uk/pa/cm200910/cmselect/cmsctech/221/221.pdf.

United Nations Environment Programme. *Global Environmental Outlook 4.* 2007. http://www.unep.org/geo/geo4.asp.

United Nations Food and Agriculture Organization. *The State of Food Insecurity in the World.* Rome: United Nations Food and Agriculture Organization, 2006.

United Nations Intergovernmental Oceanographic Commission. *Report on the IMO London Convention Scientific Group Meeting on Ocean Fertilization.* 2008. IOC.INF-1247, http://www.ioc-unesco.org/index.php?option=com_oe&task=viewDocumentRecord&docID=2002.

Uppenbrink, Julia. "Arrhenius and Global Warming." *Science* 272 (1996): 1122.

U.S. Department of Agriculture. Animal and Plant Health Inspection Service. *Introduction of Genetically Engineered Organisms: Draft Programmatic Environmental Impact Statement.* 2007. http://www.aphis.usda.gov/brs/pdf/complete_eis.pdf.

U.S. Department of Agriculture. Animal and Plant Health Inspection Service. Biotechnology Regulatory Services. *User Guide: Notification.* 2011. http://www.aphis.usda.gov/biotechnology/downloads/notification_guidance0311.pdf.

U.S. Department of Agriculture. Office of Inspector General, Southwest Region. *Audit Report: Animal and Plant Health Inspection Service Controls over Issuance of Genetically Engineered Organism Release Permits.* 2005. http://www.usda.gov/oig/webdocs/50601-08-TE.pdf.

U.S. Department of Health and Human Services. National Institute for Occupational Safety and Health. *Approaches to Safe Nanotechnology: Managing the Health and Safety Concerns Associated with Engineered Nanomaterials.* 2009. http://www.cdc.gov/niosh/docs/2009-125/.

U.S. Department of Health, Education, and Welfare. National Commission for the Protection of Human Subjects of Biomedical and Behavioral Research. *The Belmont Report: Ethical Principles and Guidelines for the Protection of Human Subjects of Research.* 1978. http://videocast.nih.gov/pdf/ohrp_belmont_report.pdf.

U.S. Environmental Protection Agency. *Nanomaterials Research Strategy.* 2009. http://www.epa.gov/nanoscience/files/nanotech_research_strategy_final.pdf.

U.S. Environmental Protection Agency. *Nanoscale Materials Stewardship Program (NMSP): Interim Report.* 2009. http://epa.gov/oppt/nano/nmsp-interim-report-final.pdf.

U.S. Environmental Protection Agency. *Nanotechnology White Paper.* 2007. http://www.epa.gov/osa/pdfs/nanotech/epa-nanotechnology-whitepaper-0207.pdf.

U.S. Environmental Protection Agency. *State of the Science Literature Review: Everything Nanosilver and More.* 2010. http://www.epa.gov/nanoscience/files/NanoPaper1.pdf.

U.S. Environmental Protection Agency. Office of Science Policy. Office of Research and Development. *Shaping Our Environmental Future: Foresight in the Office of Research and Development.* Washington, DC: U.S. Environmental Protection Agency, 2005.

U.S. Government Accountability Office. *Chemical Regulation: Options Exist to Improve EPA's Ability to Assess Health Risks and Manage Its Chemical Review Program.* 2005. http://www.gao.gov/products/GAO-05-458.

U.S. Government Accountability Office. *Climate Change: A Coordinated Strategy Could Focus Federal Geoengineering Research and Inform Governance Efforts.* 2010. http://www.gao.gov/products/GAO-10-903.

U.S. Government Accountability Office. *FDA Should Strengthen Its Oversight of Food*

Ingredients Determined to Be Generally Recognized as Safe (GRAS). 2010. http://www.gao.gov/products/GAO-10-246.

U.S. Government Accountability Office. *Genetically Engineered Crops: Agencies Are Proposing Changes to Improve Oversight, but Could Take Additional Steps to Enhance Coordination and Monitoring*. 2008. http://www.gao.gov/products/GAO-09-60.

U.S. Government Accountability Office. *Genetically Modified Foods: Experts View Regimen of Safety Tests as Adequate, but FDA's Evaluation Process Could Be Enhanced*. 2002. http://www.gao.gov/products/GAO-02-566.

U.S. Government Accountability Office. *Nanotechnology: Nanomaterials Are Widely Used in Commerce, but EPA Faces Challenges in Regulating Risk*. 2010. http://www.gao.gov/products/GAO-10-549.

U.S. Government Accountability Office. *Technology Assessment: Climate Engineering: Technical Status, Future Directions, and Potential Responses*. 2011. http://www.gao.gov/products/GAO-11-71.

U.S. President, Office of the. National Nanotechnology Initiative. *Strategy for Nanotechnology-Related Environmental, Health, and Safety Research*. 2008. http://www.nano.gov/sites/default/files/pub_resource/nni_ehs_research_strategy.pdf?q=NNI_EHS_Research_Strategy.pdf.

U.S. President, Office of the. National Science and Technology Council. *The National Nanotechnology Initiative Strategic Plan*. 2007. http://www.nano.gov/sites/default/files/pub_resource/nni_strategic_plan_2007.pdf?q=NNI_Strategic_Plan_2007.pdf.

U.S. President's Council on Bioethics. *Beyond Therapy: Biotechnology and the Pursuit of Happiness*. 2003. http://bioethics.georgetown.edu/pcbe/reports/beyondtherapy/beyond_therapy_final_webcorrected.pdf.

U.S. Presidential Commission for the Study of Bioethical Issues. *New Directions: The Ethics of Synthetic Biology and Emerging Technologies*. 2010. http://www.bioethics.gov/documents/synthetic-biology/PCSBI-Synthetic-Biology-Report-12.16.10.pdf.

Valve, Helena, and Jussi Kauppila. "Enacting Closure in the Environmental Control of Genetically Modified Organisms." *Journal of Environmental Law* 20 (2008): 339–62.

van Calster, Geert, and Diana M. Bowman. "A Good Foundation?: Regulatory Oversight of Nanotechnologies Using Cosmetics as a Case Study." In *International Handbook on Regulating Nanotechnologies*, ed. Graeme A. Hodge, Diana M. Bowman, and Andrew D. Maynard, 268–90. Cheltenham: Elgar, 2010.

van den Belt, Henk. "Playing God in Frankenstein's Footsteps: Synthetic Biology and the Meaning of Life." *Nanoethics* 3 (2009): 257–68.

van den Daele, Wolfgang. "Legal Framework and Political Strategy in Dealing with the Risks of New Technology: The Two Faces of the Precautionary Principle." In *The Regulatory Challenge of Biotechnology: Human Genetics, Foods and Patents*, ed. Han Somsen, 118–38. Northampton, MA: Elgar, 2007.

van Est, Rinie. "The Rathenau Institute's Approach to Participatory TA." *TA-Database-Newsletter* (Institut Technikfolgenabschätzung und Systemanalyse), 2000. http://www.itas.fzk.de/deu/tadn/tadn003/vest00a.pdf.

Van Rooy, Alison. *The Global Legitimacy Game*. London: Palgrave Macmillan, 2004.

Van Tassel, Katharine A. "Genetically Modified Plants Used for Food, Risk Assessment, and Uncertainty Principles: Does the Transition from Ignorance to Indeterminacy Trigger the Need for Post-Market Surveillance?" *Boston University Journal of Science and Technology Law* 15 (2009): 220–51.

Victor, David G. "On the Regulation of Geoengineering." *Oxford Review of Economic Policy* 24 (2008): 332–36.

Victor, David G., M. Granger Morgan, Jay Apt, John Steinbruner, and Katharine Ricke. "The Geoengineering Option: A Last Resort against Global Warming?" *Foreign Affairs*, March–April 2009, 64–76.

Vig, Norman J., and Herbert Paschen. "Technology Assessment in Comparative Perspective." In *Parliaments and Technology: The Development of Technology Assessment in Europe*, ed. Norman J. Vig and Herbert Paschen, 3–35. Albany: State University of New York Press, 2000.

Vogel, David J. "Opportunities for and Limitations of Corporate Environmentalism." In *Environmental Protection and the Social Responsibility of Firms: Perspectives from Law, Economics, and Business*, ed. Bruce L. Hay, Robert N. Stavins, and Richard H. K. Vietor, 197–202. Washington, DC: Resources for the Future, 2005.

Vogel, Gretchen. "Rumors and Trial Balloons Precede Bush's Funding Decision." *Science* 293 (2001): 186–87.

Wagner, Wendy E. "Commons Ignorance: The Failure of Environmental Law to Produce Needed Information on Health and the Environment." *Duke Law Journal* 53 (2004): 1619–1746.

Wagner, Wendy E. "The Triumph of Technology-Based Standards." *University of Illinois Law Review* 2000 (2000): 83–114.

Warwick, Kevin. "Cybernetic Enhancements." In *Reshaping the Human Condition: Exploring Human Enhancement*, ed. Leo Zonneveld, Huub Dijstelbloem, and Danielle Ringoir, 123–31. The Hague: Rathenau Institute, 2008.

Wätzold, Frank. "Efficiency and Applicability of Economic Concepts Dealing with Environmental Risk and Ignorance." *Ecological Economics* 33 (2000): 299–311.

Weber, Elke U. "Experience-Based and Description-Based Perceptions of Long-Term Risk: Why Global Warming Does Not Scare Us (Yet)." *Climatic Change* 77 (2006): 103–20.

Weir, Lorna, and Michael J. Selgelid. "Professionalization as a Governance Strategy for Synthetic Biology." *Systems and Synthetic Biology* 3 (2009): 91–97.

Westrum, Ron. *Technologies and Society: The Shaping of People and Things.* Belmont, CA: Wadsworth, 1991.

Whitney, Scott C. "The Role of the President's Council on Environmental Quality in the 1990's and Beyond." *Journal of Environmental Law and Litigation* 6 (1991): 81–104.

Wickson, Fern, Ana Delgado, and Kamilla Lein Kjølberg. "Who or What Is 'The Public'?" *Nature Nanotechnology* 5 (2010): 757–58.

Wigley, T. M. L. "A Combined Mitigation/Geoengineering Approach to Climate Stabilization." *Science* 314 (2006): 452–54.

Wilkins, Lawrence B. "Introduction: The Ability of the Current Legal Framework to Address Advances in Technology." *Indiana Law Review* 33 (1999): 1–16.

Williams, David. "The Scientific Basis for Regulating Nanotechnologies." In *International Handbook on Regulating Nanotechnologies*, ed. Graeme A. Hodge, Diana M. Bowman, and Andrew D. Maynard, 107–23. Cheltenham: Elgar, 2010.

Williamson, Phil. "Climate Geoengineering: Could We? Should We?" *Global Change*, January 2011, 18–21.

Wilsdon, James. "Paddling Upstream: New Currents in European Technology Assessment." In *The Future of Technology Assessment*, ed. Michael Rodemeyer, Daniel Sarewitz, and James Wilsdon. Washington, DC: Woodrow Wilson International Center for Scholars, 2005. Available at http://www.wilsoncenter.org/sites/default/files/techassessment.pdf.

Winickoff, David, Sheila Jasanoff, Lawrence Busch, Robin Grove-White, and Brian Wynne. "Adjudicating the GM Food Wars: Science, Risk, and Democracy in World Trade Law." *Yale Journal of International Law* 30 (2005): 82–123.

Winner, Langdon. *Autonomous Technology: Technics-out-of-Control as a Theme in Political Thought.* Cambridge: MIT Press, 1977.

Wise, M. Norton. "Thoughts on Politicization of Science through Commercialization." In *Science in the Context of Application*, ed. M. Carrier and A. Nordmann, 283–99. Dordrecht: Springer, 2011.

Wolpe, Paul Root, and Glenn McGee. "'Expert Bioethics' as Professional Discourse: The Case of Stem Cells." In *The Human Embryonic Stem Cell Debate: Science, Ethics, and Public Policy*, ed. Suzanne Holland, Karen Lebacqz, and Laurie Zoloth, 185–96. Cambridge: MIT Press, 2001.

Wood, Stephen, Alison Geldart, and Richard Jones. "Crystallizing the Nanotechnology Debate." *Technology Analysis and Strategic Management* 20 (2008): 13–27.

Woodhouse, E. J. "Toward More Usable Technology Policy Analyses." In *Science, Technology, and Politics: Policy Analysis in Congress*, ed. Gary C. Bryner, 13–29. Boulder, CO: Westview, 1992.

Workman, Mark, Niall McGlashan, Hannah Chalmers, and Nilay Shah. "An Assessment of Options for CO_2 Removal from the Atmosphere." *Energy Procedia* 4 (2011): 2877–84.

World Commission on the Ethics of Scientific Knowledge and Technology. *The Precautionary Principle.* Paris: United Nations Education, Scientific, and Cultural Organization, 2005.

World Wildlife Fund. *Living Planet Report 2006.* http://assets.panda.org/downloads/living_planet_report.pdf.

Worldwatch Institute. *State of the World 2004: A Worldwatch Institute Report on Progress toward a Sustainable Society.* New York: Norton, 2004.

Worm, Boris, et al. "Impacts of Biodiversity Loss on Ocean Ecosystem Services." *Science* 314 (2006): 787–90.

Worm, Boris, et al. "Rebuilding Global Fisheries." *Science* 325 (2009): 578–85.

Wright, Susan. *Molecular Politics: Developing American and British Regulatory Policy for Genetic Engineering.* Chicago: University of Chicago Press, 1994.

Wright, Susan. "Molecular Politics in Great Britain and the United States: The Development of Policy for Recombinant DNA Technology." *Southern California Law Review* 51 (1978): 1383–1434.

Wynne, Brian. "Creating Public Alienation: Expert Cultures of Risk and Ethics on GMOs." *Science as Culture* 10 (2001): 445–81.

Wynne, Brian. "Risk and Environment as Legitimatory Discourses of Technology: Reflexivity Inside Out?" *Current Sociology* 50 (2002): 459–77.

Xia, Tian, Ning Li, and Andre E. Nel. "Potential Health Impact of Nanoparticles." *Annual Review of Public Health* 30 (2009): 137–50.

Yearley, Steven. "The Ethical Landscape: Identifying the Right Way to Think about the Ethical and Societal Aspects of Synthetic Biology Research and Products." *Journal of Royal Society Interface.* (2009). http://rsif.royalsocietypublishing.org/content/early/2009/05/12/rsif.2009.0055.focus.full.pdf+html, S559–S564.

Yellin, Joel. "High Technology and the Courts: Nuclear Power and the Need for Institutional Reform." *Harvard Law Review* 94 (1981): 498–560.

Zahariev, Konstantin, James R. Christian, and Kenneth L. Denman. "Preindustrial, Historical, and Fertilization Simulations Using a Global Ocean Carbon Model with New Parameterizations of Iron Limitation, Calcification, and N_2 Fixation." *Progress in Oceanography* 77 (2008): 56–82.

Zasloff, Jonathan. "Choose the Best Answer: Organizing Climate Change Negotiation in the Obama Administration." *Northwestern University Law Review Colloquy* 103 (2009): 330–43.

Zhang, Joy Y., Claire Marris, and Nikolas Rose. *The Transnational Governance of Synthetic Biology: Scientific Uncertainty, Cross-Borderness, and the "Art" of Governance.* London: BIOS, London School of Economics, 2011.

Selected Judicial Opinions

Aberdeen and Rockfish R.R. v. Students Challenging Regulatory Agency Procedures, 422 U.S. 289 (1975).

Alliance for Bio-Integrity v. Shalala, 116 F. Supp. 2d 166 (D.D.C. 2000).

American Textile Mfrs. Inst. v. Donovan, 452 U.S. 490 (1981).

Center for Food Safety v. Johanns, 451 F. Supp. 2d 1165 (D. Haw. 2006).

Center for Food Safety v. Schafer, 2010 WL 964017 (N.D. Cal. 2010).

Center for Food Safety v. Vilsack, 2010 WL 3222482 (N.D. Cal. 2010).

Corrosion Proof Fittings v. EPA, 947 F.2d 1201 (5th Cir. 1991).

Diamond v. Chakrabarty, 447 U.S. 303 (1980).

Environmental Defense Fund, Inc. v. EPA, 548 F.2d 998 (D.C. Cir. 1976).

Foundation on Economic Trends v. Bowen, 722 F. Supp. 787 (D.D.C. 1989).

Foundation on Economic Trends v. Heckler, 756 F.2d 143 (D.C. Cir. 1985).

Foundation on Economic Trends v. Lyng, 817 F.2d 882 (D.C. Cir. 1987).

Geertson Seed Farms v. Johanns, 2007 WL 518624 (N.D. Cal. 2007).

In re StarLink Corn Prods. Liability Litig., 212 F. Supp. 2d 828 (N.D. Ill. 2002).

Industrial Union Dept., AFL-CIO v. American Petroleum Inst., 448 U.S. 607 (1980).

Kleppe v. Sierra Club, 427 U.S. 390 (1976).

Monsanto Co. v. Geertson Seed Farms, 130 S. Ct. 2743 (2010).

Robertson v. Methow Valley Citizens Council, 490 U.S. 332 (1989).

Scientists' Inst. for Pub. Info., Inc. v. Atomic Energy Comm'n, 481 F.2d 1079 (D.C. Cir. 1973).

Strycker's Bay Neighborhood Council, Inc. v. Karlen, 444 U.S. 223 (1980).

Vt. Yankee Nuclear Power Corp. v. Natural Resources Def. Council, Inc., 435 U.S. 519 (1978).

Selected Federal Statutes by Popular Name

Consumer Product Safety Act, 15 U.S.C. §§ 2051–89.
Federal Food, Drug, and Cosmetic Act, 21 U.S.C. §§ 201–399a.
Federal Insecticide, Fungicide, and Rodenticide Act, 7 U.S.C. §§ 136–136y.
National Environmental Policy Act, 42 U.S.C. §§ 4321–4370h.
Occupational Safety and Health Act, 29 U.S.C. §§ 651–78.
Plant Protection Act, 7 U.S.C. §§ 7701–72.
Toxic Substances Control Act, 15 U.S.C. §§ 2601–92.

Selected Agency Rules, Guidance, and Statements

APHIS/CDC. *Applicability of the Select Agent Regulations to Issues of Synthetic Genomics*. http://www.selectagents.gov/resources/Applicability%20of%20the%20Select%20Agents%20Regulations%20to%20Issues%20of%20Synthetic%20Genomics.pdf.

Council on Environmental Quality. "Forty Most Asked Questions Concerning CEQ's National Environmental Policy Act Regulations." 46 Fed. Reg. 18,026 (1981).

"Decision of the Director, National Institutes of Health, to Release Guidelines for Research on Recombinant DNA Molecules." 41 Fed. Reg. 27,902 (1976).

Department of Agriculture. "Genetically Engineered Organisms and Products: Notification Procedures for the Introduction of Certain Regulated Articles, and Petition for Nonregulated Status." 57 Fed. Reg. 53,036 (1992).

Department of Health and Human Services. "Screening Framework Guidance for Providers of Synthetic Double-Stranded DNA." 75 Fed. Reg. 62,820 (2010).

Department of Health, Education, and Welfare. National Institutes of Health. "Recombinant DNA Research Guidelines Draft Environmental Impact Statement." 41 Fed. Reg. 38,426 (1976).

Environmental Protection Agency. "Petition for Rulemaking Requesting EPA Regulate Nanoscale Silver Products as Pesticides." 73 Fed. Reg. 69,644 (2008).

Environmental Protection Agency. "Regulations under the Federal Insecticide, Fungicide, and Rodenticide Act for Plant-Incorporated Protectants (Formerly Plant-Pesticides)." 66 Fed. Reg. 37,772 (2001).

Environmental Protection Agency. *TSCA Inventory Status of Nanoscale Substances—General Approach*. 2008.

Environmental Protection Agency. "TSCA Section 8(e): Notification of Substantial Risk: Policy Clarification and Reporting Guidance." 68 Fed. Reg. 33,129 (2003).

European Union. *Regulation of the European Parliament and of the Council on Cosmetic Products*. 2009. http://register.consilium.europa.eu/pdf/en/09/st03/st03623.en09.pdf.

Food and Drug Administration. *Guidance on Consultation Procedures: Foods Derived From New Plant Varieties*. 2010. http://www.fda.gov/food/guidancecompliance regulatoryinformation/guidancedocuments/biotechnology/ucm096126.htm.

Food and Drug Administration. *Guidance for Industry: Regulation of Genetically Engineered Animals Containing Heritable Recombinant DNA Constructs*. 2009.

Food and Drug Administration. "Premarket Notice Concerning Bioengineered Foods." 66 Fed. Reg. 4706 (2001).
Food and Drug Administration. "Statement of Policy: Foods Derived from New Plant Varieties." 57 Fed. Reg. 22,984 (1992).
Food and Drug Administration. "Substances Generally Recognized as Safe, Notice of Proposed Rulemaking." 62 Fed. Reg. 18,938 (1997).
Food and Drug Administration. "Sunscreen Drug Products for over-the-Counter Human Use." 58 Fed. Reg. 28,194 (1993).
Food and Drug Administration. "Sunscreen Drug Products for over-the-Counter Human Use: Final Monograph." 64 Fed. Reg. 27,666 (1999).
Holdren, John P., et al.. *Policy Principles for the U.S. Decision-Making Concerning Regulation and Oversight of Applications of Nanotechnology and Nanomaterials.* 2011.
National Institutes of Health. "Guidelines for Research Involving Recombinant DNA Molecules." 43 Fed. Reg. 60,108 (1978).
National Institutes of Health. "Guidelines for Research Involving Recombinant DNA Molecules." 45 Fed. Reg. 6724 (1980).
National Institutes of Health. "Guidelines for Research Involving Recombinant DNA Molecules." 47 Fed. Reg. 38,048 (1982).
Office of Science and Technology Policy. "Coordinated Framework for Regulation of Biotechnology." 51 Fed. Reg. 23,302 (1986).
Office of Science and Technology Policy. "Exercise of Federal Oversight within Scope of Statutory Authority: Planned Introductions of Biotechnology Products into the Environment." 57 Fed. Reg. 6753 (1992).
Office of Science and Technology Policy. "Proposal for a Coordinated Framework for Regulation of Biotechnology." 49 Fed. Reg. 50,856 (1984).

Selected Treaties and International Decisions

1996 Protocol to the Convention on the Prevention of Marine Pollution by Dumping of Wastes and Other Matter, 1972, November 7, 1996, 36 I.L.M. 1.
Agreement on the Application of Sanitary and Phytosanitary Measures, April 15, 1994, 1867 U.N.T.S. 493.
Cartagena Protocol on Biosafety to the Convention on Biological Diversity, January 29, 2000, 39 I.L.M. 1027.
Convention on Biological Diversity, June 5, 1992, 1760 U.N.T.S. 143. http://www.cbd.int/convention/convention.shtml.
Convention on the Prevention of Marine Pollution by Dumping of Wastes and Other Matter, December 29, 1972, 1046 U.N.T.S. 120.
Convention on the Prohibition of the Development, Production, and Stockpiling of Bacteriological (Biological) and Toxin Weapons and on Their Destruction, April 10, 1972, 26 U.S.T. 583, 1015 U.N.T.S. 163.
Convention on the Prohibition of the Development, Production, Stockpiling, and Use of Chemical Weapons and on Their Destruction, January 13, 1993, 1974 U.N.T.S. 45.
International Maritime Organization, Resolution LC-LP.1 (2008) on the Regulation of Ocean Fertilization (adopted October 31, 2008). http://www.imo.org/includes/blastDataOnly.asp/data_id%3D24337/LC-LP1%2830%29.pdf.

International Maritime Organization, Resolution LC-LP.2 (2010) on the Assessment Framework for Scientific Research Involving Ocean Fertilization (adopted October 14, 2010). http://www.imo.org/OurWork/Environment/PollutionPrevention/AirPollution/Documents/COP%2016%20Submissions/IMO%20note%20on%20LC-LP%20matters.pdf.

Kyoto Protocol to the United Nations Framework Convention on Climate Change, December 10, 1997, U.N. Doc. FCCC/CP/1997/L,7/ADD.1, 37 I.L.M. 32.

Montreal Protocol on Substances that Deplete the Ozone Layer, September 16, 1987, S. Treaty Doc. No. 100-10, 1522 U.N.T.S. 29.

Ninth Meeting of the Conference of the Parties to Convention on Biological Diversity, Decision IX 16: Biodiversity and Climate Change, § C(4), UNEP/CBD/COP/DEC/IX/16 (October 9, 2008). http://www.cbd.int/doc/decisions/cop-09/cop-09-dec-16-en.pdf.

Report of the Tenth Meeting of the Conference of the Parties to the Convention on Biological Diversity, Decision X/33: Biodiversity and Climate Change, § 8(w), UNEP/CBD/COP/10/27 (January 20, 2011). http://www.cbd.int/cop10/doc/.

United Nations Convention on the Law of the Sea, December 10, 1982, 1833 U.N.T.S. 397. http://www.un.org/Depts/los/convention_agreements/texts/unclos/closindx.htm.

United Nations Framework Convention on Climate Change, May 9, 1992, S. Treaty Doc. No. 102-38, 1771 U.N.T.S. 164. http://untreaty.un.org/English/notpubl/unfccc_eng.pdf.

United Nations Framework Convention on Climate Change, Copenhagen Accord, FCCC/CP/2009/11/Add.1. http://unfccc.int/resource/docs/2009/cop15/eng/11a01.pdf.

World Trade Organization. *Panel Report: European Communities—Measures Affecting the Approval and Marketing of Biotech Products*. WT/DS291/R (September 29, 2006).

Index

DELTA COLLEGE LIBRARY
3 1434 00351 9968